MUTAGENESIS

Edited by **Rajnikant Mishra**

Mutagenesis
http://dx.doi.org/10.5772/2937
Edited by Rajnikant Mishra

Contributors

Chuan Tang Wang, Yue Yi Tang, Xiu Zhen Wang, Qi Wu, Hua Yuan Gao, Tong Feng, J un Wei Su, Shu Tao Yu, Xian Lan Fang, Wan Li Ni, Yan Sheng Jiang, Lang Qian, Dong Qing Hu, Fusao Hirata, George B. Corcoran, Aiko Hirata, Cherie Musgrove, Manel Camps, Marc Vermulst, Konstantin Khrapko, Jonathan Wanagat, Biljana Nikolić, Dragana Mitić-Ćulafić, Branka Vuković-Gačić, Jelena Knežević-Vukčević, Hidetaka Torigoe, Takeshi Imanishi, Sang Sun Kang, Anna Sikora, Celina Janion, Elżbieta Grzesiuk, Kristin Lampe, Siobhan Cashman, Halil Aksoylar, Kasper Hoebe, Petra Kozjak, Vladimir Meglič, María Pertusa, Hans Moldenhauer, Sebastián Brauchi, Ramón Latorre, Rodolfo Madrid, Patricio Orio

Published by InTech

Janeza Trdine 9, 51000 Rijeka, Croatia

Publishing Process Manager Danijela Duric
Technical Editor InTech DTP team
Cover InTech Design team

Additional hard copies can be obtained from orders@intechopen.com

Mutagenesis, Edited by Rajnikant Mishra
p. cm.
ISBN 978-953-51-0707-1

For EU product safety concerns:
IN TECH d.o.o., Prolaz Marije Krucifikse Kozulić 3, 51000 Rijeka, Croatia,
info@intechopen.com or visit our website at intechopen.com.

Meet the editor

The Editor is a Biochemist and Molecular Biologist with Ph.D. in Zoology from Banaras Hindu University. He is the recipient of Junior, and Senior Research Fellowships, and Senior Research Associate ship (Scientist Pool) of Council of Scientific and Industrial Research (C.S.I.R), young scientist awardees of the Department of Science & Technology, India, and Postdoctoral fellowship with honorable mention in the category of basic science research for the top 10% for the abstract scores, and ranked in the top 3 for the poster competition among trainees of 2000-2001 at U.T.-M.D. Anderson Cancer Center, Houston, Texas, USA. He has twenty years of research and teaching experience in the area of Biochemistry, Molecular Biology & Bioinformatics. His research team works to understand mechanisms of functions of paired box family transcription factors in adult neurogenesis, ocular defects, and lymphoma. He had published more than 25 research articles in International journals of repute including a book-chapter on "Gene Targeting".

Contents

Preface

Mutagenesis is a powerful method directed towards mutating genes or engineering genome for functional analysis, producing mouse models of human disease, and understanding molecular mechanism of biodiversity at different levels of biological organizations. Several insights have been revealed through studies using experimental models, or *in silico*, on natural mutations in general and site-specific mutagenesis in particular. However, unified text on recent trends in mutagenesis is poorly available within a single cover. Trust, this book would cater the need of readers who are looking for unified text on mutagenesis and its applications in research and teaching in the area of basic, agricultural or medical sciences.

I express my gratitude to authors for sharing their experience on mutagenesis. I gratefully acknowledge the time and valuable suggestions of reviewers Dr Padmanabh Dwivedi, and Dr Anju Shrivastava. I extend my sincere thanks to my dear students Miss Brij Bharti, Mr Ratnakar Tripathi, Mr Sachin Shukla and Mr Kumar Shubham for their critical help in editing this book.

Rajnikant Mishra, Ph.D
Biochemistry & Molecular Biology Laboratory
Department of Zoology
Banaras Hindu University,Varanasi-221005,
India

Mutagenesis: A Useful Tool for the Genetic Improvement of the Cultivated Peanut *(Arachis hypogaea L.)*

Chuan Tang Wang, Yue Yi Tang, Xiu Zhen Wang, Qi Wu,
Hua Yuan Gao, Tong Feng, Jun Wei Su, Shu Tao Yu, Xian Lan Fang,
Wan Li Ni, Yan Sheng Jiang, Lang Qian and Dong Qing Hu

Additional information is available at the end of the chapter

http://dx.doi.org/10.5772/50514

1. Introduction

Ranking 4th among world oilseed crops and 13th among food crops, the cultivated peanut (*Arachis hypogaea* L.) is an important cash crop currently grown in over eighty countries/regions from 40°N to 40°S across tropical and warm temperate regions [1,2]. Its seeds contain about 50% oil and 25% protein, and the crop is thus deemed as a rich source of edible oil and dietary protein. In developing countries, a large portion of peanuts are crushed for edible oil [3]. Food uses of peanut are predominant in developed nations, where high oleate, high protein and reduced fat peanuts are most preferred, as high oleate not only means better keeping quality, but also brings about multiple health benefits, for example, reduced risk of cardiovascular disease, increased sensitivity to insulin, preventative effects on tumorigenesis, and amelioration of some inflammatory diseases [3,4].

Breeding for high yield has been and will continue to be one of the most important objectives of any peanut genetic improvement programs. For quite a long time, intraspecific hybridization (IH) has been the major breeding method of peanut, through which a sizable number of peanut cultivars with high yields have been released. The narrow gene base of the peanut cultigen caused by reproductive isolation from wild *Arachis* relatives has rendered the genetic improvement of the cultivated peanut through IH more and more difficult. To obtain fertile interspecific hybrids, in most cases special technical measures have to be taken to cope with incompatible obstacles and/or ploidy difference in wide crosses [5,6]. Furthermore, backcrossing is needed to break undesirable linkages. Wide segregation range in peanut interspecific derivatives and linkage drags make the breeding process using wild *Arachis* species lengthy and tedious. In contrast, traits in mutagenized populations tend to stabilize easily. Recent years, in peanut breeding, much attention has therefore been paid to mutagenesis.

Developing peanut cultivars with improved quality has long been proposed as a major breeding objective of the crop. With the depletion of fossil fuel, interest in peanut as a source of renewable energy is growing. Peanut genotypes with both high oil and high oleate are considered most suitable for biodiesel production. Yet little progress had been made in peanut quality improvement, especially for oil and protein content during the past several decades, largely due to the limited genetic variability within the cultivated peanut gene pool as well as unavailability of simple, rapid and cost-effective selection techniques. Conventional analytical procedures for oil, protein and fatty acid determination are destructive, costly and time-consuming, unsuitable for handling large samples frequently encountered in the process of breeding [7]. Fortunately, the stagnant situation is being changed as the wide application of mutagenesis in peanut breeding and the development of near infrared spectroscopy calibration equations for peanut by several authors [8-14].

Using sun-dried peanut seed we have developed NIRS (near infrared reflectance spectroscopy)calibration equations predictive of main quality characters of bulk seed samples and intact single seeds, making it possible to screen a large number of peanut seeds for multiple quality traits rapidly, simultaneously and non-destructively [13,14].

This chapter summarized the recent progress in peanut mutagenesis for yield and quality made by scientists from our research group.

2. Development of high-yielding peanut mutants through chemical mutagenesis

2.1. Flower injection of ethyl methane sulfonate (EMS)

Through injection of 0.3% EMS into flowers of Huayu 16 at 9:00-9:30 a.m. and subsequent selection, we were able to develop a high-yielding peanut cultivar - Huayu 40 [15,16](Table 1).

Huayu 40 has an erect growth habit and sequential branching pattern. As compared with its wild type (Huayu 16), Huayu 40 possesses faster growing and darker green foliage [16]. A study conducted at flowering and pegging stage in 2009 showed that leaf water content, chlorophyll a and b content of Huayu 40 were significantly higher than those of Huayu 16 [15]. IT-ISJ (Intron-Targeted Intron-Exon Splice Conjunction) profiling using 7 primer pairs resulted in 8 bands that could differentiate Huayu 40 and Huayu 16 [15,17]. While Huayu 16 has white and yellow inner seed coat color, Huayu 40 only has white inner seed coat color. At Shandong Peanut Research Institute (SPRI), Huayu 40 showed a yield increase of about 5% over Huayu 16 [15] (Table 1). Basal pod setting around central axes and high peg strength minimizes pod losses at harvest [15,16].

The cultivar, also known as 08-test-A2, was approved for release in Anhui province in 2011 and in Jilin province in 2012 [15,16]. In summer sowing in Anhui province, Huayu 40 matured in 116 days, and produced an average pod yield of 3970.8 kg ha^{-1} at 3 locations, outyielding the local control Luhua 8 by 14.6%, ranking first among the 11 entries in the test [16]. In Jilin provincial peanut cultivar evaluation test conducted in 2010 and 2011, Huayu 40 matured in 124 days. It produced an average pod yield of 3385 kg ha^{-1}, 22.32% over the local check Baisha 1016.

Year	Action
2001	EMS injection and seed harvest
2002	Cultivation and harvest. Extremely poorly performed plants were discarded
2003-2004	Single plant selection
2005	A plant line (E7-2) with more sound mature kernels (SMK) was noticed and selected
2006-2007	Multiplication of E7-2 to obtain adequate seeds for primary yield evaluation
2008-2009	Yield evaluation in Laixi, Shandong, under the name of 08-test-A2. It showed a superiority of 24.93% (kernel yield) over Luhua 11 in 2008, 5.42% and 10.71% over Huayu 16 and Fenghua 1, respectively, in 2009.
2010-present	Continual yield evaluation and seed multiplication in Laixi, Shandong, and in Gongzhuling, Jilin. National peanut primary and secondary yield evaluation test in northern China (2010 and 2011, spring sowing), National peanut pre-release final yield evaluation test in northern China (2012, spring sowing), Jilin provincial peanut cultivar evaluation test (2010 and 2011, spring sowing), Anhui yield evaluation test (2011, summer sowing)

Table 1. Huayu 40: how it was bred

In 2010 Anhui yield evaluation test (2010, summer sowing), Jilin provincial peanut cultivar evaluation test (2010 and 2011, spring sowing), national peanut primary and secondary yield evaluation test in northern China (2010 and 2011, spring sowing), national peanut pre-release final yield evaluation test in northern China (201 2, spring sowing), Huayu 40 produced an average kernel yield of 3057.15 kg ha^{-1}, outyielding the national control Huayu 19 by 8.22%. Having passed the 2 years' regional evaluation test in northern China, Huayu 40 is eligible to enter the national peanut pre-release final yield evaluation test to be performed in May-September, 2012.

In a trial conducted in 2011 to select suitable peanut cultivars for northeast China, Huayu 40 showed high and stable yields across all locations (Prof Hua Yuan Gao, Hong Bo Yu, Shu Li Kang, unpublished data).

2.2. Seed treatment with chemical mutagens

Other lines with high productivity derived from chemical mutagenized peanut cultivars have also been bred. In primary yield evaluation test conducted in 2011 in Laixi, Shandong, three mutant lines derived from Huayu 22 (a Virginia type peanut cultivar) seeds performed well. 11-L36, a line developed through treatment of Huayu 22 peanut seeds with 0.39% sodium azide (NaN_3), outyielded the local control Fenghua 1 by 27.04% (kernel yield). 11-L39 and 11-L40, both bred through treatment of Huayu 22 peanut seeds with 0.39% diethyl sulphate (DES), had 37.60% and 22.60% more kernel yield than Fenghua 1. These promising lines are to be tested further for yield stability over years and locations.

3. Utility of mutagenesis in peanut breeding for better quality

3.1. NIRS to select quality materials from EMS mutagenized peanut populations

Previously, with the help of NIRS, a peanut plant (M_2) with elevated oleate content was selected from sodium azide mutagenized Huayu 22 seeds [14]. In autumn 2011, single seeds (M_5) with over 70% oleate content was identified by NIRS and further confirmed by GC (gas chromatography).

At SPRI, the NIRS calibration equations has also been successfully used to identify individual single peanut seeds with high oleate, high oil or high protein from EMS mutagenized peanut populations. In the experiment, 2 Virginia type peanut cultivars with desirable external traits for export, viz., LF 2 and Huayu 22 (Table 2), were chosen for mutagenic treatment with a hope to develop peanut cultivars with improved quality attributes and comparable or even higher productivity. The seeds in mesh bags (1140 seeds for each genotype) were soaked in tap water for 4 hours. Just prior to EMS treatment, 0.5%, 1.0% and 1.5% EMS solutions were prepared in 0.1 M phosphate buffered saline (PBS) (pH 7.0). The pre-soaked seeds were then treated with EMS (5ml EMS solution per seed) for 2 hours with continuous agitation. After treatment, the seeds were thoroughly washed in running water for 2 hours. Untreated seeds of LF 2 and Huayu 22 were used as controls. After 30 min of air drying, the mutagenized seeds and the untreated controls were sown (one seed per hill) in twin-row seedbeds with 80 cm bed spacing, 30 cm inter-row spacing and 16.67 cm within-row inter-plant spacing under polythene mulch on 4 May 2010 at the Nianzhitou Experimental Plots, Laixi, China. Routine cultural practices were followed [18]. Since rainfall was adequate during the crop season, no irrigation was applied. Peanut was harvested on 12 September 2010.

As shown in Table 2, EMS treatment of peanut seeds significant influenced the percentage of fertile plants, and the productivity of resultant M_1 plants as well (data not shown). With the increase of EMS concentration from 0.5 % to 1.5%, the number of fertile peanut plants harvested generally decreased; so did the number of plants producing adequate seeds as bulk seed samples for NIRS scanning (Table 2).

Oleate, protein and oil contents of sun dried peanut seeds (M_2) from individual single peanut plants (M_1) were predicted by NIRS using the calibration equations for bulk seed samples [13,14], provided that seeds were enough for the rotating sampling cup of the NIRS machine (Matrix-I, Bruker Optics, Germany). Each seed sample was measured once. Only the seeds from individual single plants predicted as with >58% oleate, >55% oil or >28%

protein were used for further analysis. Individual single seeds from selected single plants were then scanned with the same NIRS machine using a small cup for a single seed. Each seed sample was measured 3 times. Oleate, oil and protein contents were predicted by NIRS equations for single intact peanut seeds [13,14]. For comparison, seed samples (at least 30 seed samples from individual single plants and at least 30 intact single seeds for each genotype) of the untreated controls were analyzed with NIRS. The number of M_1 peanut plants with >60% oleate, >55% oil, or >27% protein predicted by NIRS calibration equations for bulk seed samples and the number of single intact M_2 seeds from these plants that had over 58% oleate, >55% oil for LF 2 and >58% oil for Huayu 22, or >28% protein, predicted by NIRS calibration equations for single intact seeds were listed in Table 3 and Table 4. A large number of single seeds selected had quality traits going far beyond the variation scope of the controls (Table 5). There were marked genotypic effects on quality of M_2 peanut seeds. For LF2-derived populations, 8, 118 and 11 M_2 seeds were predicted to have >60% oleate, >55% oil, and >28% protein by NIRS, respectively; of the 3 treatments, 1.0% EMS produced the largest number of quality materials, as far as oleate, oil and protein contents were concerned (Table 3). For Huayu 22, a total of 14, 70 and 23 M_2 seeds were predicted to have >60% oleate, >58% oil, and >28% protein by NIRS, respectively; a large portion of high oleate/ protein seeds (M_2) were from 0.5% EMS treatment, while 1.5% EMS was most suitable for induction of high oil mutations (Table 4). The frequency of high oil single seeds was generally higher than that of high oleate/protein single seeds (Table 3 and Table 4).

EMS (%)	LF2			Huayu 22		
	NST	NFPH	NPS(%)	NST	NFPH	NPS
0.5	1140	717(62.89)*	517(45.35)	1140	839(73.60)	523(45.88)
1.0	1140	740(64.91)	413(36.23)	1140	588(51.58)	488(42.81)
1.5	1140	425(37.28)	231(20.26)	1140	422(37.02)	324(28.41)

*Values in parenthesis are percentages of NST.

Table 2. No. of seeds treated with EMS (NST), no. of fertile plants harvested (NFPH) and no. of plants scanned by NIRS (NPS)

EMS (%)	No. of quality plants predicted by NIRS			No. of quality seeds predicted by NIRS		
	Oleate (>58%)	Oil (>55)	Protein (>27%)	Oleate (>60%)	Oil (>55%)	Protein (>28%)
0.5	6	2	1	1	33	5
1.0	7	7	13	6	70	6
1.5	4	1	1	1	15	0

Table 3. No. of quality plants (M_1)/seeds (M_2) identified by NIRS in mutagenized LF 2 populations

EMS (%)	No. of quality plants predicted by NIRS			No. of quality seeds predicted by NIRS		
	Oleate (>58%)	Oil (>55%)	Protein (>27%)	Oleate (>60%)	Oil (>58%)	Protein (>28%)
0.5	4	2	1	11	15	19
1.0	6	4	5	2	14	2
1.5	10	4	6	1	41	2

Table 4. No. of quality plants (M_1)/seeds (M_2) identified by NIRS in mutagenized Huayu 22 populations

Untreated control		Single plant			Single seed		
		Oleate (%)	Oil (%)	Protein (%)	Oleate (%)	Oil (%)	Protein (%)
LF 2	Min.	39.80	47.76	24.10	29.26	47.41	20.94
	Max.	51.53	51.46	26.52	39.70	53.19	27.70
	Mean	50.54	49.30	24.56	38.96	50.00	21.00
Huayu 22	Min.	49.28	47.58	23.20	30.96	51.63	20.75
	Max.	56.86	52.24	26.77	52.25	56.99	26.63
	Mean	50.12	49.97	25.90	50.30	55.50	23.82

Table 5. Variation ranges and averages of oleate, oil and protein content in bulk seeds from single plants and in single seeds of two peanut cultivars as predicted by NIRS

Some seeds (M_2) with high oleate, oil or protein content predicted by NIRS were sampled and sent to Food Supervising and Testing Centre (Wuhan), China to analyze their quality by standard methods. For quality analysis by conventional means, a small seed portion of selected single seeds, distal to embryo end, weighing no less than 100 mg, was cut off and sent. Oleate, oil and protein contents were determined using GC for determination of fatty acids, Soxlet oil extraction method and Kjeldhal nitrogen determination procedure, respectively. A conversion factor of 5.46 was used to convert the amount of nitrogen to amount of proteins in the samples.

High oleate trait of 5 M_2 seeds selected by NIRS were confirmed by GC analysis; oleate content in these seeds was not less than 60.0%, significantly higher than that in untreated seeds (ck) (Table 6 and Table 5). Similarly, several high oil/protein M_2 seeds were also analyzed with conventional methods, and all of them were found to contain over 60% oil or over 29% protein, much higher than untreated controls (Table 7 and Table 8).

EMS (%)	Cultivar	Seed serial no.	NIRS	GC
0 (ck)	LF 2	LF 2-1-15	38.1	44.2
1.0	LF 2	E2-4-256-24	60.5	62.9
1.0	LF 2	E2-4-178-18	62.4	66.4
1.0	LF 2	E2-4-83-12	63.6	64.3
0(ck)	Huayu 22	HY22-2-10	34.9	47.3
0.5	Huayu 22	E1-3-173-10	63.8	67.3

Table 6. Oleate content (%) in selected single peanut seeds (M_2)

EMS (%)	Cultivar	Seed serial no.	NIRS	Soxhlet extraction
0(ck)	Huayu 22	HY22-2-10	55.5	52.82
0.5	Huayu 22	E1-3-343-24	60.2	61.49
0.5	Huayu 22	E1-3-343-25	60.8	62.38

Table 7. Oil content (%) in selected single peanut seeds (M_2)

EMS (%)	Cultivar	Seed serial no.	NIRS	Kjeldahl nitrogen determination
0 (ck)	Huayu 22	HY22-2-10	20.8	19.5
0.5	Huayu 22	E1-3-180-6	29.0	31.35
0.5	Huayu 22	E1-3-180-10	28.6	30.49

Table 8. Protein content (%) in selected single peanut seeds (M_2)

M_2 seeds that were identified as with high oil, protein or oleate were sown and resultant M_3 seeds were harvested. NIRS analysis of seeds (M_3) from some of the single plants showed that a portion of the single plants kept the quality trait(s), while others did not. For example, Q3-5-23 (plant no. of 2011) was grown from an M_2 seed (seed no. of 2010: E1-3-343-8) with 57.78% oil, and 23 seeds of the plant were analyzed by NIRS, of which 18 contained higher than 55% oil, with the highest being 58.71%. Q3-12-24 (plant no. of 2011), grown from a seed with 62.97% oleate (seed no. of 2010: E1-2-11-16), still had 60.55% oleate in M_3 generation.

To summarize, the present study demonstrated the successful use of NIRS in selecting a limited number of peanut breeding materials with desired altered quality characters from large populations of M_2. Some of the single seeds with high oil, protein or oleate content lost the quality trait(s) in subsequent generation; these quality traits appeared to be caused by physiological abnormalities rather than genetic changes. However, there were other M_3 seeds with quality traits inherited from previous generation, which could be ascribed to mutations in related genes. We have reported a high oleate EMS mutant of LF 2 (2010 seed no.: E2-4-83-12), an output of the present study, that had dysfunctional mutated *FAD2A* (G448A in the coding region) and *FAD2B* (C313T in the coding region) [4]. Despite the uncertainty in quality traits, selection in M_2 seeds is still necessary as it helps to reduce the population size of M_2 plant and hence M_3 seed generations.

In a separate study, we identified several peanut induced/natural mutants with over 70% oleate content, and used them in a hybridization program [3]. Thus far, a great number of high oleate lines including those with double "high"- high oleate and high yield, or high oleate and high shelling percentage, have been tentatively developed. Notably, some of the large seeded lines had a high shelling outturn when planted in Sanya, Hainan province (located in tropical zone), and some lines consistently exhibited high yield when planted in Laixi (located in temperate zone) and Sanya.

3.2. A bold-seeded peanut natural mutant of *A. duranensis* with high oil content

Breeding high oil peanut cultivars especially those with large seeds is most challenging, and variations in oil between years and locations may be quite high. In fact, in China, several high oil peanut cultivars have been released in Hebei, Hubei and Henan, but all of these peanut cultivars were found to only contain less than 54% seed oil when planted in Shandong peninsula. Availability of high oil genetic resources is of vital importance to the genetic improvement of the cultivated peanut in this region and the like. High oil peanut wild species have been reported by several research groups, but none of them possessed large seeds. A variant of a wild peanut species, *Arachis duranensis* PI 262133, was identified for the first time as with high oil content and large seeds at SPRI.

Cloning and sequencing of the rDNA ITS internal transcribed spacer (ITS) sequences from *A. duranensis* and the bold-seeded genotype detected no difference in their ITS sequences, suggesting that the variant was unlikely to be a result of natural hybridization.

The mutant was identified in a separate square cement block allotted to *A. duranensis* in the wild peanut nursery at SPRI Experiment Station, Laixi. The plant was grown from a seed of similar size to that of *A. duranensis*, but was found to possess thicker branches, and larger leaflets, pods and seeds (Figure 1, Table 9). The size and weight of pods and seeds of wild and mutant type *A. duranensis* significantly differed (P<0.01).

Figure 1. Pods and seeds of mutant type (top 2 rows) and wild type (bottom 2 rows) of *A. duranensis*

Trait		Mean	SD
Pod weight (g)	Mutant type	0.93	0.24
	Wild type	0.24	0.05
Pod length (cm)	Mutant type	2.12	0.23
	Wild type	1.28	0.10
Pod thickness (cm)	Mutant type	1.30	0.16
	Wild type	0.75	0.06
Pod width (cm)	Mutant type	1.11	0.12
	Wild type	0.70	0.07
Seed weight (g)	Mutant type	0.67	0.17
	Wild type	0.19	0.05
Seed length (cm)	Mutant type	1.67	0.20
	Wild type	1.03	0.09
Seed thickness (cm)	Mutant type	0.92	0.13
	Wild type	0.57	0.07
Seed width (cm)	Mutant type	0.70	0.08
	Wild type	0.55	0.06

Table 9. Difference in pod and seed size and weight between wild and mutant type *A. duranensis*

Totally 110 seeds were analyzed by NIRS to predict their oil content (Figure 2). Of them, 40 contained more than 55% oil, with 60.33% being the highest. The average oil content of the mutant was 54.23%, as against 52.62% in the wild type *A. duranensis*.

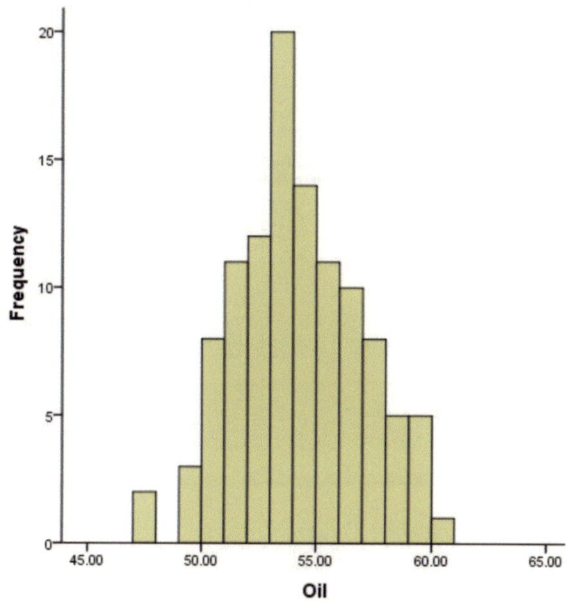

Figure 2. Frequency distribution of seed oil content (%) in 110 seeds of the *A. duranensis* mutant

According to latest classification proposed by Krapovickas and Gregory (1994), the genus *Arachis* is devided into 9 sections, consisting of more than 80 species [19]. Of them, *A. hypogaea* L. is widely cultivated for oil extraction and food uses. Other *Arachis* species of economic importance include *A. glabrata, A. pintoi,* and *A. duranensis*, which are being utilized as forage and/or ground cover [20]. Wild species have higher genetic diversity than the cultivated peanut, providing a desirable source for stress resistance, high oil, protein, oleate, amino acid content as well as high yield factors [20].

Whether the mutant was resulting from chromosome doubling or gene mutation is still unknown. Chromosome counting is absolutely necessary. Anyhow, the mutant reported here is of relevance both to the genetic improvement of the cultivated peanut and to its direct utilization as forage and/or ground cover.

4. Conclusion

Our study on peanut mutagenesis clearly demonstrated that it was possible to induce mutations in productivity, oleate, oil and protein content in the cultivated peanut. NIRS may facilitate identification of induced/natural mutants with improved quality characters. Although not all of the variations in quality traits of M_2 seeds are inheritable, selection in this generation may help to reduce the size of M_3 seed population.

Author details

Chuan Tang Wang*, Yue Yi Tang, Xiu Zhen Wang and Qi Wu
Shandong Peanut Research Institute (SPRI), Qingdao, China

Hua Yuan Gao and Tong Feng
Institute of Economic Plants, Jilin Academy of Agricultural Sciences, Gongzhuling, China

Jun Wei Su and Shu Tao Yu
Sandy Land Amelioration and Utilization Research Institute of Liaoning,Fuxin, China

Xian Lan Fang
South Jiangxi Academy of Sciences, Ganzhou, China

Wan Li Ni
Institute of Crop Plants, Anhui Academy of Agricultural Sciences, Hefei, China

Yan Sheng Jiang
Weifang Academy of Agricultural Sciences, Weifang, China

Lang Qian
Dalian Academy of Agricultural Sciences, Dalian, China

Dong Qing Hu
Qingdao Entry-Exit Inspection and Quarantine Bureau, Qingdao, China

Acknowledgement

The authors thank the financial support from China Agricultural Research System (Grant No. CARS-14), Ministry of Agriculture, China, Qingdao Science & Technology Support Program (Grant No. 10-3-3-20-nsh, Grant No. 09-1-3-67-jch), and Jilin Natural Science Foundation (Grant No. 20101580).).

5. References

[1] McGill JF. Economic importance of peanuts. In: Peanut-Culture and uses, a symposium. American Peanut Research and Education Association. 1973. p.3-16.

[2] Rao VR. Chapter 3. Botany. In: Reddy PS. (ed.) Groundnut. Indian Council of Agricultural Research (ICAR). 1988. p. 24-64.

[3] Wang CT, Hu DQ, Ding YF, Yu HT, Tang YY, Wang XZ, Zhang JC, Chen DX. A new set of allelespecific PCR primers for identification of true hybrids in normal oleate × high oleate crosses in groundnut. Journal of SAT Agricultural Research 2011; 9. ejournal.icrisat.org/Volume9/Groundnut/A_new_set.pdf (accessed 6 June 2012)

* Corresponding Author

[4] Fang CQ, Wang CT, Wang PW, Tang YY, Wang XZ, Cui FG, Yu SL. Identification of a novel mutation in *FAD2B* from a peanut EMS mutant with elevated oleate content. Journal of Oleo Science 2012; 61(3):143-148.

[5] Hammons RO. Peanuts: genetic vulnerability and breeding strategy. Crop Science 1975; 19(4): 527-530.

[6] Wang CT, Yang XD, Xu JZ, Liu GZ. Differences in pod characters among L7-1 peanut cultivar and its chemical mutants. International *Arachis* Newsletter 2006; 26:12-14.

[7] Zhang JC, Wang CT, Tang YY, Wang XZ. Effects of grading on the main quality attributes of peanut kernels. Frontiers of Agriculture in China 2009; 3(3):291-293.

[8] Fox GP, Cruickshank A. Near infrared reflectance as a rapid and inexpensive surrogate measure for fatty acid composition and oil content in peanuts (*Arachis hypogaea* L.). Journal of Near Infrared Spectroscopy 2005; 5, 287-291.

[9] Yu SL, Zhu YJ, Min P, Liu H, Cao YL. NIRS to determine protein and oil contents in peanut kernels. Journal of Peanut Science 2003; 32 (Supplement):138-143.

[10] Yu SL, Zhu YJ, Min P, Liu H, Cao YL, Wang CT, Zhang CS, Liu X, Zhou XQ. NIRS to determine major fatty acids contents in peanut kernels. In: Wan SB. (ed.) High Quality and High Output Peanuts: Principle and techniques for Production. Beijing: China Agricultural Science & Technology Press; 2003. p. 344-349.

[11] Tillman B, Gorbet D, Person G. Predicting oleic and linoleic acid content of single peanut seeds using near-infrared reflectance spectroscopy. Crop Science 2006; 5:2121-2126.

[12] Sundaram J, Kandala CV, Butts CL. Application of near infrared spectroscopy to peanut grading and quality analysis: overview. Sensing and Instrumentation for Food Quality and Safety 2009; 3,156-164.

[13] Wang CT, Yu SL, Zhang SW, Wang XZ, Tang YY, Zhang JC, Chen DX. Novel protocol to identify true hybrids in normal oleate x high oleate crosses in peanut. Electronic Journal of Biotechnology 2010; 13(5). http://dx.doi.org/10.2225/vol13-issue5-fulltext-18 (accessed 6 June 2012)

[14] Wang CT, Tang YY, Wang XZ, Zhang SW, Li GJ, Zhang JC, Yu SL. Sodium azide mutagenesis resulted in a peanut plant with elevated oleate content. Electronic Journal of Biotechnology. 2011; 14(2). http://dx.doi.org/10.2225/vol14-issue2-fulltext-4 (accessed 6 June 2012)

[15] Wang CT, Wang XZ, Tang YY, Chen DX, Zhang JC, Cui FG, Yu SL. High yielding mutants achieved by injecting EMS into peanut flower organs. Journal of Nuclear Agricultural Sciences 2010; 24(2):239–242.

[16] Wang CT, Wang XZ, Tang YY, Zhang JC, Chen DX, Xu JZ, Yang XD, Song GS, Cui FG. Huayu 40, a groundnut cultivar developed through EMS mutagenesis. Journal of SAT Agricultural Research 2011; 9. ejournal.icrisat.org/Volume9/Groundnut/Huayu40.pdf (accessed 6 June 2012)

[17] Zheng J, Zhang Z, Chen L, Wan Q, Hu M, Wang W, Zhang K, Liu D, Chen X, Wei X. IT-ISJ marker and its application in construction of upland cotton linkage map. Scientia Agricultura Sinica 2008; 41(8):2241-2248.

[18] Wan SB, editor. Peanut cultivation Science in China. Shanghai: Shanghai Science and Technology Press. 2003. 647pp.

[19] Krapovikas A, Gregory WC. Taxonomia del genero Arachis (Leguminosae). Bonplandia 1994; 8:1-86.

[20] Yu SL, Wang CT, Yang QL, Zhang DX, Zhang XY, Cao YL, Liang XQ, Liao BS, editors. Peanut Genetics and Breeding in China. Shanghai: Shanghai Science and Technology Press; 2011. 565pp.

Chapter 2

Mono-Ubiquitination of Nuclear Annexin A1 and Mutagenesis

Fusao Hirata, George B. Corcoran and Aiko Hirata

Additional information is available at the end of the chapter

http://dx.doi.org/10.5772/50800

1. Introduction

Annexin A1, a protein previously termed as lipomodulin and lipocortin, is a member of the protein family that binds to phospholipids in a Ca^{2+} dependent manner (Hirata, 1998; Gerke & Moss, 2002; Lim & Pervaiz, 2007). This protein was first discovered as a phospholipase A_2 inhibitory protein, and from its chemical nature was thought to be closely associated with membrane functions such as membrane organization, trafficking and metabolism (Hirata, 1998; Gerke & Moss, 2002; Lim & Pervaiz, 2007). On the other hand, annexin A1 is a major substrate of oncogenic kinases such as c-met and c-src, and is thus, proposed to be involved in signal transduction of growth factors and mitogens (Hirata et al., 1984; Skouteris & Schröder, 1996). Therefore, this protein is thought to have some regulatory roles in cancer development. Indeed, certain types of cancers such as hepatoma and pancreas cancers have higher levels of annexin A1 (Lim & Pervaiz, 2007). However, transfection of cDNA encoding annexin A1 often results in apoptosis of cells or interference of cell proliferation, consistent with tumor suppressing functions (Debret et al., 2003; Hsiang et al., 2006). In keeping with this interpretation, some types of cancers such as esophageal carcinoma and prostate cancer have decreased levels of annexin A1 (Lim & Pervaiz, 2007). However, recent pathohistochemical evidence with esophageal carcinoma and neck squamous carcinoma suggests that such down-regulation of annexin A1 is partially attributed to nuclear translocation, and the nuclear translocation of annexin A1 is facilitated by tyrosine and/or serine phosphorylation and Ca^{2+} signals as well as by oxidative stress (Rhee et al., 2000; Kim et al., 2003; Liu et al., 2003; Cui et al., 2007; Lin et al., 2008). The presence of annexin A1 in nuclei is now proposed to be a poor prognostic marker of squamous cancer or to be associated with malignancy of gastric carcinoma, while changes in cellular expression of annexin A1 may not be involved in tumorigenesis (Lin et al., 2008; Zhu et al., 2010). Therefore, nuclear annexin A1 is thought to play an important role in cell proliferation and/or cell transformation. Since this protein is reported to reside on DNA synthesomes within nuclei (Lin et al., 1997), it is likely that nuclear annexin A1 is involved in DNA replication, especially DNA damage induced gene mutation, since DNA damage induced mutagenesis plays an important role in tumorigenesis.

Mutagenesis is largely the outcome of insults to DNA by environmental agents including alkylating agents and by endogenous metabolic oxidative metabolites such as reactive oxygen, and plays an important role in initiation, progression, and ultimately formation of cancer (Wang, 2001). Heavy metals such as As^{3+} are known to promote the mutagenic action of another DNA damaging agent including reactive oxygen, while they alone are weakly- or non-mutagenic (Sekowski et al., 1997; Maier et al., 2002). Such promotion by heavy metals is attributed not only to inhibition of DNA repair systems such as mismatch repair but also to relaxation of the semi-conservative replication machinery for translesion DNA synthesis that bypasses sites of damage (Calsou et al., 1996; Jin et al., 2003). Translesion DNA synthesis is catalyzed by error-prone DNA polymerases, and exchange of DNA polymerases is promoted by ubiquitination of nuclear proteins such as proliferating cell nuclear antigen (PCNA) (Ulrich, 2005). Translesion DNA synthesis is thought to be the major cause of mutagenesis rather than incorrect repair of damaged DNA (Kunz et al., 2000). Accordingly, we have investigated how annexin A1 in nuclei stimulates DNA damage-induced mutagenesis.

2. Modifications of annexin A1 in nuclei with ubiquitin and ubiquitin-like proteins

Ubiquitin and ubiquitin-like modification systems are related pathways that covalently attach a protein modifier to a lysine residue of a target protein. Ubiquitin classically marks proteins for proteosomal destruction typically when polymeric chains (longer than four ubiquitin subunits) assemble *via* ubiquitin-ubiquitin isopeptide-linkages (Gill, 2004; Chen & Sun, 2009). However, other functions of ubiquitin have been recently discovered that do not involve the proteosome (Hicke, 2001; Chen & Sun, 2009). The small ubiquitin-related modifier, SUMO, post-translationally modifies many proteins with diverse functions including regulation of transcription, chromatin structure and DNA repair, and facilitates their nuclear translocation (Gill, 2004). In addition, repair of and tolerance to DNA damage are regulated by modifications with ubiquitin and SUMO (Ulrich, 2005), and their modifications are shown to have antagonistic effects on functions of the target proteins (Hilgarth et al., 2004; Huang & D'Andrea, 2006).

Since annexin A1 contains the consensus sequence, ψKxE/D, for SUMOylation, purified bovine annexin A1 was incubated with human recombinant Ubc9 (E2) and SAE I/SAE II (E1) in the presence of SUMO 1, 2 or 3 to test whether annexin A1 can be SUMOylated. When the reaction mixtures were analyzed by Western blots, monospecific anti-SUMO antibodies stained two proteins with apparent molecular weights around 38,000 and 34,000 Da (F. Hirata et al., 2010). Anti-annexin A1 antibody detected a broad single protein band with an apparent molecular weight around 37,000 Da but not a protein band with a molecular weight of 34,000 Da. The protein band with a molecular weight of 34,000 was identified, using anti-Ubc9 antibody as SUMOylated Ubc9 (F. Hirata et al., 2010). With SUMO 2 or 3 as a substrate, SUMOylation of annexin A1 was apparently facilitated. Rates of annexin A1 SUMOylation with SUMO 1, 2 or 3 were approximately 1:3:5, providing that the amounts of SUMOylated Ubc9 formed with SUMO 1, 2, or 3 were essentially the same. Ca^{2+} was required for the maximal modification, and increased SUMOylation by 2.6 fold *vs.* without Ca^{2+}. Therefore, we concluded that annexin A1 was conjugated with SUMOs under these conditions. While the SUMOylation barely altered the molecular weight of bovine annexin A1 (37,000 Da) as detected by anti-annexin A1 antibody, the protein band with an apparent molecular weight around 38,000 Da could not be detected by anti-SUMO antibody in the absence of E1 or E2 for SUMOylation or in the absence of annexin A1. To further confirm that annexin A1 is covalently modified with SUMO, the reaction mixtures in the absence and presence of annexin A1 for SUMO 3 modification were scaled up by 5 fold, and the incubation was continued overnight. Then, the reaction mixtures were analyzed using FLPC (Amersham Biosciences) with a MiniQ column. Native and SUMOylated annexin A1 were separately eluted as measured by conductance, suggesting that charges in native and SUMOylated annexin A1 are distinct. To establish this contention, 2D electrophoresis was performed. Native annexin A1 was detected at pI 6.4. The reaction mixture containing SUMOylated annexin A1 showed a new annexin A1 location in the pI 6.1 area as detected by anti-annexin A1 antibody, while its mobility on SDS gel electrophoresis was barely shifted. This new protein with pI 6.1 was stained with anti-SUMO antibody as well as with anti-annexin antibody, and its density increased, when the incubation was prolonged. Omission of SUMO or ATP from the reaction mixture resulted in no protein around pI 6.1. These observations supported the conclusion that annexin A1 was covalently modified with SUMO, even though no significant mobility shift on SDS electrophoresis was detected after the modification of annexin A1 with SUMOs. Since ^{160}LKRD in the annexin repeat domain II is conserved among mammalian annexin A1 proteins (Gerke & Moss, 2002), it is likely that SUMOylation takes place in the core domain II. Mutation experiments by mutating K to R will be essential for determination of the site of SUMOylation.

On the other hand, the molecular weight of annexin A1 shifted from 37,000 Da to approximately 45,000 Da after ubiquitination. The ubiquitination of annexin A1 required UbcH2A (Rad6 homologue) together with HeLa S100 lysate that contained E3 ubiquitin ligases. Since HeLa S100 lysate pretreated with anti-Rad 18 antibody did not catalyze ubiquitination of annexin A1, it is most likely that Rad18 is an E3 ligase for ubiquitination of annexin A1. Ca^{2+} was required for the maximal ubiquitination, but its stimulation was not as much as seen with SUMOylation, when amounts of ubiquitinated annexin A1 were adjusted with the total amounts of annexin A1 (free and ubiquitinated annexin A1). These observations suggest that the modification site with ubiquitin is distinct from that with SUMO. HeLa S100 lysate also contained annexin A1 as detected by anti-annexin A1 antibody, but under the present experimental conditions, no significant ubiquitination of endogenous annexin A1 was detected in the presence of ubiquitin and an ATP generating system without UbcH2A. UbcH2A could be equally replaced by its related enzyme, UbcH2B but not by E2-25K. These observations suggest but do not necessarily prove that annexin A1 is ubiquitinated by the Rad6–Rad18 system which is closely associated with response to DNA damage (Kunz et al., 2000; Ulrich, 2005). The difference in stimulation of SUMO and ubiquitin conjugation by Ca^{2+} is apparently attributed to sites of modification and Ca^{2+} induced conformational changes, in which the N-terminal domain is exposed and flexibility of the core domain residues are increased by Ca^{2+} (Shesham et al., 2008). Therefore, we suggest that SUMOylation takes place in the core domain regions, while ubiquitination takes place outside the core domain regions.

3. DNA damage and modification of annexin A1

Among post-translational modification systems, ubiquitin and ubiquitin-like molecules including SUMO are a unique family of protein modifiers that play pivotal roles in regulation of protein stability and function (Hicke, 2001; Gill, 2004; Hilgarth et al., 2004; Huang & D'Andrea, 2006; Chen & Sun, 2009). SUMOylation is involved in protein stabilization, nucleo-cytoplasmic trafficking, cell cycle regulation, maintenance of genome integrity and transcription. Indeed, annexin A1 present in nuclei is mostly modified with SUMO or ubiquitin (F. Hirata et al., 2010). SUMO 1-modified annexin A1 resulted in enhanced helicase activity, while SUMO2/3 were the better substrates for *in vitro* annexin A1 conjugation, suggesting that SUMOylated annexin A1 might be involved in cell proliferation and differentiation (Yang & Paschen, 2009). On the other hand, mono-ubiquitination is thought to enable the modified proteins to interact and form complexes with other proteins via ubiquitin binding proteins and ubiquitin receptors. Mono-ubiquitination changes subcellular localization and alters certain structural and targeting properties, while polyubiquitination targets proteins for the degradation pathway (Huang & D'Andrea, 2006). Mono-ubiquitination by the Rad6-Rad18 system is proposed to play an important role in DNA damage response (Ulrich, 2005).

To ask whether mono-ubiquitination of annexin A1 is, indeed, involved in DNA damage response, we investigated if nuclear annexin A1 is modified with SUMO or ubiquitin in mouse L5178Y *tk*(+/-) lymphoma cells treated with DNA damaging agents, 15 μM MMS or 3 μM $AsCl_3$ for 3 or 6 hr. Under these conditions, the mutation rate of the thymidine kinase gene increased from 23 x 10^{-6} (vehicle control) to 67 x 10^{-6} and 104 x 10^{-6} with $AsCl_3$ or MMS, respectively, as measured by the number of colonies with trifluorothymidine resistance according to the method described by Honma et al. (1999). These observations suggest that DNA damage was induced under these conditions. After 3 hr treatments with 15 μM MMS or 3 μM As^{3+}, nuclear and cytoplasmic extracts were isolated and were analyzed by Western blots with anti-annexin A1 antibody. Nuclear annexin A1 was increased by the treatments, while cytoplasmic annexin A1 was decreased. These observations suggest nuclear translocation of annexin A1 following DNA damage signaling. Nuclear annexin A1 exhibited apparent molecular weights around 38,000 and 45,000 Da, whereas the molecular weights of cytoplasmic annexin A1 were 37,000, 30,000 and 27,000 Da (F. Hirata et al., 2010). Annexin A1 with the molecular weights of 30,000 and 27,000 Da are reported as products of N-terminal cleavage (Kim et al., 2003; Sakaguchi et al., 2007). Such cleavage was also detected in nuclear annexin A1. Polyubiquitination of annexin A1 has recently been shown to be catalyzed by E6AP in the presence of Ca^{2+}, and polyubiquitinated annexin A1 is degraded by proteosomes (Shimoji et al., 2009). Although poly-ubiquitination of proteins is proposed to be a major pathway of protein degradation (Hilgarth et al., 2004; Huang & D'Andrea, 2006; Wilkinson et al., 2008), poly-ubiquitinated annexin A1 was not detected in either cytosol or nuclei. The direct cleavage of annexin A1 was thought to be its major degradation pathway under our experimental conditions. The Western analysis of the extracts with anti-SUMO 1 and anti-ubiquitin antibodies suggested that majority of nuclear annexin A1 might be modified, while cytoplasmic annexin A1 was not. MMS increased ubiquitination of annexin A1 in nuclei by 2 fold, whereas these treatments decreased its SUMOylation by 70%. $AsCl_3$ alone was less effective but was more than additive, when MMS was present. Cytosolic annexin A1 was not stained with anti-SUMO 1 or anti-ubiquitin antibodies (data not shown). Accordingly, we concluded that the modifications of annexin A1 with SUMO or ubiquitin facilitate its nuclear translocation and that ubiquitinated annexin A1 is involved in DNA damage response, in which the Rad6-Rad18 system plays an important role (Ulrich, 2005).

4. DNA helicase activity of nuclear annexin A1

Annexins have a common internal structure comprising 4 or 8 repeats of a conserved 70 amino acid domain, and differ primarily in the length and composition of the amino-terminal domains (F. Hirata, 1998; Gerke and Moss, 2002; Lim and Pervaiz, 2007). Since this amino-terminal domain contains the sites for phosphorylation and glycosylation, it is considered a regulatory domain. A defining feature of annexins is their ability to bind, in a Ca^{2+}-dependent manner, to negatively charged phospholipids such as phosphatidylserine (PtdSer). This functional feature is attributed to the conserved C-terminal domain, and is essential for biological functions of the annexins. The cell-cycle dependent existence of annexin A1 and A2 in nuclei suggests a close association with nuclear functions, while they are major substrates of the oncogenic tyrosine kinases, met and src (Katoh et al., 1995; Rydal et al., 1992). Thus, annexin A1 and A2 are proposed to be a biological marker of proliferating cells (cancer cells) (Masaki et al., 1994). In accord with this notion, the treatments of A347 and HeLa cells with antisense annexin A1 and A2 oligonucleotides reduce the synthesis and subsequent phosphorylation at tyrosine of the annexins, thereby inhibiting cell proliferation (Kumble et al., 1992; Skouteris & Schröder, 1996). Translocation of annexins from cytosol to nuclei apparently requires their phosphorylation at tyrosine and Ca^{2+} signaling (Mohiti et al., 1997). Purified annexin A1 and A2 can stimulate DNA synthesis in cell free systems of HeLa cells, Xenopus oocytes and rat hepatocytes (Vishwanatha & Kumble, 1992; Vishwanatha et al., 1992; Tavokoli-Nezhad et al., 1998). In addition, annexin A1 is present in DNA synthesomes and annexin A2 is located in nuclear matrix (Kumble et al, 1992; Lin et al., 1997). These observations strongly suggest that nuclear annexins regulate DNA replication.

Annexin A2 functions in DNA replication as a primer recognition factor for Pol α (Vishwanatha & Kumble, 1993), while this protein is also reported to be an RNA binding protein, interacting with c-myc (Fillipenko et al., 2004). The binding of annexin A2 to RNA and DNA requires Ca^{2+}. Accordingly, it was thought that DNA and RNA bind to acidic phospholipid binding sites via ionic interaction. Our laboratory has investigated details of annexin A1 binding to RNA and DNA (Hirata & Hirata, 1999; 2002). Annexin A1 purified from rat liver nuclei binds to purine clusters in RNA, while it preferentially binds to pyrimidine clusters in DNA. The size of maximal recognition for binding was around 20-25 nt. Since phospholipids, especially acidic phospholipids such a phosphatidylserine, enhanced DNA and RNA binding, the binding of annexin A1 to RNA and DNA was not due to simple ionic interactions, and the sites for binding to phospholipids and DNA/RNA were distinct. Indeed, the RNA binding site of annexin A2 was reported to be C-D helices of Domain IV (Aukrust et al., 2007), while the consensus sequence for phosphatidylserine binding is proposed to be (R/K)XXXK-(B-C helices)-(R/K)XXXXDXXS(D/E) in Domain I and II (Montaville et al., 2002).

Annexins are also reported to interact with ATP and GTP (Bandorowicsz-Pikula & Pikula, 1998), although they do not have consensus sequences for typical ATP binding sites such as Walker A. As seen with annexin A7 that forms ion channels in lipid bilayers, GTP and other nucleotides are thought to regulate Ca^{2+} gating, and/or Ca^{2+} dependent membrane trafficking such as exocytosis (Caohuy et al., 1996). Cotton fiber annexin and N-terminal deleted annexin A1 can hydrolyze GTP in the presence of Mg^{2+}, and Ca^{2+} is not required for this hydrolysis (Hyun et al., 2000; Shin & Brown, 1999). Since annexin A1 can bind not only ssDNA but also dsDNA in the presence of both Mg^{2+} and Ca^{2+}, we examined effects of various DNAs on ATP hydrolysis by annexin A1 (Hirata & Hirata, 2002). dsDNA such as calf thymus DNA and annealed M13mp18 but not ssDNA stimulated ATP hydrolysis by annexin A1. When DNA was analyzed, dsDNA was unwound to form ssDNA, suggesting that annexin A1 has DNA helicase activity (Hirata & Hirata, 2002). Interestingly, its annealing reaction did not require Mg^{2+} nor ATP, but Ca^{2+} was necessary. Therefore, binding of dsDNA requires Mg^{2+}, and that of ssDNA takes place in the presence of Ca^{2+} (Fig. 1).

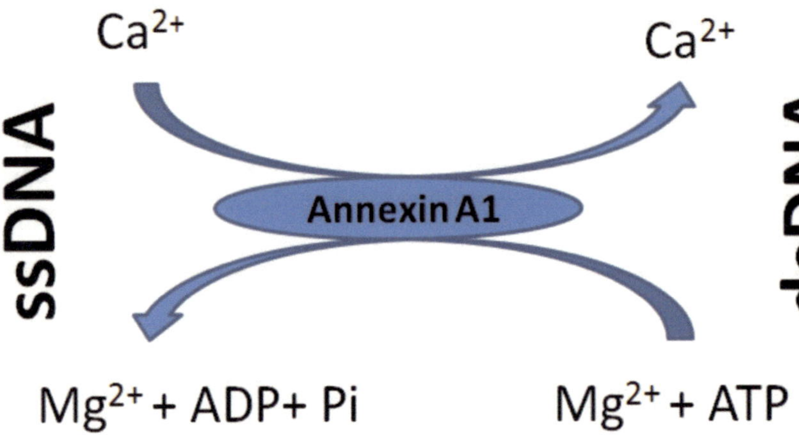

Figure 1. Helicase activity of annexin A1

5. Modification with SUMOylation and ubiquitination and annexin helicase activity

Purified rat nuclear annexin A1 had an apparent molecular weight of approximately 92,000 ± 2,000 Da, and was ubiquitinated as detected by anti-ubiquitin antibody (A. Hirata et al., 2010). Under reducing conditions, its molecular weight was approximately 45,000 ± 1,000 Da on SDS-PAGE, suggesting that high molecular weight annexin A1 is a homodimer of mono-ubiquitinated annexin A1 rather than a heterodimer complex with S100 as previously thought. Since this homodimer with an apparent molecular weight of 92,000 Da exhibited a Ca^{2+}- and Mg^{2+}-regulated helicase activity, we performed helicase assays of the reaction mixtures for ubiquitination and SUMOylation of annexin A1. The reaction mixtures containing annexin A1 modified with SUMO1 and native annexin A1 showed helicase activity as measured by unwinding of dsDNA . While purified native annexin A1 also exhibited low helicase activity, SUMOylated annexin A1 exhibited much higher activity. Since under the present conditions, approximately 10% and 20% of annexin A1 were modified in the absence and presence of Ca^{2+} with SUMO1, we calculated that SUMOylation stimulated the helicase activity of annexin A1 by approximately 3.5 fold (F. Hirata et al., 2010).

HeLa S100 lysate that was required for ubiquitination reactions contained other types of helicases beside annexin A1, and thus increased control activity. Despite this challenge, the reaction mixtures containing mono-ubiquitinated annexin A1 demonstrated clearly enhanced helicase activity (F. Hirata et al., 2010). Assuming that annexin A1 in HeLa S100 exhibits helicase activity equal to purified bovine annexin A1, mono-ubiquitination of annexin A1 results in an approximately 6-fold activation as compared with native annexin A1. Since purified rat nuclear annexin A1 shows 9 fold higher activity than native rat annexin A1 (A. Hirata et al., 2010), these observations suggest that the conjugation of annexin A1 with ubiquitin or SUMO enhances its helicase activity.

6. Heavy metals and annexin A1

Annexin A1 has 3 different types of Ca^{2+} binding sites, type II, type III and type III (AB) (Weng et al., 1993). Type II calcium binding sites have the highest affinity for Ca^{2+}, and are found only at AB loops. The coordination of the type II sites is octahedral. It consists of three peptide oxygens from the AB loops with the (K, R)-(G, R)-X-G-T sequence and bidentate ligands from the acidic groups of either an aspartate or a glutamate residue downstream in the sequence. The remaining two calcium coordinating sites show electro-density for water molecules. The calcium ions at the type III sites coordinate to two backbone carbonyl oxygens and one nearby acidic side chain. Water molecules have been found at most of the remaining three coordinating sites to complete the six-ligand octahedral coordination. The type III sites correspond to the two minor calcium sites labeled by lanthanum in annexin A5.

As seen with EF band calcium binding proteins that have type I Ca^{2+} binding sites, phospholipid aggregation experiments suggest that other divalent metals such as Pb^{2+} and Zn^{2+} can replace Ca^{2+} in annexin A1 (Mel'gunov et al., 2000). Since type III Ca^{2+} binding sites can be labeled with La^{3+}, we tested not only Pb^{2+} and Cd^{2+} but also As^{3+} and Cr^{6+} for DNA helicase assays (A. Hirata et al., 2010). Pb^{2+} alone stimulated the DNA binding activity of purified mono-ubiquitinated nuclear annexin A1 in the absence of Mg^{2+} and Ca^{2+}. Heat denatured nuclear annexin A1 did not exert DNA binding even with these metals. Therefore, we thought that Pb^{2+} and Ca^{2+} were acting in essentially similar manners on mono-ubiquitinated annexin A1. Similar results were observed with Cd^{2+}, suggesting that these divalent metals were able to replace Ca^{2+} for DNA binding activity of annexin A1 as previously shown for its phospholipid binding activity (Mel'gunov et al., 2000). To clarify whether carcinogenic heavy metals such as As^{3+} and Cr^{6+} can promote or block the DNA binding activity of nuclear annexin A1, we tested the effects of As^{3+} on binding of nuclear annexin A1 to P0G, a 80mer polynucleotide that is complimentary to the *ori* of M13mp18 (see below). As^{3+} was synergistic with Ca^{2+} and with Mg^{2+}, yet As^{3+} alone apparently promoted the binding of annexin A1 to ssDNA. With As^{3+}, formation of the nuclear annexin A1-P0G complex increased in a concentration-dependent manner. The concentration of As^{3+} for half-maximal activation was 2.2 μM in the absence of phospholipids. Similar concentration-dependent activation was observed with Pb^{2+}, Cd^{2+}, and Cr^{6+}. Half-maximal binding in the absence of phospholipids was observed at 30, 0.1, and 2 μM for Pb^{2+}, Cd^{2+}, and Cr^{6+}, respectively. Phospholipids increased complex affinity for heavy metals by approximately 10 fold, as seen previously with Ca^{2+} for phospholipid binding (Gerke & Moss, 2002). These observations suggest that carcinogenic divalent and some multivalent heavy metal cations are able to replace Ca^{2+} in the DNA binding activity of mono-ubiquitinated annexin A1. It is noted that Cd^{2+} at concentrations higher than 5 μM caused potent inhibition, possibly due to thiol oxidation of the annexin A1 molecule.

Ca^{2+} facilitates the annealing of C_{20}-P0G to M13mp18 by nuclear annexin A1 (Hirata & Hirata, 2002). Although poly(dC)$_{20}$ was added at the 3′- or at the 5′-end of P0G as a binding site for annexin A1, which demonstrates a higher affinity for poly(dC)$_{20}$ (Hirata & Hirata, 2002), nuclear annexin A1 bound to P0G without a poly(dC)$_{20}$ tail. To test the interpretation that As^{3+} or Cr^{6+} can replace Ca^{2+}, the annealing activity of nuclear annexin A1 was measured in the presence of As^{3+} or Cr^{6+} and the absence of Ca^{2+} or Mg^{2+}. As expected, As^{3+} promoted the annealing of P0G to M13m18 by nuclear annexin A1 in a concentration-dependent manner. Half-maximal stimulation was observed at 1.4 μM $AsCl_3$. This concentration was consistent with that required for half maximal binding of ssDNA. Similar results with DNA annealing were obtained with CrO_3, $PbCl_2$ and $CdCl_2$, with concentrations for half-maximal annealing approximately 3,30 and 0.1 μM, respectively. Since heat denatured nuclear annexin A1 did not promote DNA annealing even with heavy metals under the present experimental conditions (data not shown), it was concluded that carcinogenic heavy metals As^{3+} and Cr^{6+} and divalent metals Pb^{2+} and Cd^{2+} can replace Ca^{2+} for the ssDNA binding and DNA annealing activities of nuclear annexin A1 (A. Hirata et al., 2010).

7. DNA damage and mono-ubiquitinated annexin A1

Mono-ubiquitination of nuclear proteins is mainly involved in tolerance of DNA damage, while SUMOylated nuclear proteins generally function in repair of damaged DNA (Ulrich,

2005). Therefore, we tested nuclear annexin A1 for binding to damaged DNA. We synthesized the 80mer, 5′-GTCCACTATTAAAGAACGTGGACTCCAACGTCAAAGGGC-GAAAAACCGTCTATCAGGGCGATGGCCCACTACGTGAACCA-3′ (P0G), and 3 additional 80mers, each with a selected single G in position 14, 30 or 37 replaced by 8-oxo-guanosine (8-oxo-G) to model DNA damaged at a specific site by oxidation (A. Hirata et al., 2011). These damaged DNAs were designated as P14G, P30G, and P37G. We previously demonstrated that nuclear annexin A1 binds to ssDNA in a Ca^{2+}-dependent manner, and binding to dsDNA or aggregated DNA occurs in a Mg^{2+}-dependent manner (Hirata & Hirata, 1999; 2002). In the presence of 50 μM Ca^{2+}, mono-ubiquitinated annexin A1 purified from rat liver exhibited higher affinity for damaged DNA, while SUMOylated annexin A1 did not show much of difference in preference. In the presence of 50 μM Ca^{2+}, *Km* values for P30G, P37G and P14G were 0.20, 0.28, and 0.44 nM. All showed significantly higher affinity than P0G which had a *Km* value of 0.62 nM. Guanosine damaged by oxidation in the middle of the polynucleotide rather than at its ends appears to be readily tolerated by mono-ubiquitinated annexin A1.

Ca^{2+} induces annexin A1 conformational changes (Shesham et al. 2008). Since As^{3+} and Cr^{6+} also bind to the Ca^{2+} binding sites of annexin A1, we tested As^{3+} and Cr^{6+} for damaged DNA binding of mono-ubiquitinated annexin A1. The carcinogenic heavy metals, As^{3+} and Cr^{6+}, increased the affinity of nuclear annexin A1 for the oxidatively damaged DNA, P30G, but not for undamaged P0G. The *Km* value in the presence of 50 μM Ca^{2+} for G30P (0.20 nM) was significantly changed to 0.12 and 0.10 nM in the presence of 10 μM Cr^{6+} and 30 μM As^{3+}, respectively. However, maximal binding did not appear to be significantly altered (0.42 and 0.39 mol ssDNA/mol protein for G0P and G30P, respectively). Pb^{2+} and Cd^{2+} also increased affinity, but effects were much smaller. Ca^{2+} and heavy metals promote annealing of the damaged DNAs to M13mp18 by nuclear annexin A1 (Hirata & Hirata, 2002; A. Hirata et al., 2010). K_m values for the annealing reaction were essentially the same as values for the binding reaction. The affinity of nuclear annexin A1 for oxidatively damaged DNA was much higher than that for undamaged P0G and amounts of the DNA annealed with oxidatively damaged P14G were higher than with undamaged P0G. The specificity of nuclear annexin A1 for various oxidatively damaged DNAs was not substantially altered in the presence of heavy metals.

The damaged oligonucleotide-M13mp18 duplexes were also unwound in the presence of Mg^{2+} and ATP by mono-ubiquitinated annexin A1. ATP was hydrolyzed under these conditions. Unwinding velocities appeared similar for undamaged and damaged DNA. The unwinding of damaged polynucleotide-M13mp18 duplexes was inhibited by Ca^{2+} and heavy metals as reported previously (Hirata & Hirata, 2002). *Ki* values for heavy metal inhibition of the unwinding reaction were essentially the same with the *Ka* values for binding and annealing reactions. These heavy metals did not inhibit but rather stimulated dsDNA-dependent ATPase activity. Therefore, the apparent inhibition of unwinding by heavy metals most likely resulted from metal-induced increases in the annealing reaction, which in turn supplied the substrate (A. Hirata et al., 2010).

8. Translesion DNA synthesis by mono-ubiquitinated annexin A1

Helicases and DNA binding proteins are among the first proteins to encounter sites of DNA damage during transcription and DNA replication. The Werner syndrome protein enhances DNA synthesis during strand replacement of damaged DNA through its helicase activity (Harrigan et al., 2003). To test if mono-ubiquitinated annexin A1 stimulates translesion DNA synthesis, Pol β was used as an error-prone DNA polymerase that bypasses 8-oxo-guanine during DNA replication (Avkin & Livneh, 2002). DNA synthesis was measured by extension of the primer, 5'-TGGTTCACGTAG-3' annealed to P0G or oxidatively damaged DNA oligonucleotide (P30G) templates. Mono-ubiquitinated annexin A1 and Ca^{2+} increased DNA replication by approximately 2.6 fold as measured by the synthesis of 80mer, a full size of the template, G30G. When DNA synthesis was terminated at the damaged site, the size of DNA newly synthesized should be around 50 mer. Because ATP is not required for the maximal activation, it was conceivable that mono-ubiquitinated annexin A1 promoted annealing of the primer or stabilized the ssDNA template by binding rather than promoting unwinding. Translesion DNA synthesis was greatly enhanced by mono-ubiquitinated annexin A1, when primer was added separately at low concentrations. The amount of primer required for half maximal DNA synthesis in the presence of nuclear annexin A1 and Ca^{2+} decreased significantly to 1 pmol from 5 pmol in the absence of annexin A1. These observations suggest that mono-ubiquitinated annexin A1 promoted annealing of primer to template, while this primer was not necessarily the best substrate for nuclear annexin A1 with regards to length and base composition (Hirata & Hirata, 1999). It is noteworthy that even when the primer was annealed to the template prior to the experiments, annexin A1 was able to enhance translesion DNA synthesis by Pol β, suggesting that annexin A1 stabilizes the ssDNA region of the template. Taken together, these observations suggest that mono-ubiquitinated annexin A1 acts as a primer-recognizing protein that anneals primers to templates, a ssDNA binding protein, or both (Jindal & Vishwanatha, 1990). Notably, the amount of mono-ubiquitinated annexin A1 required for maximal DNA synthesis was not a function of Pol β, indicating that annexin A1 did not directly interact with this polymerase.

Given that binding of annexin A1 to damaged DNA was dependent on heavy metals, the effects of heavy metals on translesion DNA synthesis by Pol β was examined directly. The concentrations of metals required for half-maximal activation of DNA synthesis equaled those for half-maximal ssDNA binding. With P30G, Ka values for Ca^{2+}, Pb^{2+}, Cd^{2+}, As^{3+} and Cr^{6+} were 12.5, 30, 0.1, 1.4, and 3 μM, respectively, without phosphatidylserine. These values are essentially the same as those observed with these heavy metals for helicase activity.

9. Annexin A1-dependent promotion of translesion DNA synthesis by L lymphoma nuclear extracts

L lymphoma, L5178Y *tk*(+/-) cells contain annexin A1 with an apparent molecular weight 45,000 Da in nuclei (F. Hirata et al., 2010). The majority of nuclear annexin A1 was present in DNA synthesomes (Lin et al., 1997). Heavy metals inhibited DNA unwinding and stimulated DNA annealing by the nuclear extracts, and these metal effects were blocked by anti-annexin A1 antibody but not by anti-IgG (A. Hirata et al., 2010). To establish if nuclear annexin A1 plays a role in translesion DNA synthesis, we tested nuclear extracts for DNA synthesis using the P30G oligonucleotide as a template. Synthesis of 80mer DNA was considered as full translesion synthesis. Translesion DNA synthesis from damaged DNA templates was stimulated only in the presence of Ca^{2+} or heavy metals. Heavy metal-stimulated translesion DNA synthesis was partially blocked by monospecific anti-annexin A1 antibody but not by anti-IgGs, indicating that translesion DNA synthesis was dependent on annexin A1. However, it is noted that some degree of replication by nuclear extracts with damaged DNA templates took place in the absence of heavy metals or Ca^{2+}. Therefore, it cannot be ruled out that nuclear annexin A1 can function in recruiting error-prone DNA polymerases to the site of DNA synthesis. A recent study by O'Brien and colleagues (2009) demonstrates that Pol τ plays an important role in Cr^{6+}-induced mutation. It is also possible that the type of error prone polymerase that is recruited depends on the specific type of DNA damage.

When L5178Y *tk*(+/-) lymphoma cells were exposed to 20 μM MMS and 3 μM $AsCl_3$, mutation of the *tk* gene was induced as measured by the number of clones resistant to 5-fluoro-thymidine. Such mutation was suppressed by pretreatment of L5178Y *tk*(+/-) cells with an annexin A1 anti-sense oligonucleotide (unpublished data by F. Hirata). These observations suggest that DNA damage induced mutagenesis is mediated through annexin A1.

10. Summary

Cellular contents of annexin A1 increase in a variety of cancer cells including pancreatic cancer, glioma and hepatoma (Lim & Pervaiz, 2007). Not only precancerous hepatocytes but also proliferative hepatocytes following liver damage express annexin A1, while normal hepatocytes do not. Therefore, annexin A1 is proposed to be a biomarker of cell transformation and/or hyperproliferative state (Masaki et al., 1994). In contrast, some cancer cells such as prostate, breast and esophageal cancers demonstrate downregulation of annexin A1 expression (Lim & Pervaiz, 2007). Transfection of cDNA encoding annexin A1 into some cells alters the MAP kinase pathway *via* interaction of annexin A1 with Grb2, thereby delaying the cell cycle (Lim & Pervaiz, 2007). These observations lead to the proposal that annexin A1 is a suppressor gene that promotes cellular differentiation rather than an oncogene that promotes cancer. However, recent studies have shown that the down-regulation of annexins in certain types of malignant cancers is mainly attributable to epigenetic regulation by DNA methylation and histone acetylation, and to deletion of the annexin genes due to chromosomal instability following malignant transformation (Lim & Pervaiz, 2007). In esophageal epithelial carcinoma cells, nuclear annexin A1 is increased by its translocation from the cytosol, and cytosolic annexin A1 is consequently decreased (Liu et al., 2008). Therefore, it is proposed that nuclear annexin A1 rather than cytosolic annexin A1 is more closely associated with cell transformation, namely, cancer initiation, while changes of annexin A1 expression may not be involved in tumorigenesis.

Annexin A1 contains sequence and structural motifs for binding of nucleotides (non-Walker A type), binding of Ca^{2+} and heavy metals (non-EF hand type II and III) and DNA and/or RNA (helix-loop or turn-helix). Further, this protein is present in DNA synthesomes, suggesting some roles in the replication machinery. Nuclear annexin A1 is modified with ubiquitin and ubiquitin like molecules, and mono-ubiquitination of annexin A1 increases in response to DNA damage. Mono-ubiquitinated annexin A1 exhibits a helicase activity which has higher affinity for damaged DNA. Since helicases are among the first proteins that encounter damaged sites of DNA, this protein may regulate assembly of proteins required for repair of and tolerance to DNA damage. Mono-ubiquitinated but not SUMOylated nor native annexin A1 stimulates error-prone DNA polymerases such as Pol β with damaged DNA as a template *via* its helicase activity. In keeping with this interpretation that annexin A1 plays an important role in error prone lesion bypassing DNA synthesis in response to DNA damage, nuclear extracts of L5178Y *tk*(+/-) lymphoma cells promote translesion DNA synthesis in an annexin A1 dependent manner. Carcinogenic heavy metals such as As^{3+} and Cr^{6+} bind to the Ca^{2+} binding sites of annexin A1, and promote translesion DNA synthesis via annexin A1. DNA damage induced mutagenesis was found to be annexin A1 dependent both *in vivo* and *in vitro* (Fig. 2).

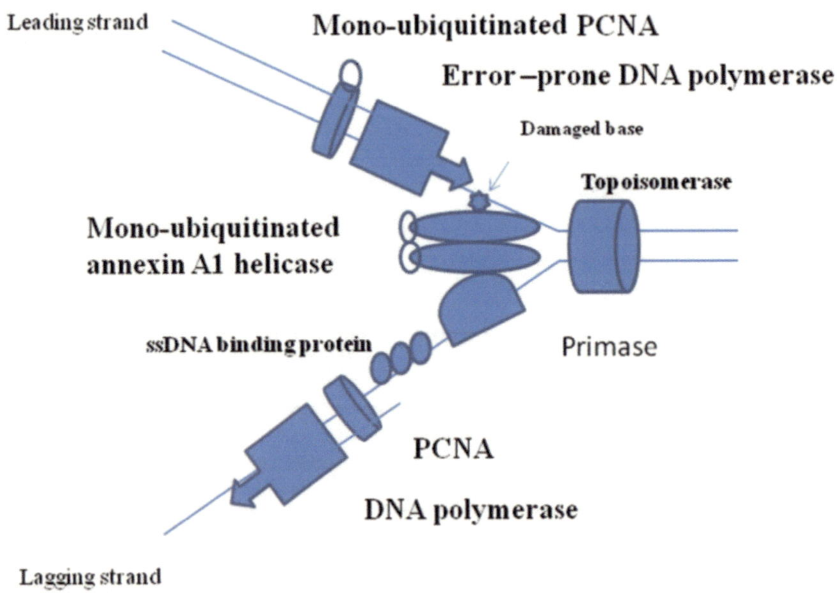

Figure 2. Annexin A1 mediated translesion DNA synthesis

Author details

Fusao Hirata, George B. Corcoran and Aiko Hirata
Department of Pharmaceutical Sciences, Eugene Applebaum College of Pharmacy and Health Sciences, Wayne State University, Detroit, MI, USA

11. References

Aukrust, I., Hollås, H., Strand, E., Eversen L., Travé, G., Flatmark, T. & Vedeler, A. (2007) The mRNA-binding site of annexin A2 resides in helices C-D of its domain IV. *Journal of Molecular Biology*, Vol.368, No.5, (May 2007, EPub Mar 2007), pp. 1367-1378, ISSN 0022-2836

Avkin, S. & Livneh, Z. (2002) Efficiency, specificity and DNA polymerase-dependence of translesion replication across the oxidative DNA lesion 8-oxoguanine in human cells. *Mutation Research*, Vol. 510, No.1-2, (December 2002), pp.81-90, ISSN 0027-5107

Bandorowicz-Pikula, J. & Pikula, S. (1998) Annexins and ATP in membrane traffic: a comparison with membrane fusion machinery. *Acta Biochimica Polonica,* Vol.45, No.3, pp. 721-733, eISSN 1734-154X

Calsou, P., Frit, P., Bozzato, C. & Salles, B. (1996) Negative interference of metal (II) ions with nucleotide excision repair in human cell-free extracts. *Carcinogenesis*, Vol.17, No.12, pp. 2779-2782, ISSN 0143-3334, eISSN 1460-2180

Caohuy, H., Srivastava, M. & Pollard, H.B. (1996) Membrane fusion protein synexin (annexin VII) as a Ca^{2+}/GTP sensor in exocytotic secretion. *Proceedings of the National Academy of Sciences USA*, Vol.93, No.20, pp. 10792-17802, eISSN 1091-6490

Chen, Z.J. & Sun, L.J. (2009) Nonproteolytic functions of ubiquitin in cell signaling. *Molecular Cell*, Vol.33, No.3, (February 2009) pp. 275-286, doi: 10.1016/j.molcel.2009.01.014

Cheung, C.W., Gibbons, N., Johnson, D.W. & Nicol, D.L. (2010) Silibinin-a promising new treatment for cancer. *Anti-Cancer Agents in Medicinal Chemistry*, Vol.10, No.3, (March 2010), pp. 186-195, ISSN 1871-5206

Cui, L., Wang, Y., Shi, Y., Zhang, Z., Xia, Y., Sun, H., Wang, S., Chen, J., Zhang, W., Lu, Q., Song, L., Wei, Q., Zhang, R. & Wang, X. (2007) Overexpression of annexin A1 induced by terephthalic acid calculi in rat bladder cancer. *Proteomics*, Vol.7, No.22, (November 2007), pp. 4192-4202, eISSN 1615-9861

Debret, R., El Btaouri, H., Duca, L., Rahman, I., Radke, S., Haye, B., Sallenave, J.M. & Antonicelli, F. (2003) Annexin A1 processing is associated with caspase-dependent apoptosis in BZR cells, *FEBS Letters*, Vol.546, No.2-3, pp.195-202, ISSN 0014-5793

Filipenko, N.R., MacLeod, T.J., Yoon C.S. & Waisman, D.M. (2004) Annexin A2 is a novel RNA-binding protein. *The Journal of Biological Chemistry*, Vol.279, No.10, (March 2004), pp. 8723-8731, ISSN 0021-9258, eISSN 1083-351X

Gerke, V. & Moss, S.E. (2002) Annexins: from structure to function. *Physiological Reviews*, Vol.82, No.2, (April 2002), pp. 331-371, ISSN 0031-9333, eISSN 1522-1210

Gill, G. (2004) SUMO and ubiquitin in the nucleus: different functions, similar mechanisms? *Genes & Development*, Vol.18, No.17, (September 2004), pp. 2046-2059, ISSN 0890-9369, eISSN 1549-5477

Harrigan, J.A., Opresko, P.L., von Kobbe, C., Kedar, P.S., Prasad, R., Wilson, S.H. & Bohr, V.A. (2003) The Werner syndrome protein stimulates DNA polymerase beta strand displacement synthesis via its helicase activity. *The Journal of Biological Chemistry*, Vol.

278, No.25, (June 2003, EPub Mar 2003), pp. 22686-22695, ISSN 0021-9258, eISSN 1083-351X

Hicke, L. (2001) Protein regulation by monoubiquitin. *Nature Reviews/Molecular Cell Biology,* Vol.2, No.3, (March 2001), pp. 195-201, ISSN 1471-0072

Hilgarth, R.S., Murphy, L.A., Skaggs, H.S., Wilkerson, D.C., Xing, H. & Sarge, K.D. (2004) Regulation and function of SUMO modification. *The Journal of Biological Chemistry*, Vol. 279, No.52, (December 2004, EPub Sept 2004), pp. 53899-53902, ISSN 0021-9258, eISSN 1083-351X

Hirata, A., Corcoran, G.B. & Hirata, F. (2010) Carcinogenic heavy metals replace Ca^{2+} for DNA binding and annealing activities of mono-ubiquitinated annexin A1 homodimer. *Toxicology & Applied Pharmacology,* Vol.248, No.1, (October 2010, EPub Jul 2010) pp. 45-51, ISSN 0041-008X

Hirata, A., Corcoran, G.B. & Hirata, F. (2011) Carcinogenic heavy metals, As^{3+} and Cr^{6+}, increase affinity of nuclear mono-ubiquitinated annexin A1 for DNA containing 8-oxo-guanosine, and promote translesion DNA synthesis. *Toxicology & Applied Pharmacology,* Vol.252, No.2, (April 2011, EPub Feb 2011), pp. 159-164, ISSN 0041-008X

Hirata, A. & Hirata, F. (1999) Lipocortin (annexin) I heterotetramer binds to purine RNA and pyridine DNA. *Biochemical & Biophysical Research Communications,* Vol.265, No.1, (November 1999), pp. 200-204, ISSN 0006-291X

Hirata, A. & Hirata, F. (2002) DNA chain unwinding and annealing reaction of lipocortin (annexin) I heterotetramer: regulation by Ca^{2+} and Mg^{2+}. *Biochemical & Biophysical Research Communications,* Vol.291, No.2, (February 2002), pp. 205-209, ISSN 0006-291X

Hirata, F. (1998) Annexins (Lipocortins), In: *Encyclopedia of Immunology, 2nd ed.*, Roitt, I.M. and Delves, P.J. Eds., pp. 111-115, Academic Press Ltd., ISBN 978-0-12-226765-9, New York

Hirata, F., Matsuda, K., Notsu, Y., Hattori, T. & DelCarmine, R. (1984) Phosphorylation at tyrosine residue of lipomodulin in mitogen-stimulated murine thymocytes, *Proceeding of the National Academy of Sciences USA,* Vol.81, No.15, (August 1984), pp. 4717-4721, eISSN 1091-6490

Hirata, F., Thibodeau, L.M. & Hirata, A. (2010) Ubiquitination and SUMOylation of annexin A1 and its helicase activity. *Biochimica et Biophysica Acta,* Vol.1800, No.9, (September 2010, EPub Mar 2010), pp. 899-905, ISSN 0304-4165

Honma, M., Hayashi, M., Shimada, H., Tanaka, N., Wakuri, S., Awogi, T., Yamamoto, K.I., Kodani, N.K., Nishi, Y., Nakadate, M. & Sofuni, T. (1999) Evaluation of the mouse lymphoma *tk* assay (microwell method) as an alternative to the in vitro chromosomal aberration test. *Mutagenesis,* Vol.14, No.1, (January 1999), pp. 5-22, ISSN 0267-8357, eISSN 1464-3804

Hsiang, C.H., Tunoda, T., Whang, Y.E., Tyson, D.R. & Ornstein, D.K. (2006) The impact of altered annexin I protein levels on apoptosis and signal transduction pathways in prostate cancer cells. *The Prostate,* Vol.66, No.13, (September 2006), pp. 1413-1424, eISSN 1097-0045

Huang, T.T. & D'Andrea, A.D. (2006) Regulation of DNA repair by ubiquitylation. *Nature Reviews/Molecular Cell Biology*, Vol.7, No.5, (May 2006), pp. 323-334, ISSN 1471-0072

Hyun, Y-L., Park, Y. M. & Na, D.S. (2000) ATP and GTP hydrolytic function of N-terminally deleted annexin I. *Journal of Biochemistry & Molecular Biology*, Vol.33, No.4, pp. 289-293, ISSN 2152-4114

Jin, Y.H., Clark, A.B., Slebos, R.J., Al-Refai, H., Taylor, J.A., Kunkel, T.A., Resnick, M.A. & Gordenin, D.A. (2003) Cadmium is a mutagen that acts by inhibiting mismatch repair. *Nature Genetics*, Vol.34, No.3, (July 2003), pp. 326-329, ISSN 1061-4036

Jindal, H.K. & Vishwanatha, J.K. (1990) Purification and characterization of primer recognition proteins from HeLa cells. *Biochemistry*, Vol.29, No.20, (May 1990), pp. 4767-4773, ISSN 0006-2960, eISSN 1520-4995

Katoh, N., Suzuki, T., Yuasa, A. & Miyamoto, T. (1995) Distribution of annexin I, II and IV in bovine mammary gland. *Journal of Dairy Science*, Vol.78, No.11, (November 1995), pp. 2382-2387, ISSN 0022-0302

Kim, Y.S., Ko, J., Kim, I.S., Jang, SW., Sung, H.J., Lee, H.J., Kim, Y. & Na, D.S. (2003) PKC delta-dependent cleavage and nuclear translocation of annexin A1 by phorbol 12-myristate 13-acetate. *European Journal of Biochemistry*, Vol.270, No.20, October 2003), pp. 4089-4094, ISSN 0022-0302

Kumble, K.D., Iversen, P.L. & Vishwanatha, J.K. (1992) The role of primer recognition proteins in DNA replication: inhibition of cellular proliferation by antisense oligonucleotides. *Journal of Cell Science*, Vol.101, No.Pt 1, (January 1992), pp. 35-41, ISSN 0021-9533, eISSN 1477-9137

Kunz, B.A., Straffon, A.F. & Vonarx, E.J. (2000) DNA damage-induced mutation: tolerance via translesion synthesis. *Mutation Research*, Vol.451, No.1-2, (June 2000) pp. 169-185, ISSN 0027-5107

Lim, L.H. & Pervaiz, S. (2007) Annexin 1: the new face of an old molecule. *The FASEB Journal*, Vol.21, No.4, (April 2007, Epub Jan 2007), pp.968–975, ISSN 0014-5793

Lin, C.Y., Jeng, Y.M., Chou, H.Y., Hsu, H.C., Yuan, R.H., Chiang, C.P. & Kuo, M.Y. (2008) Nuclear localization of annexin A1 is a prognostic factor in oral squamous cell carcinoma. *Journal of Surgical Oncology*, Vol.97, No.6, (May 2008), pp. 544-550, eISSN 1096-9098

Lin, S., Hickey, R. & Malkas, L. (1997) The biochemical status of the DNA synthesome can distinguish between permanent and temporary cell growth arrest. *Cell Growth & Differentiation*, Vol.8, No. , pp. 1359-1369, ISSN 1541-7786, eISSN 1557-3125

Lin, S., Hickey, R.J. & Malkas, L.H. (1997) The isolation of a DNA synthesome from human leukemia cells. *Leukemia Research*, Vol.21, No.6, (June 1997), pp. 501-512, ISSN 0145-2126

Liu, Y., Wang, H.X., Lu, N., Mao, Y.S., Liu, F., Wang, Y., Zhang, H.R., Wang, K., Wu, M. & Zhao, X.H. (2003) Translocation of annexin I from cellular membrane to the nuclear membrane in human esophageal squamous cell carcinoma. *World Journal Gastroenterology*, Vol.9, No.4, pp. 645-649, ISSN 1007-9327

Liu, Y.C., Yang, Z.Y., Du, J., Yao, X.J., Lei, R.X., Zheng, X.D., Liu, J.N., Hu, H.S. & Li, H. (2008) Study on the interactions of kaempferol and quercetin with intravenous immunoglobulin by fluorescence quenching, fourier transformation infrared spectroscopy and circular dichroism spectroscopy. *Chemical & Pharmaceutical Bulletin,* Vol.56, No.4, (April 2008), pp. 443-451, ISSN 0009 2363, eISSN 1347-5223

Maier, A., Schumann, B.L., Chang, X., Talaska, G. & Puga, A. (2002) Arsenic co-exposure potentiates benzo[a]pyrene genotoxicity. *Mutation Research,* Vol.517, No.1-2, (May 2002), pp. 101-111, ISSN 1383-5718

Masaki, T., Tokuda, M., Ohnishi, M., Watanabe, S., Fujishima, T., Miyamoto, K., Itano, T., Matsui, H., Arima, K., Shirai, M., Maeba, T., Sogawa, K., Konishi, R., Taniguchi, K., Hatanaka, Y., Hatase, O. & Nishioka, M. (1996) Enhanced expression of the protein kinase substrate annexin in human hepatocellular carcinoma. *Journal of Hepatology,* Vol.24, No.1, (July 1996), pp. 72-81, ISSN 0168-8278

Masaki, T., Tokuda, M., Fujimura, T., Ohnishi, M., Tai, Y., Miyamoto, K., Itano, T., Matsui, H., Watanabe, S., Sogawa, K., Yamada, T., Konishi, R., Nishioka, M. & Hatase, O. (1994) Involvement of annexin I and annexin II in hepatocyte proliferation: can annexins I and II be markers for proliferative hepatocytes? *Journal of Hepatology,* Vol.20, No.2, (August 1994), pp. 425-435, ISSN 0168-8278

Mel'gunov, V.I., Akimova, E.I. & Krasavchenko, K.S. (2000) Effect of divalent metal ions on annexin-mediated aggregation of asolectin liposomes. *Acta Biochimica Polonica,* Vol.47, No.3, pp. 675-683, eISSN 1734-154X

Mohiti, J., Caswell, A.M. & Walker, J.H. (1997) The nuclear location of annexin V in the human osteosarcoma cell line MG-63 depends on serum factors and tyrosine kinase signaling pathways. *Experimental Cell Research,* Vol.234, No.1, (July 1997), pp. 98-104, ISSN 0014-4827

Montaville, P., Neumann, J.M., Russo-Marrie, F., Ochsenbein, F. & Sanson, A. (2002) A new consensus sequence for phosphatidylserine recognition by annexins. *The Journal of Biological Chemistry,* Vol.277, No.27, (July 2002), pp. 24684-24693, ISSN 0021-9258, eISSN 1083-351X

O'Brien, T.J., Witcher, P., Brooks, B. & Patierno, S.R. (2009) DNA polymerase zeta is essential for hexavalent chromium-induced mutagenesis. *Mutation Research,* Vol.663, No.1-2, (April 2009), pp. 77-83, ISSN 0027-5107

Raydal, P., van Bergen en Henegouwen, P.M., Hullin, F., Ragab-Thomas, J.M., Fauvel, J., Verleij, A. & Chap, H. (1992) Morphological and biochemical evidence for partial nuclear localization of annexin 1 in endothelial cells. *Biochemical & Biophysical Research Communications,* Vol.186, No.1, (July 1992), pp. 432-439, ISSN 0006-291X

Rhee, H.J., Kim, G.Y., Huh, J.W., Kim, S.W. & Na, D.S. (2000) Annexin I is a stress protein induced by heat, oxidative stress and a sulfhydryl-reactive agent. *European Journal of Biochemistry,* Vol.267, No.11, (June 2000), pp. 3220-3225, eISSN 1742-4658

Sakaguchi, M., Murata, H., Sonegawa, H., Sakaguchi, Y., Futami, J., Kitazoe, M., Yamada, H., & Huh, N.H. (2007) Truncation of annexin A1 is a regulatory lever for linking epidermal growth factor signaling with cytosolic phospholipase A2 in normal and malignant squamous epithelial cells. *The Journal of Biological Chemistry,* Vol.282, No.49, (December 2007, Epub Oct 2007), pp. 35679-35686, ISSN 0021-9258, eISSN 1083-351X

Sekowski, J.W., Malkas, L.H., Wei, Y. & Hickey, R.J. (1997) Mercuric ion inhibits the activity and fidelity of the human cell DNA synthesome. *Toxicology & Applied Pharmacology,* Vol.145, No.2, (August 1997), pp. 268-276, ISSN 0041-008X

Shesham, R.D., Bartolotti, L.J. & Li, Y. (2008) Molecular dynamics simulation studies on Ca^{2+} -induced conformational changes of annexin I. *Protein Engineering Design & Selection,* Vol.21, No.2, (February 2008), pp. 115-120, ISSN 1741-0126, eISSN 1741-0134

Shimoji, T., Murakami, K., Sugiyama, Y., Matsuda, M., Inubushi, S., Nasu, J., Shirakura, M., Suzuki, T., Wakita, T., Kishino, T., Hotta, H., Miyamura, T. & Shoji, I. (2009) Identification of annexin A1 as a novel substrate for E6AP-mediated ubiquitylation. *Journal of Cellular Biochemistry,* Vol.106, No.6, (April 2009), pp. 1123-1135, eISSN 1097-4644

Shin, H. & Brown, R.M. Jr. (1999) GTPase activity and biochemical characterization of a recombinant cotton fiber annexin. *Plant Physiology,* Vol.119, No.3, (March 1999), pp. 925-934, ISSN 0032-0889, eISSN 1532-2548

Skouteris, G.G. & Schröder, C.H. (1996) The hepatocyte growth factor receptor kinase-mediated phosphorylation of lipocortin-1 transduces the proliferating signal of the hepatocyte growth factor. *The Journal of Biological Chemistry,* Vol.271, No.44, (November 1996), pp. 27266-27273, ISSN 0021-9258, eISSN 1083-351X

Tavokoli-Nezhad, M., Hirata, A. & Hirata, F. (1998) Lipocortin I stimulates the synthesis of the DNA lagging strand. *The FASEB Journal,* Vol.12, A1354, ISSN 0014-5793

Ulrich, H.D. (2005) The RAD6 pathway: control of DNA damage bypass and mutagenesis by ubiquitin and SUMO. *Chembiochem,* Vol.6, No.10, (October 2005), pp. 1735-1743, eISSN 1439-7633

Vishwanatha, J., Jindal, H.K. & Davis, R.G. (1992) The role of primer recognition proteins in DNA replication: association with nuclear matrix in HeLa cells. *Journal of Cell Science,* Vol.101, No.Pt 1, (January 1992), pp. 25-34, ISSN 0021-9533, eISSN 1477-9137

Vishwanantha, J.K. & Kumble, S. (1993) Involvement of annexin II in DNA replication: evidence from cell-free extracts of xenopus eggs. *Journal of Cell Science,*Vol.105, No.Pt 2, (June 1993), pp. 533-540, ISSN 0021-9533, eISSN 1477-9137

Wang, Y., Serfass, L., Roy, M.O., Wong, J., Bonneau, A.M. & Georges, E. (2004) Annexin-I expression modulates drug resistance in tumor cells. *Biochemical & Biophysical Research Communications,* Vol.314, No.2, (February 2004), pp. 565-570, ISSN 0006-291X

Wang, Z. (2001) DNA damage-induced mutagenesis: a novel target for cancer prevention. *Molecular Interventions,* Vol.1, No.5, (December 2001), pp. 269-281, ISSN 1534-0384, eISSN 1543-2548

Weng, X., Lueke, H., Song, I.S., Kung, D.S., Kim, S.H. & Huber, R. (1993) Crystal structure of human annexin I at 2.5 A resolution. *Protein Science,* Vol.2, No.3, (December 2008), pp. 448-458, eISSN 1469-896X

Wilkinson, K.A., Nishimune, A. & Henley, J.M. (2008) Analysis of SUMO-1 modification of neuronal proteins containing consensus SUMOylation motifs. *Neuroscience Letters,* Vol.436, No.2, (May 2008), pp. 239-244, ISSN 0304-3940

Yang, W. & Paschen, W. (2009) Gene expression and cell growth are modified by silencing SUMO2 and SUMO3 expression. *Biochemical & Biophysical Research Communications*, Vol. 382, No.1, (April 2009), pp. 215-218, ISSN 0006-291X

Zhu, F., Xu, C., Jiang, Z., Jin, M., Wang, L., Zeng, S., Teng, L. & Cao, J. (2010) Nuclear localization of annexin A1 correlates with advanced disease and peritoneal dissemination in patients with gastric carcinoma. *The Anatomical Record,* Vol.293, No.8, (August 2010), pp. 1310-1314, eISSN 1932-8494

Chapter 3

Models for Detection of Genotoxicity *in vivo*: Present and Future

Cherie Musgrove and Manel Camps

Additional information is available at the end of the chapter

http://dx.doi.org/10.5772/50554

1. Introduction

DNA damage is toxic to the cell, both acutely (perturbing the cell cycle and inducing apoptosis), and in the longer term (accelerating senescence and causing cancer and genetic disease) (1,2). Therefore it is of great interest for public health to determine the potential of anthropogenic chemicals and other compounds found in the environment to cause DNA damage as likely toxicants, carcinogens and teratogens (3).

This chapter will review methods that use *in vivo* models (*i.e.* living organisms and cell lines) for detection of genotoxic damage caused by exposure to chemicals. The reason that living models play such a prominent role in mutagenesis detection is two-fold:

1. **Extremely low frequency of mutation:** the mutation frequency induced by exogenous agents is extremely low (in the range of 1 mutation in 10^6 to 10^7 nucleotides). The ability of living organisms to amplify these rare events through positive selection is the basis for a number of these model systems.
2. **Modulation by metabolism:** metabolism has a dual role for activation (bioactivation) and for detoxification of genotoxic compounds. Therefore, metabolism needs to be taken into account by models of genotoxic exposure. Living organisms incorporate metabolic activity into the equation, although they only approximate human metabolism to various degrees.

In vivo models fall into two broad categories according to how they detect genotoxicity: direct or indirect genotoxicity detection methods. Direct measurement detects alterations in DNA either by sequencing, by the generation of a phenotype linked to specific mutations, or by visualization of the DNA damage such as micronucleus formation, detection of aberrant chromosomes, or an increase in the number of DNA breaks (4-6) (**Fig. 1**).

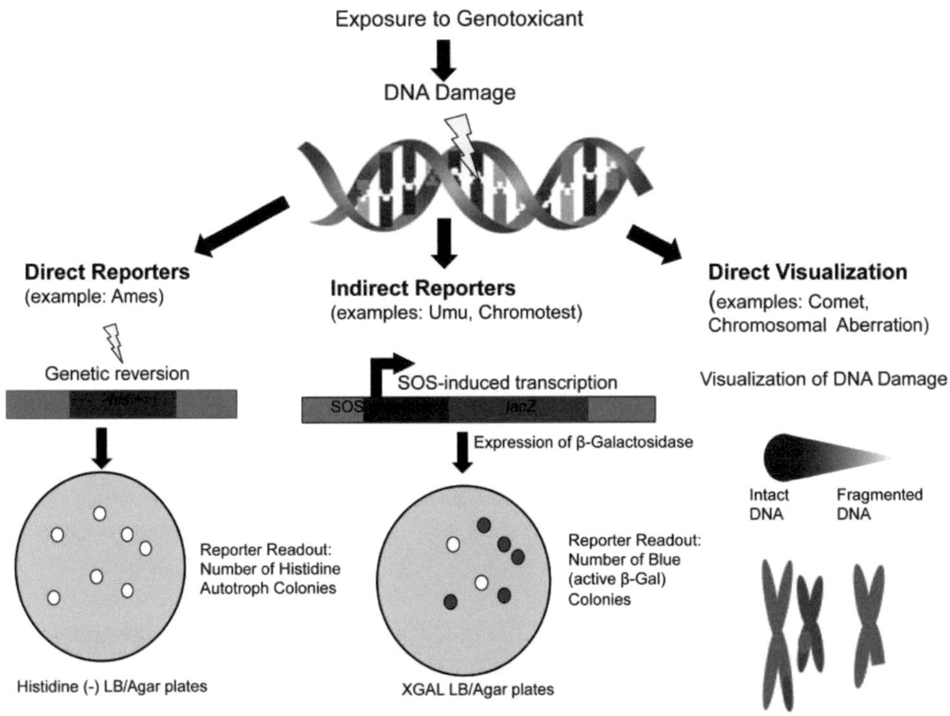

Figure 1.

Phenotypic detection of DNA damage is based on loss-of-function (*forward*), or gain-of-function (*reverse*) reporters. Forward mutation reporters are based on the loss of a phenotypically-detectable trait such as color or sensitivity to a metabolic poison. Therefore they can detect a range of mutations (miss-sense, transcriptional termination, frameshift, indels, etc.) along a sizeable target sequence, which increases the overall frequency of detectable events, allowing in many cases for direct screening. Forward mutation reporters also provide a representation of the range of genotoxic effects induced by the relevant compound, although biased for changes that lead to functional inactivation. Reversion reporters, by contrast, are based on reversion of a specific mutation inactivating a selectable marker. Therefore, reversion markers report very specific mutations at pre-determined sites, which may not be representative of the range of lesions introduced into DNA. Also reversion events are exceedingly rare due to the small size of the target, and can only be identified by positive selection.

Phenotypic detection methods generally produce binary readouts, with the presence of growth on limiting solid or liquid media, or changes in color as primary readouts. This means that the generation of a single data point requires fine-tuning the dose and of the dilution to obtain countable colonies (on solid plates) or a number positive wells that follows a Poisson distribution (in liquid culture) (7).

Genotoxic potential can also be detected by indirect measurement methods, usually based on transcriptional fusion of a reporter gene to a promoter responsive to DNA damage (specific examples are discussed below). Indirect detection methods provide an indication that the cell has sensed genotoxic stress, but the accuracy of each indirect reporting system depends on the range of lesions inducing the relevant transcriptional response and on the specificity of the relevant promoter for DNA damage relative to other types of stress. The reporter can be colorimetric, fluorescent or luminescent. Examples include: *lacZ* (beta-galactosidase), GFP (green fluorescent protein), luciferase, and *phoA* (alkaline phosphatase). Some of these markers (GFP and luciferase for example) have a wide dynamic range and are proportional to the amount of damage, greatly facilitating quantification (8,9). The use of GFP as a reporter, rather than reliance on an enzymatic reaction, produces a measurable response to DNA damage in a shorter time frame.

In general, indirect assays are better suited for high-throughput analysis because they can produce quantitative signals. Surface markers such as CD59 also provide forward mutation reporters that are quantitative, *i.e.* whose loss in a cell population following chemical exposure is directly proportional to the amount of genotoxic damage induced (10,11). Quantification of surface markers in large cell populations is made possible by the use of FACS (Fluorescence-Activated Cell Sorting) analysis, which is a high-throughput method that detects the presence or absence of the relevant marker in individual cells. The availability of quantitative reporters for mutagenesis substantially reduces the amount of test sample required. On the other hand, direct mutagenesis detection assays are more labor-intensive but more specific because they detect alterations in DNA.

DNA-damaging chemicals are frequently generated from precursors as reactive metabolic intermediates (12,13). The precursors are known as procarcinogens even though carcinogenicity has not in all cases been demonstrated. Procarcinogens include most genotoxic natural products and environmental agents since they would be expected to react with other molecules before reaching DNA. Chemotherapeutic drugs and other anthropogenic chemicals or contaminants, on the other hand, are frequently direct-acting. **Table 1** provides some examples of direct-acting and of bioactivation-dependent genotoxicants.

Xenobiotic metabolism, which is designed to solubilize lipophilic compounds as a way to facilitate their excretion, contributes prominently to bioactivation (12,13). The liver is the primary site of metabolic bioactivation, given its large metabolic capacity as well as its anatomical position as the gateway for compounds absorbed in the GI tract. Bioactivation by metabolism also occur in other tissues, including skin, lung, bone marrow, and GI tract. Bioactivation in the GI tract can also result from the action of the intestinal flora, or due to the drastic pH changes that occur as food moves through the tract (12).

Xenobiotic metabolism often involves two reactions, an oxidative one and a conjugative one. Oxygenation is typically carried out by members of the cytochrome P450 family (CYPs). CYPs are membrane-bound heme-proteins that require an effective reductase system to provide electrons (14). These enzymes tend to exhibit low catalytic efficiency and broad substrate specificity. Humans bear over 50 different CYP genes, which have some overlapping substrate specificities. Conjugation on the other hand transfers N- or O- acetyl groups (acetyl transferases), sulfates (sulfotransferases) or glutathione (glutathione-S transferases) to electrophilic substrates. Metabolism genes (particularly CYPs) are extraordinarily polymorphic, explaining the presence of wide interindividual differences in response to xenobiotics (13). Liver metabolism can be mimicked in testing paradigms by adding primary hepatocytes, liver slices, or various organ extract fractions to tester cell cultures, or by liver perfusion (15). The standard fraction is known as S9 fraction, which combines microsomes (containing CYPs) and cytosol (enriched for transferases) from the liver of rodents whose metabolism has been activated through xenobiotic pre-treatment (15,16).

Type of damage	Direct-acting Chemical class	Bioactivation-dependent Chemical class	
Alkylating	Diazo compounds	Triazenes	
	Nitrosamines	Azoxy compounds	
	Nitrosoureas	N-alkylnitrosamines	
	Aziridines	Aromatics	Polyaromatics
	Halogenated methanes, ethanes		Heteroaromatics
	Mustards		Nitroaromatic amines*
	Sulfates, sulfonates	Some proximate mustards	
Oxidizing	Simple epoxides	Quinones	
	Thiiranes, oxiranes		
	Simple peroxides		

Table 1. Examples of direct-acting and bioactivation-dependent genotoxic agents

In the context of drug discovery, *in vivo* methods are second-line assays performed to support the safety of a compound that is in the pipeline for clinical development (17,18).

Lead compounds are typically first prioritized by structure-activity relationship (SAR) analysis. This is a computational method that links specific chemical features of a given compound to individual biological activities, including genotoxicity. Due to its correlative nature, the predictive value of SAR analysis largely depends on how well represented the relevant class of compounds is in the database (19). Lead compounds that have made this first cut are then tested *in vivo* by direct mutagenesis detection methods for regulatory compliance. These include at least one prokaryotic phenotypic mutagenesis assay, one eukaryotic cell culture assay, and one animal test visualizing DNA damage (20,21). The standard test battery includes the Ames Test, the Mouse Lymphoma TK Assay, and the Micronucleus Test, respectively (see below). Transcriptional reporter-based assays can also be used for pre-screening prior to direct detection tests. Finally, the most reliable method to determine the potential carcinogenicity of a compound is testing it in a mammalian animal model (rat or mouse). This is done last, given the high cost of rodent carcinogenicity assays.

Here we discuss different *in vivo* models reporting on the ability chemicals to induce DNA damage, flagging these compounds as potential hazards to public health. This includes a variety of detection methods in prokaryotic, eukaryotic, tissue culture, whole-animal, and transgenic animal models. We finish by highlighting active areas of technology development and briefly speculate on the impact that next generation sequencing will likely to have in the field.

2. Prokaryotic reporter systems

Prokaryotes are useful for assessing DNA damage because they are haploid, reproduce quickly, and are easily grown in culture. Their use as a model for testing genotoxicity in humans is based on the universal nature of chemical mechanisms of DNA modification, as well as on the strong conservation of mechanisms of DNA repair between bacteria and humans (with the important exception of nucleotide-excision repair) (1). Genetic alterations are frequently used to enhance the sensitivity of prokaryotic reporter systems to DNA damage. Examples include mutations that increase membrane permeability (*rfaE, tolC*) and deficiencies in DNA repair (*uvrB, uvrA, umuD*). Most B- and K-derived laboratory strains of *E. coli* already exhibit increased permeability to xenobiotics as a result of loss of LPS selected during the long passage of these strains in culture (D. Josephy, personal communication). *E. coli* tends to be more sensitive to chemical mutagens than *Salmonella*, particularly to oxidizing mutagens, cross-linking agents and hydrazines (22). On the other hand, *Salmonella* facilitates detection of aromatic amines and of nitroaromatic compounds because of substantial endogenous bacterial nitroreductase (NR) and O-acetyltransferase (O-AT) metabolic activity.

2.1. Phenotypic reporter systems

Phenotypic reporters in prokaryotes are based on reversion of an auxotrophic marker. The *Ames Test* was the first of these assays to be developed, revolutionizing the field of genetic toxicology for its low cost and simplicity. This assay is based on reversion of a mutation preventing the biosynthesis of histidine. Reversion is detected by growth of colonies on solid agar in the presence of trace amounts of histidine (23). Growth on solid agar requires a large amount of test sample (~1 mg) but allows testing of non-water soluble compounds. A set of six strains have been developed to detect a broad range of point mutations and frameshifts (24). The Ames Test is still by far the most widely-used prokaryotic testing method, in part because it is mandatory for regulatory compliance.

Two variations have been developed to facilitate high-throughput formatting and to reduce the amount of sample needed: *Mini-Ames* and the *Ames Fluctuation Test*. Mini-Ames (also known as *Mini-Mutagenicity Test*) follows the standard Ames Test protocol, except at 1/5 the size. This reduces the amount of sample required to 300 mg of compound for the whole set of 6 reporter strains (25). Despite these advantages, Mini-Ames is still not widely used. The Ames Fluctuation Test is a variation of the Ames Test that is performed in liquid culture, using a chromophore as a binary indicator of growth (26). This assay has been adapted to a microtiter format (*Ames II Test*) (27). This format can incorporate microsomes, S9 fraction, or hepatocytes for bioactivation. Commercially available, The Ames II Test has comparable accuracy relative to the traditional Ames Test for most compounds (and even higher accuracy for low-potency liquid mixtures) (28), and its use is overtaking that of the traditional Ames Test.

The *AraD Test* is an alternative assay that detects forward mutations in the arabinose operon. The cells used in this assay have a mutation in the *araD* gene, which leads to accumulation of a toxic intermediate when arabinose is present. Mutations that inactivate the operon prevent the metabolism of arabinose, allowing cells to grow on arabinose (29,30). The *AraD* Test exhibits a different sensitivity profile than Ames, although being a forward mutation assay it has two advantages over Ames: more sensitive to point mutations (larger target for mutagenic action) and producing more accurate spectrum of mutation than the Ames test (since mutations are not limited to a single site). However, in practice this assay does not represent a significant alternative to the Ames Test or other cell-based mutagenesis assays (17).

2.2. Transcriptional reporter systems

Transcriptional reporter systems are based on the fusion of reporter genes to promoters of the SOS regulon, which includes a battery of ~40 genes involved in the response to DNA damage (31). This regulon is under the control of the *lexA* repressor, which upon genotoxic stress is cleaved by RecA, relieving repression (31). Two systems enjoy widespread use: the *UmuC Test* and the *SOS Chromotest.* Both systems are based on transcriptional fusions of DNA damage-inducible promoters (*umuC* and *sfiA*, respectively) to *lacZ* (32,33).

SfiA detects a broader range of genotoxic damage than umuC. On the other hand, the host for the UmuC Assay is a *Salmonella* strain (NM3009), making this assay particularly suited for detection of nitroarenes, such as those found in combustion products. The UmuC Test has been adapted to micro-titer plate format. High throughput, fully automated microtiter plate versions are also available for the SOS-Chromotest as well as numerous commercially available kits for testing specific sample types. Thus, the SOS-Chromotest provides easily quantifiable, reproducible and customizable ways to measure genotoxicity in a variety of samples, from wastewater to blood serum (33). An additional system, based on a fusion between the SOS-inducible gene *sulA* and the alkaline phosphatase-encoding gene phoA has also been recently described (34).

3. Eukaryotic reporter systems

Eukaryotic systems are also extensively used for detection of genotoxic activity. They have the advantage of having a DNA repair machinery that is even closer to that of humans (with homologous nucleotide excision repair machinery and more translesion synthesis polymerases for example) and of having comparable replication machinery, allowing detection of genotoxicants that interfere with mitosis. Higher eukaryotes also have metabolic systems that are much closer to our own, although even rodents show marked differences in metabolism relative to humans (12,13). Another advantage is the larger size of the cell nucleus and genome, which facilitates detection of rearrangements and other genomic abnormalities. Disadvantages include higher cost (particularly rodent systems), the presence of efflux pumps that can prevent accumulation of xenobiotics, and a diploid genome, which masks the phenotypic effects of heterozygous recessive mutations. Different strategies for improved detection of recessive mutations have been devised. These include: sporulation (in yeast), heterologous expression of single human chromosomes (in Chinese hamster cells), and selection of X-linked and heterozygous loci as markers (which become dominant or homozygous recessive with only one mutation).

As in prokaryotes, genetic modifications in yeast enhance sensitivity to DNA damage. Examples include deletion of efflux pumps (35), removing *mag1* (a N3mA DNA glycosylase), hindering base excision repair (36), and deletion of *mre11*, preventing both homologous recombination repair and non-homologous end joining pathways (36).

We group eukaryotic models into three sections: 1) phenotypic detection; 2) transcriptional detection; and 3) direct visualization of DNA damage.

3.1. Phenotypic reporter systems

Phenotypic detection of genotoxic damage in eukaryotes follows principles of forward and reverse mutation analogous to prokaryotic systems. While some of these systems are as old as the Ames' test, others are still being actively developed. Below we discuss two yeast mutagenesis reporter systems (the *DEL Assay* and the *Mitotic Gene-Conversion Assay*) and four mammalian cell-based ones (HPRT, TK, *The Human-Hamster Hybrid (A(L))* and the *PigA assay)*.

3.1.1. Yeast DEL Assay

This assay detects chemical induction of recombination events by reversion of a *his3* locus that has been interrupted by short repeats. Reversion to his+ can be measured by plating (37), or more recently in microtiter plate format, using a colorimetric readout (38,39). This assay has thus far proven to be very accurate, discriminating between carcinogens and non-carcinogens of the same chemical class, and showing a 92% correlation with two prokaryotic genotoxicity assays (Ames and UmuC) (39).

3.1.2. Yeast Mitotic-Gene-Conversion Assay

The Mitotic Gene-Conversion Assay uses a combination of heteroallelic (*ade2-40/ade2-119 and trp5-12/trp5-27*) and homoallelic (*ilv1-92/ilv1-92)* gene loci to detect induction of mitotic crossing over, mitotic gene conversion and reverse mutation (40). The original heteroallelic condition *ade2-40/ade2-119* forms white colonies. Mitotic crossing over can be detected visually as pink and red twin sectored colonies due to the formation of homozygous cells of the genotype *ade240/ade240* (deep red) and *ade-2-119/ade2-119* (pink). Mitotic gene conversion can be detected by the loss of auxotrophy for adenine (*ade2* locus) or tryptophan (*trp5* locus). Mutation induction can be followed by the appearance of isoleucine non-requiring colonies on selective media. Detecting both reversions and repair-associated recombination events is a unique feature of this assay that increases the sensitivity to genotoxicity. This assay is widely used and included in the Code of Federal regulations of the United States of America. However, the need to assess mitotic cross-over by screening for changes of color makes full automation of this assay very difficult.

3.1.3. HPRT Assay

This assay measures inactivating mutations at the *hprt* locus, which encodes the salvage-pathway enzyme hypoxanthine-guanine phosphoribosyl transferase (HPRT). HPRT catalyzes the formation of inosine or guanosine monophosphate from hypoxanthine or guanine, respectively. Treatment with 6-thioguanine generates 6-thioguanine monophosphate (6-TGM), which is highly cytotoxic to wild-type cells (41). Inactivating mutations in the *hprt* gene are dominant because this gene is carried on the X chromosome and is subject to X-inactivation (42-44). The standard cells for use in this assay are CHO (Chinese hamster ovary) cells, V79 (Chinese hamster lung cells), G12 or G10 cells (V79-derived cells). A variation on this assay is the expression of bacterial *gpt* gene (the functional homolog of HPRT) in an HPRT- background (42).

3.1.4. Cell Line TK Assay (mouse or human lymphoma cells)

The cells used for the assay are mouse lymphoma cells L5178Y, which are heterozygous at the thymidine kinase locus (*tk1*) on chromosome 11. Inactivating the WT allele induces trifluorothymidine (TFT) resistance, and tk–/– mutants can be selected for in a background of tk+/– non-mutant cells (45-47). Colony size is an indicator of the type of mutation involved: large colonies typically correspond to small TK-inactivating mutations, while small colonies often indicate clastogenic damage. This test (which is mandatory for regulatory compliance) is the most favored of the cell-line based assays because of its sensitivity to mutagens (48,49). However, this assay is also very susceptible to false positives (48,49).

Two tests use surface proteins as forward mutation markers, the Human-Hamster Hybrid (A(L)) Cell Mutagenesis System and the PigA Assay. Surface markers offer several advantages over drug-dependent readouts. Results are quantitative, producing not only a binary (yes/no) result but an indication of the potency of the chemical tested. In addition, with these assays cells do not need to be lysed for analysis, which enables tracking of the phenotype over time, as well as testing multiple or constant low-level exposures.

3.1.5. The Human-Hamster Hybrid (A(L)) Cell Mutagenesis System

Human-hamster hybrid (A(L)) cells were generated containing a single human chromosome 11 in addition to a standard set of CHO chromosomes. This human chromosome expresses CD59, CD44 and CD 90 surface antigens. The presence of CD59 on the cells' surface makes them sensitive to binding by a polyclonal antibody known as E7. Upon binding of the E7 antibody, incubation with serum stimulates the complement cascade, which lyses the cells.

The yield of CD59- mutants can also be detected by immunofluorescence and quantified using flow cytometry, providing a quantitative readout for mutagenesis (50,51). Detection of CD59- mutants exhibited a linear correlation with clastogen (gamma-radiation) and point mutagen (MNNG) dose, confirming the quantitative nature of the assay (50).

3.1.6. PigA Assay

The phosphatidylinositol glycan complementation group A (Pig-A) gene encoded on the X-chromosome is essential for attaching GPI-anchored proteins to the cell surface. The PigA assay detects the loss of CD59 (incidentally the same marker used in the Human-hamster hybrid (A(L)) assay described above) in red blood cells as and indicator of loss-of-function mutations at the endogenous Pig-A locus. Anti-CD59-PE is used to stain blood cells, and individual cell fluorescence is monitored by FACS analysis (10,52). Thiazole orange is used to differentiate between mature erythrocytes, reticulocytes (RETs), and leukocytes; and anti-CD61 to resolve platelets (10). The assay has been adapted for monkeys, mice, rats and humans (11). In rats, phenotypes can be detected earlier in reticulocytes than in erythrocytes (2 weeks versus 2 months following exposure, respectively).

3.2. Transcription reporter systems

As in the case of prokaryotes, eukaryotic transcription reporter systems are transcriptional fusions to genes that are specifically induced in response to DNA damage.

In yeast, one of the promoters of choice is that of ribonucleotide reductase 3 (rnr3), which encodes a form of the large subunit of ribonucleotide reductase. This gene is transcribed in response to low levels of damage, discriminates between DNA damage and other forms of stress, and its expression reaches higher levels than other DNA damage-responsive genes (53). The *rnr3* promoter is therefore ideal as a reporter. An assay was developed with *rnr3* driving *lacZ* expression (*rnr3-lacZ*) (35), and has had a modest impact following its initial description (10 PubMed citations, and 260 Google Scholar hits). A promising variation was developed that uses secreted *Cypridina* luciferase as a reporter (*rnr3-luciferase*) in a DNA repair-deficient yeast strain (54,55). Secretion of luciferase into the culture medium facilitates sequential measurements of DNA damage because cells don't need to be collected. This allows the detection of chronic effects, *i.e.* accumulated damage due to chronic low-level exposue over an extended period of time. It is easy to envision a fully automated adaptation of this assay, which would be cheap, and would not require specialized technicians.

Another promoter that has been used in yeast is that of HUG1 (Hydroxyurea- UV- and Gamma radiation-induced). The HUG1 promoter is used to drive expression of GFP (56). While the specific function of HUG1 is unknown, it is a part of the Mec1p kinase pathway, a signal transduction cascade that has a pivotal role in DNA damage-sensing in yeast. The sensitivity of the initial strain was enhanced by two deletions: that of *mag1* and that of *mre11*. These changes increased the sensitivity of the assay to alkylating agents and to inducers of strand-breaks, but did not change the sensitivity to other forms of DNA damage (36). However, in order for this system to have a clear advantage over other luminescence-based reporters, the sensitivity still needs to improve considerably.

In human cell lines, the promoter of choice is that of the Growth Arrest and DNA Damage 45 (GADD45) gene, which is a sensor for genotoxic stress in mammalian cells. GADD45α is induced upon exposure to clastogens, aneugens, and mutagens. The *GreenScreen Assay*, which uses a transcriptional fusion with GFP transformed into human lymphoblastic TK6 primary cells as a reporter, showed high specificity for carcinogens that do not require metabolic activation (100% accuracy with 75 chemicals tested) (57), as well as for procarcinogens (91% accuracy with 23 chemicals tested) (58). This assay has since undergone extensive validation with more than 8,000 compounds. The overall specificity to genotoxins has remained quite high at 95% (59). The GreenScreen Assay has been commercialized by the company Gentronix, including a 96-well plate version, and is becoming increasingly popular.

3.3. Direct visualization of DNA damage

A sensitive way to visualize DNA damage in eukaryotic cells is the COMET assay. This assay detects DNA fragmentation, which can result from a wide range of lesions including double strand breaks (DSBs), single strand breaks (SSBs), alkali labile (abasic) sites, oxidative DNA base damage, and DNA-DNA/DNA-protein/DNA-drug crosslinking. Cells are embedded in a thin layer of agarose, which is mounted on a microscope slide. The slide is immersed in an ionic running buffer (usually TBE or TAE) and the cells are electrophoresed through the agarose. DNA fragments will travel faster than the intact parts of the nucleus, and will run in front of the nucleus. When the DNA is stained and observed with a microscope, the fragments form what looks like a comet's tail, and the nucleus forms the comet's head (17,18) (**Fig1. direct visualization**). The COMET assay does not test for a specific end-point and can therefore be used to monitor both the genotoxic effects of chemical exposure and the kinetics of DNA repair. The use of the COMET assay as a readout for genotoxicity is increasing. Full automation has recently been achieved (60,61), which will greatly facilitate standardization and use of the assay as a screening tool. In addition, the sensitivity of this assay is being improved through combinations with other visualization methods, such as FISH (6,62).

3.3.1. Sister Chromatid Exchange Assay

The Sister Chromatid Exchange (SCE) Assay detects reciprocal exchanges between two sister chromatids of a replicating chromosome, apparently involving homologous loci (63). The DNA is labeled for two cell cycles (for example with bromodeoxyuridine) and visualized by fluorescence microscopy. It can be performed on a variety of cells, including cells from sentinel species like mussels and fish, which makes this assay extremely useful for environmental monitoring (64). This assay has also been used in humans as a marker for genotoxic exposure (65). While this assay does not detect DNA damage *per se*, SCE is an indication of ongoing DNA repair and therefore a genuine indicator of genotoxicity.

Two other visualization methods (the *Micronucleus Test* and the *Chromosomal Aberration Test*) are largely aimed at the detection of clastogens (agents that produce alterations affecting more than a few contiguous bases) and of aneugens (agents that alter the number of chromosomes). Prokaryotic systems have poor sensitivity for clastogens and aneugens because prokaryotes do not have multiple chromosomes, and their replication shares little mechanistic homology with mitosis. In addition, the larger size of the genome and of the nucleus in eukaryotes greatly facilitates the direct visualization of large aberrations. Therefore, these two assays complement prokaryotic reporter systems and are required for regulatory compliance.

3.3.2. Micronucleus Test (MN)

Micronuclei (MN) are broken fragments of daughter chromosomes that did not make it into the nucleus during mitosis. MN formation is therefore diagnostic for chromosomal DNA damage. It is detected by staining for DNA, cell membrane and nuclear membranes, followed by observation of individual cells with microscopy. The approved method for scoring micronucleus induction is to image stained cells and count those with MN. Cells can be harvested from a live animal or from tissue culture. The presence of MN is best visualized in erythrocytes (because they are anucleate) but it can also be used with other cell types. The success of this test relies on proper cell harvesting and culturing techniques, as the integrity of the cell and nuclear membrane are vital. It also depends on careful scoring of cells, since the nucleus must be clearly defined in order to determine the occlusion of MN from the nucleus (5,66). Flow cytometry can be used to quantify cells with MN induction (67), although careful microscopy controls are recommended. Recently, micronuclei induction in TK6 cells by a battery of reference compounds was determined using both microscopy and flow cytometry (68). This study produced a good correlation between the two readouts, suggesting that MN assay by flow cytometry may become one of the methods of choice for routine genotoxicity testing in the near future, particularly in the pharmaceutical industry.

3.3.3. Chromosomal Aberration (CA) Assay

The Chromosomal Aberration Assay detects large-scale damage of chromosomes, including structural aberrations (fragmentation or intercalations) and numerical aberrations (aneuploidy and polyploidy). Numerical aberrations are most frequently the result of unequal segregation of homologous chromosomes during cell division, which can be caused by interference with cohesion during mitosis (69). The test is most commonly carried out *in vitro* by exposing cell cultures to the test substance, and then treating the cells with a compound that stops mitosis in metaphase (colcemid). Following staining, the chromosomes are analyzed microscopically for aberrations. FISH-staining techniques have been used to increase the sensitivity of CA, allowing each chromosome to be differentially stained to reveal chromosomal rearrangements not detectable with conventional staining techniques (70,71).

3.4. Transgenic animal models

Transgenic animals represent one of the pillars of toxicological analysis, because they combine exposure in a whole organism with efficient detection in microbial systems. Every cell of the transgenic animal carries a chromosomally-integrated vector-reporter fusion gene that is not expressed and is therefore free to accumulate mutations. The vector is either a bacteriophage or a bacterial plasmid. Following exposure of live animals to a test chemical, transgenes are recovered from the genomic DNA and placed in the appropriate bacteria for readout of mutational frequency. Mutants are identified through the use of phenotypic reporters and their mutational spectrum can be determined by sequencing (72). Transgene models are ideal for study of the effects of chronic and repeated exposure, given the genetic neutrality of the transgenic reporter in the live animal. When the goal is to obtain mutation spectrum information, prolonged and/or repeated genotoxic exposure maximizes the number of independent mutational events obtained.

3.4.1. Muta™Mouse

The Muta™Mouse was the first transgenic rodent system to be introduced (73). In this system, the transgenic mice have, on their third chromosome, the λ g10 bacteriophage vector linked to a single *lacZ*. Following treatment, the genomic DNA is extracted from the tissues of interest, packaged into a lambda vector and transfected into *lacZ- E. coli*. Mutations result in while plaques in the presence of X-gal. This system is not nearly as popular as the LacZ Plasmid Mouse and the Big ® Blue Mouse because of poorer yield of transgenic DNA (72).

3.4.2. Big ® Blue Assay

Another bacteriophage-based assay is the Big ® Blue Assay, which exists in both mouse and rat backgrounds (74,75), and is available from several different companies (72). The original mouse version uses *LacI*, the lac repressor, as the reporter that is linked to the λLIZα phage vector. Multiple copies of the transgene (40 for the mouse model and 30 for the rat) are integrated in the chromosome, arranged head-to-tail. Transformation of recovered vector into *E. coli* followed by plating on X-gal medium allows the identification of blue plaques (with inactive mutant *lacI*) on a background of white plaques. The Big Blue system was greatly improved by the use of the *cII* gene as a reporter. The *cII λ* gene is responsible for transition from the lytic to the lysogenic phase at low temperatures, inducing expression of the CI repressor. Inactivating mutations sustained at the *cII* locus confer phages with the ability to form plaques, making detection of mutations a simple positive selection (76,77).

3.4.3. LacZ Plasmid Mouse

The LacZ Plasmid Mouse has 20 copies of the pUR288 plasmid per haploid genome (a total of 40 copies) integrated into multiple chromosomes (78). Genomic DNA is recovered, and digested by the *HindIII* restriction enzyme, releasing single copies of linearized plasmid. Magnetic beads coated in Lac repressor protein are used to isolate plasmid DNA from the digest, and the isolated vector is then recircularized into individual plasmids. These plasmids are transformed in *E. coli* and mutant frequency can be determined by scoring white colonies in the presence of X-gal. Compared with bacteriophage-based transgenic systems, plasmids can be isolated more efficiently and are more tolerant of deletions (both internal and involving flanking sequence) (72). A significant improvement to the test was the introduction of P-gal (which generates a toxic product when broken down by β-gal) as a positive selection for β-galactosidase loss-of- function (79).

4. Conclusions

Mutagenesis detection *in vivo* is key for testing the genotoxic potential of anthropogenic chemicals produced for industrial or medical applications as well as of products present in our environment. Both prokaryotic and eukaryotic models are useful, and complement each other. Prokaryotic models are simple, inexpensive, and frequently amenable to high-throughput formatting but detection is largely restricted to mutagens that induce point mutations and frameshifts. Eukaryotic models, by contrast, are more labor-intensive and time-consuming but are more sensitive to clastogenic and aneugenic activity and facilitate visualization of DNA damage (nicks, breaks, abasic sites, etc.). Indirect methods are cost-effective and easily amenable to automation, while visual or phenotypic detection is more specific because it reports DNA damage or genetic alterations caused by DNA damage but is generally more expensive and labor-intensive.

Accurate reproduction of human metabolism in model systems of genotoxicity remains one of the most urgent challenges in the field. As mentioned in the introduction, bacterial strains used for genotoxicity testing exhibit some metabolic activities. However, they lack cytochrome p450 activity completely, making them poor models for human bioactivation. Individual CYP proteins have been expressed in *E. coli (22)*. However, expressing active CYP proteins in *E. coli* is not trivial, as it requires special media and co-expression of a reductase system as electron donor. More importantly, only a few CYP alleles can be expressed at a time, so it will be extremely difficult to reproduce the complex patterns of CYP expression occurring in liver cells. In the classic Ames Test, mammalian xenobiotic metabolism is mimicked through the addition of post-mitochondrial hepatic rodent extract (S9 fractions). While this *in vitro* metabolic model allows detection of a range of pro-carcinogens, it misses short-lived metabolites that fail to cross the bacterial cell wall and suffers from low reproducibility because of the variable composition of the extracts (15% inter-laboratory variability) (48).

Whole-animal models are still the most sensitive systems available for detection of procarcinogens. Fish were proposed as a model organism early on due to their enhanced liver metabolism relative to humans and to the easy exposure to xenobiotics in the water or in the trophic chain (80). Transgenic reporters analogous to the ones created in mice were developed in fish (80,81), although their use is not yet widespread, possibly due to the need for specialized labor and facilities. *Drosophila melanogaster* is also likely to become more prominent in the future as a model for genotoxicity because it complements in many ways bacteria or yeast-based models. It is a whole organism, but extremely cheap and easy to maintain. Like fish, *Drosophila* produce large numbers of testable offspring (high n), and have metabolic and DNA repair systems that are highly homologous to human systems. Assays for genetic damage in germ cells, mostly in males (*Sex-Linked Recessive Lethal Test (SLRLT)*, and *Reciprocal Recombination Test)* were the initially developed (82). Recombination assays were later devised in somatic cells for improved sensitivity (82). These assays rely on endogenous forward mutation markers, with visible developmental abnormalities in wings, eye morphology, or bristle shape as readouts. Flies can be exposed to test chemicals in early stages of development (larvae), further increasing the sensitivity of the assay. Larvae are very actively metabolically and have been shown to be sensitive to teratogenic effects of pro-carcinogens (83). The large number of endogenous targets, the suitability for early exposure, and its active metabolism make *Drosophila* possibly the most sensitive phenotypic detection model available and a very promising model for detection of genotoxic and teratogenic effects (83).

New molecular technologies are likely to enhance our ability to detect the presence of mutations at very low frequencies, as illustrated by the *Random Mutation Capture Assay* (84). This technique detects the loss of a specific restriction site in chromosomal or mitochondrial genomes using multiplex PCR amplification (65,84) and has enabled establishing spontaneous mutation rates in tumors (85), and in a mouse model of aging (86). Importantly, by limiting dilution of the template, this technique has the ability to detect mutations from single DNA molecules templates, identifying non-clonal mutations in a heterogeneous population (85).

High-throughput sequencing technology will also likely allow the determination of genotoxic effects in the near future with an unprecedented level of resolution. Next-generation sequencing is based on massive, parallel amplification of templates (87). While DNA amplification is PCR-based, and therefore susceptible to the error-rate of the polymerase, mutations present in the template can still be detected through redundant coverage (typically in the 30-fold range). The accuracy of coverage information can be ensured through adequate design of bar-coded primers for amplification. Because, given the structure of the human genome, most random mutations in a cell are expected to be neutral, they should occur randomly and increase the genetic diversity in exposed the population over a period of time. In the absence of positive selection, sequencing of clonal mutations (*i.e.* mutations that are present in a significant fraction of the population) would miss this underlying genetic diversity (88). Therefore, obtaining an adequate representation of chemically-induced mutations would require sequencing DNA from individual cells.

As these new models and molecular tools become established in the field of genetic toxicity, they will need to be incorporated into the regulatory process for approval of new chemicals or for reassessment of chemicals currently in use.

Author details

Cherie Musgrove and Manel Camps
Department of Microbiology and Environmental Toxicology (METX),
University of California at Santa Cruz, Santa Cruz, CA, USA

Acknowledgement

The authors would like to thank a Special Research Grant from the UCSC Academic Senate for support for this work and Dr. Jason Bielas (Fred Hutchinson Cancer Center) for his input on this manuscript.

5. References

[1] Hoeijmakers, J.H. (2001) Genome maintenance mechanisms for preventing cancer. *Nature*, 411, 366-374.

[2] Loeb, L.A. (2011) Human cancers express mutator phenotypes: origin, consequences and targeting. *Nat Rev Cancer*, 11, 450-457.

[3] Loeb, L.A. and Harris, C.C. (2008) Advances in chemical carcinogenesis: a historical review and prospective. *Cancer Res*, 68, 6863-6872.

[4] Ishidate, M., Jr., Miura, K.F. and Sofuni, T. (1998) Chromosome aberration assays in genetic toxicology testing in vitro. *Mutat Res*, 404, 167-172.

[5] Kirsch-Volders, M., Plas, G., Elhajouji, A., Lukamowicz, M., Gonzalez, L., Vande Loock, K. and Decordier, I. (2011) The in vitro MN assay in 2011: origin and fate, biological significance, protocols, high throughput methodologies and toxicological relevance. *Arch Toxicol*, 85, 873-899.

[6] Spivak, G., Cox, R.A. and Hanawalt, P.C. (2009) New applications of the Comet assay: Comet-FISH and transcription-coupled DNA repair. *Mutat Res*, 681, 44-50.

[7] Thompson, K., Pine, P.S. and Rosenzweig, B. (2008) In Sahu, S. C. (ed.), *Toxicogenomics: A Powerful Tool for Toxicity Assesment*. John Wiley & Sons, Ltd., pp. 101-109.

[8] Cui, C., Wani, M.A., Wight, D., Kopchick, J. and Stambrook, P.J. (1994) Reporter genes in transgenic mice. *Transgenic Res*, 3, 182-194.

[9] Jia, X. and Xiao, W. (2004) In Yan, Z. and Caldwell, G. W. (eds.), *Optimization in Drug Discovery: In Vitro Methods*. Humana Press, Inc, Totowa, NJ, pp. 315-324.

[10] Bryce, S.M., Bemis, J.C. and Dertinger, S.D. (2008) In vivo mutation assay based on the endogenous Pig-a locus. *Environ Mol Mutagen*, 49, 256-264.

[11] Dobrovolsky, V.N., Miura, D., Heflich, R.H. and Dertinger, S.D. (2010) The in vivo Pig-a gene mutation assay, a potential tool for regulatory safety assessment. *Environ Mol Mutagen*, 51, 825-835.

[12] Guengerich, F.P. (2000) Metabolism of chemical carcinogens. *Carcinogenesis*, 21, 345-351.

[13] Guengerich, F.P. (2008) Cytochrome p450 and chemical toxicology. *Chem Res Toxicol*, 21, 70-83.

[14] Shimada, T., Oda, Y., Gillam, E.M., Guengerich, F.P. and Inoue, K. (2001) Metabolic activation of polycyclic aromatic hydrocarbons and other procarcinogens by cytochromes P450 1A1 and P450 1B1 allelic variants and other human cytochromes P450 in Salmonella typhimurium NM2009. *Drug Metab Dispos*, 29, 1176-1182.

[15] Brandon, E.F., Raap, C.D., Meijerman, I., Beijnen, J.H. and Schellens, J.H. (2003) An update on in vitro test methods in human hepatic drug biotransformation research: pros and cons. *Toxicology and applied pharmacology*, 189, 233-246.

[16] Cantelli-Forti, G., Hrelia, P. and Paolini, M. (1998) The pitfall of detoxifying enzymes. *Mutat Res*, 402, 179-183.

[17] Krewski, D., Acosta, D., Jr., Andersen, M., Anderson, H., Bailar, J.C., 3rd, Boekelheide, K., Brent, R., Charnley, G., Cheung, V.G., Green, S., Jr. *et al.* (2010) Toxicity testing in the 21st century: a vision and a strategy. *J Toxicol Environ Health B Crit Rev*, 13, 51-138.

[18] Lynch, A.M., Sasaki, J.C., Elespuru, R., Jacobson-Kram, D., Thybaud, V., De Boeck, M., Aardema, M.J., Aubrecht, J., Benz, R.D., Dertinger, S.D. *et al.* (2011) New and emerging technologies for genetic toxicity testing. *Environ Mol Mutagen*, 52, 205-223.

[19] Benigni, R. and Giuliani, A. (1996) Quantitative structure--activity relationship (QSAR) studies of mutagens and carcinogens. *Med Res Rev*, 16, 267-284.

[20] Gallagher, K., Goodsaid, F., Dix, D.J., Euling, S., Kramer, M., McCarroll, N., Preston, J., Sayre, P.G., Banalata, S., Wolf, D.C. *et al.* (2011) In Boverhof, D. R. and Gollapudi, B. B. (eds.), *Applications of Toxicogenomics in Safety Evaluation and Risk Assessment*. First ed. John Wiley & Sons, Inc., pp. 293-317.

[21] Hartung, T. and Daston, G. (2009) Are in vitro tests suitable for regulatory use? *Toxicological sciences : an official journal of the Society of Toxicology*, 111, 233-237.

[22] Josephy, P.D. (2002) Genetically-engineered bacteria expressing human enzymes and their use in the study of mutagens and mutagenesis. *Toxicology*, 181-182, 255-260.

[23] Mortelmans, K. and Zeiger, E. (2000) The Ames Salmonella/microsome mutagenicity assay. *Mutat Res*, 455, 29-60.

[24] Maron, D.M. and Ames, B.N. (1983) Revised methods for the Salmonella mutagenicity test. *Mutat Res*, 113, 173-215.

[25] Flamand, N., Meunier, J., Meunier, P. and Agapakis-Causse, C. (2001) Mini mutagenicity test: a miniaturized version of the Ames test used in a prescreening assay for point mutagenesis assessment. *Toxicol In Vitro*, 15, 105-114.

[26] Bridges, B.A. (1980) The fluctuation test. *Arch Toxicol*, 46, 41-44.

[27] Kamber, M., Fluckiger-Isler, S., Engelhardt, G., Jaeckh, R. and Zeiger, E. (2009) Comparison of the Ames II and traditional Ames test responses with respect to mutagenicity, strain specificities, need for metabolism and correlation with rodent carcinogenicity. *Mutagenesis*, 24, 359-366.

[28] Umbuzeiro, G.D., Rech, C.M., Correia, S., Bergamasco, A.M., Cardenette, G.H., Fluckiger-Isler, S. and Kamber, M. (2009) Comparison of the Salmonella/microsome microsuspension assay with the new microplate fluctuation protocol for testing the mutagenicity of environmental samples. *Environ Mol Mutagen*.

[29] Ruiz-Vazquez, R., Pueyo, C. and Cerda-Olmedo, E. (1978) A mutagen assay detecting forward mutations in an arabinose-sensitive strain of Salmonella typhimurium. *Mutat Res*, 54, 121-129.

[30] Whong, W.Z., Stewart, J. and Ong, T. (1981) Use of the improved arabinose-resistant assay system of Salmonella typhimurium for mutagenesis testing. *Environ Mutagen*, 3, 95-99.

[31] Janion, C. (2008) Inducible SOS response system of DNA repair and mutagenesis in Escherichia coli. *International journal of biological sciences*, 4, 338-344.

[32] Oda, Y., Yamazaki, H., Watanabe, M., Nohmi, T. and Shimada, T. (1993) Highly sensitive umu test system for the detection of mutagenic nitroarenes in Salmonella typhimurium NM3009 having high O-acetyltransferase and nitroreductase activities. *Environ Mol Mutagen*, 21, 357-364.

[33] Quillardet, P. and Hofnung, M. (1993) The SOS chromotest: a review. *Mutat Res*, 297, 235-279.

[34] Biran, A., Ben Yoav, H., Yagur-Kroll, S., Pedahzur, R., Buchinger, S., Shacham-Diamand, Y., Reifferscheid, G. and Belkin, S. (2011) Microbial genotoxicity bioreporters based on sulA activation. *Anal Bioanal Chem*, 400, 3013-3024.

[35] Zhang, M., Hanna, M., Li, J., Butcher, S., Dai, H. and Xiao, W. (2009) Creation of a hyperpermeable yeast strain to genotoxic agents through combined inactivation of PDR and CWP genes. *Toxicological sciences : an official journal of the Society of Toxicology*, 113, 401-411.

[36] Benton, M.G., Glasser, N.R. and Palecek, S.P. (2008) Deletion of MAG1 and MRE11 enhances the sensitivity of the Saccharomyces cerevisiae HUG1P-GFP promoter-reporter construct to genotoxicity. *Biosens Bioelectron*, 24, 736-741.

[37] Ku, W.W., Aubrecht, J., Mauthe, R.J., Schiestl, R.H. and Fornace, A.J., Jr. (2007) Genetic toxicity assessment: employing the best science for human safety evaluation Part VII: Why not start with a single test: a transformational alternative to genotoxicity hazard and risk assessment. *Toxicological sciences : an official journal of the Society of Toxicology*, 99, 20-25.

[38] Hafer, K., Rivina, Y. and Schiestl, R.H. (2010) Yeast DEL assay detects protection against radiation-induced cytotoxicity and genotoxicity: adaptation of a microtiter plate version. *Radiat Res*, 174, 719-726.

[39] Hontzeas, N., Hafer, K. and Schiestl, R.H. (2007) Development of a microtiter plate version of the yeast DEL assay amenable to high-throughput toxicity screening of chemical libraries. *Mutat Res*, 634, 228-234.

[40] Zimmermann, F., Kern, R. and Rasenberger, H. (1975) A Yeast Strain for Simultaneous Detection of Induced Mitotic Crossing Over, Mitotic Gene Conversion, and Reverse Mutation. *Mutation Research*, 28, 381-388.

[41] Caskey, C.T. and Kruh, G.D. (1979) The HPRT locus. *Cell*, 16, 1-9.

[42] Klein, C.B. and Rossman, T.G. (1990) Transgenic Chinese hamster V79 cell lines which exhibit variable levels of gpt mutagenesis. *Environ Mol Mutagen*, 16, 1-12.

[43] Klein, C.B., Su, L., Rossman, T.G. and Snow, E.T. (1994) Transgenic gpt+ V79 cell lines differ in their mutagenic response to clastogens. *Mutat Res*, 304, 217-228.

[44] Klein, C.B., Su, L., Singh, J. and Snow, E.T. (1997) Characterization of gpt deletion mutations in transgenic Chinese hamster cell lines. *Environ Mol Mutagen*, 30, 418-428.

[45] Clements, J. (2000) The mouse lymphoma assay. *Mutat Res*, 455, 97-110.

[46] Honma, M., Hayashi, M., Shimada, H., Tanaka, N., Wakuri, S., Awogi, T., Yamamoto, K.I., Kodani, N., Nishi, Y., Nakadate, M. *et al.* (1999) Evaluation of the mouse lymphoma tk assay (microwell method) as an alternative to the in vitro chromosomal aberration test. *Mutagenesis*, 14, 5-22.

[47] Lloyd, M. and Kidd, D. (2012) The mouse lymphoma assay. *Methods Mol Biol*, 817, 35-54.

[48] Kirkland, D., Aardema, M., Henderson, L. and Muller, L. (2005) Evaluation of the ability of a battery of three in vitro genotoxicity tests to discriminate rodent carcinogens and non-carcinogens I. Sensitivity, specificity and relative predictivity. *Mutat Res*, 584, 1-256.

[49] Seifried, H.E., Seifried, R.M., Clarke, J.J., Junghans, T.B. and San, R.H. (2006) A compilation of two decades of mutagenicity test results with the Ames Salmonella typhimurium and L5178Y mouse lymphoma cell mutation assays. *Chem Res Toxicol*, 19, 627-644.

[50] Ross, C.D., Lim, C.U. and Fox, M.H. (2005) Assay to measure CD59 mutations in CHO A(L) cells using flow cytometry. *Cytometry. Part A : the journal of the International Society for Analytical Cytology*, 66, 85-90.

[51] Zhou, H., Xu, A., Gillispie, J.A., Waldren, C.A. and Hei, T.K. (2006) Quantification of CD59- mutants in human-hamster hybrid (AL) cells by flow cytometry. *Mutat Res*, 594, 113-119.

[52] Miura, D., Dobrovolsky, V.N., Kasahara, Y., Katsuura, Y. and Heflich, R.H. (2008) Development of an in vivo gene mutation assay using the endogenous Pig-A gene: I. Flow cytometric detection of CD59-negative peripheral red blood cells and CD48-negative spleen T-cells from the rat. *Environ Mol Mutagen*, 49, 614-621.

[53] Elledge, S.J., Zhou, Z. and Allen, J.B. (1992) Ribonucleotide reductase: regulation, regulation, regulation. *Trends Biochem Sci*, 17, 119-123.

[54] Wright, J.H., Modjeski, K.L., Bielas, J.H., Preston, B.D., Fausto, N., Loeb, L.A. and Campbell, J.S. (2011) A random mutation capture assay to detect genomic point mutations in mouse tissue. *Nucleic Acids Res*, 39, e73.

[55] Bielaszewska, M., Middendorf, B., Tarr, P.I., Zhang, W., Prager, R., Aldick, T., Dobrindt, U., Karch, H. and Mellmann, A. (2011) Chromosomal instability in enterohaemorrhagic Escherichia coli O157:H7: impact on adherence, tellurite resistance and colony phenotype. *Mol Microbiol*, 79, 1024-1044.

[56] Benton, M.G., Glasser, N.R. and Palecek, S.P. (2007) The utilization of a Saccharomyces cerevisiae HUG1P-GFP promoter-reporter construct for the selective detection of DNA damage. *Mutat Res*, 633, 21-34.

[57] Hastwell, P.W., Chai, L.L., Roberts, K.J., Webster, T.W., Harvey, J.S., Rees, R.W. and Walmsley, R.M. (2006) High-specificity and high-sensitivity genotoxicity assessment in a human cell line: validation of the GreenScreen HC GADD45a-GFP genotoxicity assay. *Mutat Res*, 607, 160-175.

[58] Jagger, C., Tate, M., Cahill, P.A., Hughes, C., Knight, A.W., Billinton, N. and Walmsley, R.M. (2009) Assessment of the genotoxicity of S9-generated metabolites using the GreenScreen HC GADD45a-GFP assay. *Mutagenesis*, 24, 35-50.

[59] Walmsley, R.M. and Tate, M. (2012) The GADD45a-GFP GreenScreen HC assay. *Methods Mol Biol*, 817, 231-250.

[60] Dehon, G., Catoire, L., Duez, P., Bogaerts, P. and Dubois, J. (2008) Validation of an automatic comet assay analysis system integrating the curve fitting of combined comet intensity profiles. *Mutat Res*, 650, 87-95.

[61] Rosenberger, A., Rossler, U., Hornhardt, S., Sauter, W., Bickeboller, H., Wichmann, H.E. and Gomolka, M. (2011) Validation of a fully automated COMET assay: 1.75 million single cells measured over a 5 year period. *DNA Repair (Amst)*, 10, 322-337.

[62] Shaposhnikov, S., Thomsen, P.D. and Collins, A.R. (2011) Combining fluorescent in situ hybridization with the comet assay for targeted examination of DNA damage and repair. *Methods Mol Biol*, 682, 115-132.

[63] Kamb, A. (2011) Next-generation sequencing and its potential impact. *Chem Res Toxicol*, 24, 1163-1168.

[64] Brockmeyer, J., Spelten, S., Kuczius, T., Bielaszewska, M. and Karch, H. (2009) Structure and function relationship of the autotransport and proteolytic activity of EspP from Shiga toxin-producing Escherichia coli. *PloS one*, 4, e6100.

[65] Vermulst, M., Bielas, J.H. and Loeb, L.A. (2008) Quantification of random mutations in the mitochondrial genome. *Methods*, 46, 263-268.

[66] Madle, E., Korte, A. and Beek, B. (1986) Species differences in mutagenicity testing: I. Micronucleus and SCE tests in rats, mice, and Chinese hamsters with aflatoxin B1. *Teratog Carcinog Mutagen*, 6, 1-13.

[67] Hayashi, M., MacGregor, J.T., Gatehouse, D.G., Adler, I.D., Blakey, D.H., Dertinger, S.D., Krishna, G., Morita, T., Russo, A. and Sutou, S. (2000) In vivo rodent erythrocyte micronucleus assay. II. Some aspects of protocol design including repeated treatments, integration with toxicity testing, and automated scoring. *Environ Mol Mutagen*, 35, 234-252.

[68] Lukamowicz, M., Woodward, K., Kirsch-Volders, M., Suter, W. and Elhajouji, A. (2011) A flow cytometry based in vitro micronucleus assay in TK6 cells--validation using early stage pharmaceutical development compounds. *Environ Mol Mutagen*, 52, 363-372.

[69] Thompson, S.L., Bakhoum, S.F. and Compton, D.A. (2010) Mechanisms of chromosomal instability. *Current biology : CB*, 20, R285-295.

[70] Masood, F., Anjum, R., Ahmad, M. and Malik, A. (2012) In Malik, A. and Grohmann, E. (eds.), *Environmental Protection Strategies for Sustainable Development*. 1st ed. Springer Netherlands, Dorcrecht, pp. 229-260.

[71] Wang, G., Zhao, J. and Vasquez, K.M. (2009) Methods to determine DNA structural alterations and genetic instability. *Methods*, 48, 54-62.

[72] Boverhof, D.R., Chamberlain, M.P., Elcombe, C.R., Gonzalez, F.J., Heflich, R.H., Hernandez, L.G., Jacobs, A.C., Jacobson-Kram, D., Luijten, M., Maggi, A. *et al.* (2011) Transgenic animal models in toxicology: historical perspectives and future outlook. *Toxicological sciences : an official journal of the Society of Toxicology*, 121, 207-233.

[73] Gossen, J.A., de Leeuw, W.J., Molijn, A.C. and Vijg, J. (1993) Plasmid rescue from transgenic mouse DNA using LacI repressor protein conjugated to magnetic beads. *Biotechniques*, 14, 624-629.

[74] Stiegler, G.L. and Stillwell, L.C. (1993) Big Blue transgenic mouse lacI mutation analysis. *Environ Mol Mutagen*, 22, 127-129.

[75] Wyborski, D.L., Malkhosyan, S., Moores, J., Perucho, M. and Short, J.M. (1995) Development of a rat cell line containing stably integrated copies of a lambda/lacI shuttle vector. *Mutat Res*, 334, 161-165.

[76] Jakubczak, J.L., Merlino, G., French, J.E., Muller, W.J., Paul, B., Adhya, S. and Garges, S. (1996) Analysis of genetic instability during mammary tumor progression using a novel selection-based assay for in vivo mutations in a bacteriophage lambda transgene target. *Proc Natl Acad Sci U S A*, 93, 9073-9078.

[77] Swiger, R.R. (2001) Just how does the cII selection system work in Muta Mouse? *Environ Mol Mutagen*, 37, 290-296.

[78] Vijg, J., Dolle, M.E., Martus, H.J. and Boerrigter, M.E. (1997) Transgenic mouse models for studying mutations in vivo: applications in aging research. *Mechanisms of ageing and development*, 98, 189-202.

[79] Gossen, J.A., Martus, H.J., Wei, J.Y. and Vijg, J. (1995) Spontaneous and X-ray-induced deletion mutations in a LacZ plasmid-based transgenic mouse model. *Mutat Res*, 331, 89-97.

[80] Winn, R.N., Norris, M.B., Brayer, K.J., Torres, C. and Muller, S.L. (2000) Detection of mutations in transgenic fish carrying a bacteriophage lambda cII transgene target. *Proc Natl Acad Sci U S A*, 97, 12655-12660.

[81] Winn, R.N., Norris, M., Muller, S., Torres, C. and Brayer, K. (2001) Bacteriophage lambda and plasmid pUR288 transgenic fish models for detecting in vivo mutations. *Mar Biotechnol (NY)*, 3, S185-195.

[82] Vogel, E.W., Graf, U., Frei, H.J. and Nivard, M.M. (1999) The results of assays in Drosophila as indicators of exposure to carcinogens. *IARC scientific publications*, 427-470.

[83] Foureman, P., Mason, J.M., Valencia, R. and Zimmering, S. (1994) Chemical mutagenesis testing in Drosophila. IX. Results of 50 coded compounds tested for the National Toxicology Program. *Environ Mol Mutagen*, 23, 51-63.

[84] Bielas, J.H. and Loeb, L.A. (2005) Quantification of random genomic mutations. *Nature methods*, 2, 285-290.

[85] Bielas, J.H., Loeb, K.R., Rubin, B.P., True, L.D. and Loeb, L.A. (2006) Human cancers express a mutator phenotype. *Proc Natl Acad Sci U S A*, 103, 18238-18242.

[86] Vermulst, M., Wanagat, J., Kujoth, G.C., Bielas, J.H., Rabinovitch, P.S., Prolla, T.A. and Loeb, L.A. (2008) DNA deletions and clonal mutations drive premature aging in mitochondrial mutator mice. *Nature genetics*, 40, 392-394.

[87] Niedringhaus, T.P., Milanova, D., Kerby, M.B., Snyder, M.P. and Barron, A.E. (2011) Landscape of next-generation sequencing technologies. *Anal Chem*, 83, 4327-4341.

[88] Loeb, L.A., Bielas, J.H. and Beckman, R.A. (2008) Cancers exhibit a mutator phenotype: clinical implications. *Cancer Res*, 68, 3551-3557; discussion 3557.

Chapter 4

Mitochondrial Mutagenesis in Aging and Disease

Marc Vermulst, Konstantin Khrapko and Jonathan Wanagat

Additional information is available at the end of the chapter

http://dx.doi.org/10.5772/51099

1. Introduction

Shortly after DNA was discovered to be the carrier of hereditary information, a number of peculiar observations were made. It was found that some traits, like the ability of yeast cells to use non-fermentable sources of energy, were inherited through a cytoplasmic, and not a nuclear mechanism[1]. This observation was rather alarming at the time, since DNA was only known to be present in the nucleus. As a result, some researchers started to doubt whether DNA was truly the sole carrier of hereditary information. Could it be that the original hypothesis about DNA was incorrect? Or did the cytoplasm carry DNA molecules that were yet to be discovered? Ultimately, this conflict was resolved when clever mating experiments, biochemical tests and precise electron microscopy culminated in the discovery of a new DNA molecule, present inside mitochondria[2-4].

Now, approximately 50 years later, we can truly appreciate the historic nature of this discovery. Not only do we realize how essential mitochondrial DNA (mtDNA) is to life itself, but mtDNA has also helped us understand our origin as a species. As our ancestors spread across the globe, they acquired mutations in their mitochondrial genome, which genetically marked the humans that colonized a certain geographical location. Expansive sequencing projects have now documented these mutations, and used them to trace the origin of mankind back to central Africa[5, 6]. In addition to their historical significance, mtDNA mutations are also important in a medical context. Most diseases that are currently endemic in western society have an mtDNA component, including cancer, diabetes and Parkinson disease. As a result, mitochondrial mutagenesis is now one of the focal points of modern biomedical research[7].

This renewed interest in mitochondrial genetics has led to an enormous influx of new researchers, ideas and technology. Because of this enthusiasm new mouse models have been generated, improved mutation detection assays have been developed, and new roles for mitochondria have been discovered in immunity, signal transduction, development, and countless other biological processes that are essential to human health[8]. As a result, we are starting to expose the molecular mechanisms that underpin mitochondrial mutagenesis, and we are learning how intimately mutagenesis is related to mitochondrial genetics. The following sections of this chapter will describe this relationship in greater detail.

Feature	Nuclear DNA	Mitochondrial DNA
Length	3 billion bases	16.5kb
Shape	Linear	Circular
Copies per cell	2 chromosomes/cell	100-10.000 copies/cell
Packaging	Histones	Nucleoid
Inheritance	Paternal/maternal	Maternal
Introns	Many	None
Geneticcode	AGA = R, AGG = R AUA = I, UGA = Ter	AGA = Ter, AGG = Ter AUA = M, UGA = W
Replication	Symmetrical	Strand-displacement*
Transcription	Gene specific	Multi-cistronic

Table 1. Structural and genetic differences between the nuclear and mitochondrial genome.

2. The structure of mtDNA

Researchers who are unfamiliar with mitochondrial genetics may find mtDNA to be a peculiar molecule, as it differs from nuclear DNA (nDNA) in almost every way. Some of the most important differences between mtDNA and nDNA are summarized in table 1. First, mitochondrial DNA is a very short DNA molecule, which is approximately 16.5kb in length in most mammalian species[9, 10]. And on top of that, it is circular in nature, not unlike a plasmid. However, even though mtDNA is relatively short, it still makes up a significant portion of the genetic content of a cell because it is typically present in hundreds, or even thousands of copies per cell. These molecules are more or less randomly dispersed across the mitochondrial population of a cell, so that each mitochondrion contains approximately 2-10 mtDNA molecules. This is especially important in the context of mutagenesis, because this multiplicity ensures that, if a mutation occurs in one of these copies, multiple WT molecules can complement the acquired defect.

Remarkably, this complementation is not limited to a single mitochondrion, but encompasses the entire cell. The expansion of this safety net is made possible by the dynamic nature of mitochondrial biology. Mitochondria are highly mobile organelles, which frequently change their position inside a cell. To do this, they travel along the cytoskeleton with the help of numerous proteins, of which Milton and MIRO are two key members. These proteins form a complex with a third protein, Kinesin-1 heavy chain, to mediate mitochondrial transport[11]. Because mitochondria can travel in either direction, they frequently collide in an end-to-end fashion. During these collisions, mitochondria fuse together to form a single, continuous organelle and mix their contents in the process. This newly formed mitochondrion will eventually undergo a fission event, splitting it back into two smaller organelles. Since all the bio-molecules are randomly distributed across the emerging organelles, each mitochondrion contains a completely new mixture of DNA, RNA and proteins. In fact, each mitochondrion contains an "average" of all the contents that were present in the two mitochondria that initiated the fusion event. Because mitochondria undergo constant cycles of fusion and fission, mtDNA is constantly shared across the entire mitochondrial population. A WT genome can therefore always complement a mutation, even if it is harbored by a mitochondrion that is present elsewhere in the cell[12]. In a sense, you might even say that an "individual mitochondrion" does not really exist. Each mitochondrion is part of a fluid, interconnected network, whose components are in a constant state of flux[13].

Because of this cell-wide safety net, most mutations are thought to be harmless. As long as mutant genomes are outnumbered by WT genomes, mitochondrial fusion and fission ensures that any deficiencies are consistently complemented. However, if a mutation is replicated extensively inside a cell, it can eventually become very harmful. This occurs once a mutation is present in the majority of a cell's mtDNA, and insufficient WT genomes left for complementation. This "threshold" is typically reached when 60-90% of the mtDNA molecules carry an identical mutation[14]. Such mutations are said to be heteroplasmic (figure 7), because they are present in only a fraction of the mitochondrial genomes in a cell. A mutation that is present in all genomes it is said to be homoplasmic. One of the most remarkable things about mitochondrial mutagenesis is that all the mutations that have been found thus far require clonal expansion before they cause cellular dysfunction. This means that every pathological mutation in mtDNA is recessive. One might argue then, that if we understood this expansion process in greater detail, and learned to control it, we could combat every mtDNA disease using a single intervention.

Another feature of mitochondrial genetics that is important in the context of mutagenesis is the density of the mitochondrial genome[9] (figure 1). MtDNA encodes 13 proteins that are all essential components of the electron transport chain (ETC). In addition, mtDNA encodes 2 rRNA molecules and 22 tRNA particles that help express these proteins. This leaves room for only a short region of non-coding mtDNA. This region is called the displacement loop, or D-loop for short, which was named after its peculiar triple helix structure with one "displaced" DNA strand. The D-loop contains an origin of replication and multiple sites for transcription initiation, which makes it the most important sequence for mtDNA metabolism. When you combine the lack of non-coding regions in mtDNA, with its short length and abundance of genes, it becomes clear how compact the mitochondrial genome is. It is so compact in fact, that some genes, like the ATP6 gene, do not even encode a complete stop codon. Instead, the final base of this codon, an adenine, is added to the mRNA during polyadenylation[9]. This gene density means that almost every mutation is bound to impact mtDNA in a non-trivial way.

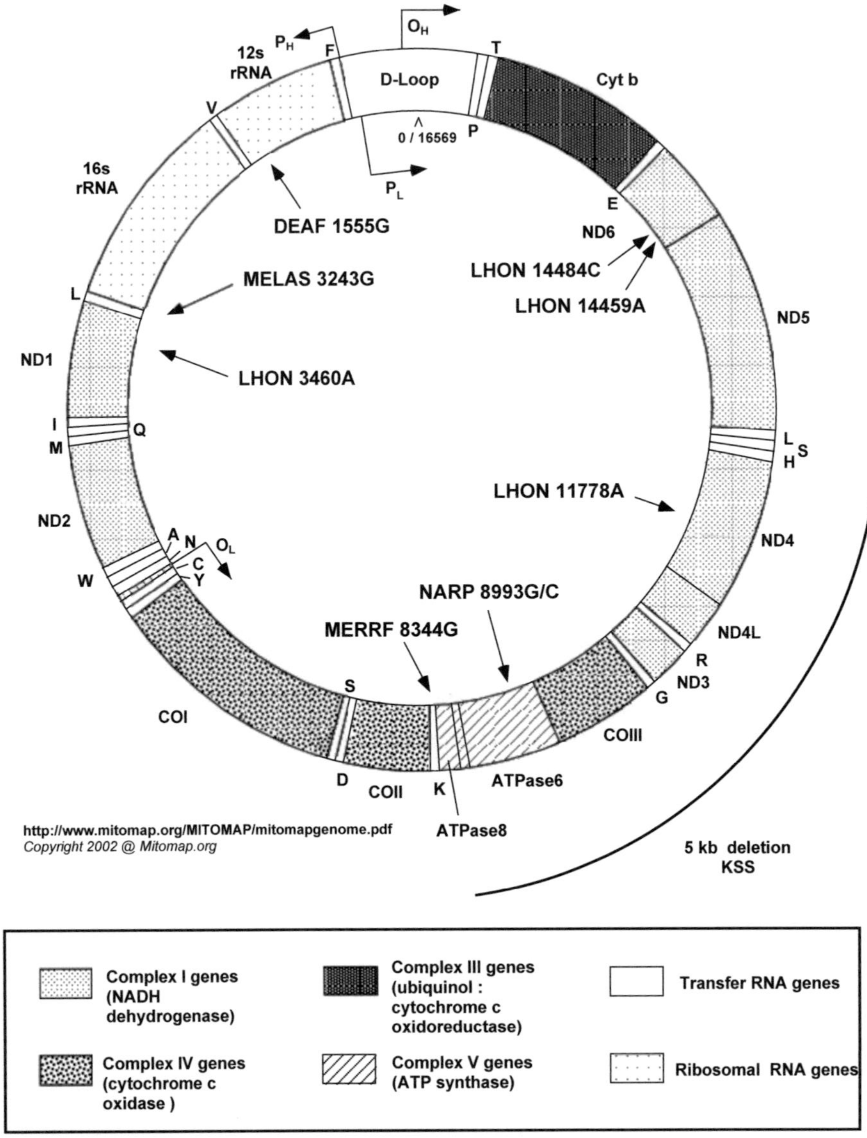

Figure 1. Organization of the mitochondrial genome. The mitochondrial genome contains 13 protein coding genes, 2 rRNA genes, and 22 tRNA genes. The mitochondrially encoded proteins are components of complex I, III, IV, or V. Notice how the tRNA genes punctuate the rRNA genes and proteins, which marks the sites where the multicistronic transcript is processed. Several mutations that are responsible for inherited mitochondrial diseases are shown.

The gene composition of mtDNA also affects mutagenesis in a less obvious way: because mtDNA only encodes proteins that are involved in energy production, mitochondria must rely on proteins that are encoded by the nucleus for DNA maintenance. These proteins are transcribed in the nucleus, translated in the cytoplasm and ultimately distributed over the mitochondrial population. The proper distribution of these proteins is a daunting task, because each cell contains hundreds of mitochondria, each of which should receive its fair share of DNA repair proteins. Ultimately, mitochondria solve this problem by undergoing continuous cycles of fusion and fission, which allows them to share DNA repair proteins and homogenize them across the mitochondrial population. Thus, mitochondrial fusion and fission promote mitochondrial function by enabling mitochondria to share proteins derived from mtDNA *and* nDNA (figure 8, 9).

Another feature of mitochondrial genetics that is important for our understanding of mutagenesis is the maternal inheritance of mtDNA[16]. This has highly complex consequences for human health. One of the more obvious of these consequences is seen in the clinic, where patients with mitochondrial disease receive genetic counseling for future pregnancies. Clearly, if mtDNA is inherited only through the maternal germ line, mutations are inherited maternally as well. Moreover, certain mutations, like large DNA deletions, are difficult to transmit, and counseling needs to take all of these factors in consideration. The more complex consequences go beyond a single human generation though, and affect us on much longer timescale.

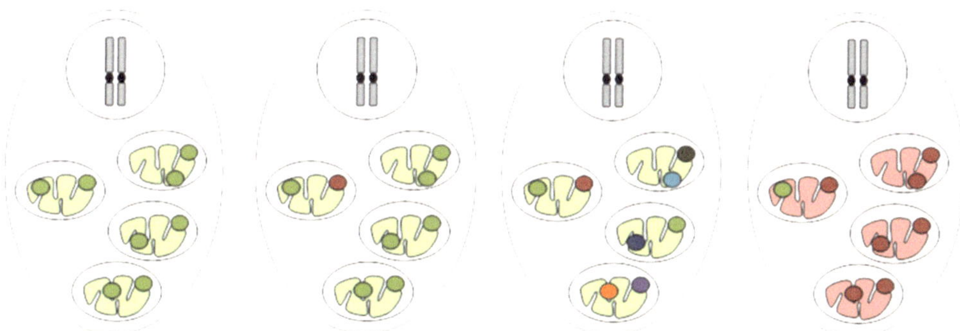

Figure 2. Heteroplasmy and homoplasmy of mtDNA in single cells. Four cartoons of cells are depicted. Each cell contains 1 nucleus (circle with 2 chromosomes), and 4 mitochondria (ovals with light green coloring) that carry 2 genomes, which are either WT (dark green circle) or mutant (colored circles). The cell on the far left contains no mutant mtDNA, and is therefore homoplasmic for WT mtDNA. The second cell from the left contains 1 mutant molecule and is therefore 12.5% heteroplasmic. The third cell contains 7 unique mtDNA mutations in 7 molecules, and is therefore 12.5% heteroplasmic for each mutation. The cell on the right contains 7 identical mutations in 7 genomes, and is therefore 87.5% heterplasmic. Although it contains strictly speaking the same amount of mutations as the third cell, only this fourth cell will display mitochondrial dysfunction.

One long-term consequence is that uniparental inheritance surrenders the benefits of sexual recombination. Sexual recombination gives evolution the opportunity to select against, and remove detrimental mutations from the germline. Without sexual recombination, detrimental mutations would simply accumulate in the germline and degrade our DNA until life becomes unviable, a process called Muller's ratchet. In the absence of sexual recombination, mtDNA must evade Muller's ratchet by some other mechanism. For instance, it is possible that some form of purifying selection occurs in the germline, which selects against oocytes that carry detrimental mtDNA mutations. Until recently though, no direct experimental evidence had been found for this hypothesis. However, two recent studies have now documented powerful proof of such a mechanism in mice. In the first study, a mouse model was generated that exhibits a greatly increased mtDNA mutation rate. The authors discovered that newborn pups of these mutator mice carried significantly more synonymous than non-synonymous mutations than could be expected from random chance[17]. Thus, some form of selection against non-synonymous mutations must be occurring in the germ line. A second study generated two mouse models, which carry either a benign, or a severe mtDNA mutation in their germ line. These authors found that the severe mtDNA mutation was quickly removed from the germ line (in four generations), whereas the benign mutation lingered on for several more generations[18]. These experiments demonstrate that a form of purifying selection protects us from detrimental mtDNA mutations in our germ line, and that the strength of the selection depends directly on the severity of the mutation. The precise mechanism by which this purifying selection operates remains obscure.

A second potential consequence revolves around the co-evolution of the nuclear and the mitochondrial genome. The electron transport chain is composed of approximately a hundred proteins that are divided into 5 major complexes. Thirteen of these proteins are encoded by mtDNA, while the remaining proteins are encoded by nDNA. Because of this shared responsibility, mitochondrial and nuclear DNA have evolved together to optimize energy production. We now know though, that mtDNA differs slightly from person to person, based on their geographical heritage. Thus, the possibility arises that the fine-tuning between mtDNA and nDNA is geographically dependent as well. If so, the co-operation between mtDNA and nDNA may be disrupted by placing mtDNA from one region into the nuclear background of another region. This might cause subtle, but significant metabolic complications. This possibility is supported by experiments in *Drosophila*. Fruit flies are an outstanding animal model for this type of experiment, because different strains of *Drosophila* carry unique mitochondrial genomes. By mating female flies of one strain, with male flies from another strain (and extensively backcrossing their daughters into the nuclear background of their father), the mtDNA of the female fly can be placed in the nuclear background of the male fly. This breaks the evolutionary link between the mtDNA and nDNA of these strains. When mitochondrial function was examined in the experimental flies, it was found that new combinations of mtDNA and nDNA are indeed suboptimal when they are compared to control flies[19, 20]. Thankfully, this is unlikely to be applicable to humans, because even mtDNA from chimpanzees or gorillas function normally in cells with a human nucleus. It is not until mtDNA from animals with a greater evolutionary distance from humans is placed in human cells, like orangutans, that mitochondrial function becomes suboptimal[21].

Other features of mitochondrial genetics that may affect mutagenesis are mtDNA replication and transcription. For over 30 years, it was thought that mtDNA replication occurs through a unique, strand-displacement model[22]. This model stands in stark contrast to the symmetrical model of nuclear DNA replication, which involves coordinated leading and a lagging strand synthesis. If the asymmetric model of mtDNA replication is correct, mtDNA must be partially single stranded for extended periods of time, which would affect the rate at which mtDNA spontaneously deaminates[23-25], or suffers single-strand and double strand breaks[25, 26].

It is important to note though, that a second model has been suggested for mtDNA replication[27, 28], which is more consistent with traditional leading and lagging strand synthesis. According to this model, enormous amounts of RNA are incorporated into the lagging strand during DNA synthesis[29]. If this is the case, then the ribonucleotide incorporation into mtDNA could have a substantial impact on mitochondrial mutagenesis and potentially explain the observed strand-asymmetry in mutagenesis[30]. Unfortunately, it is currently unclear which model is correct, and an important discussion about these models is currently ongoing[27, 28, 31-34]. Regardless of which model is correct, it is clear that the outcome would have an enormous impact on our understanding of mtDNA mutagenesis. After all, DNA replication is an important source for DNA lesions and the tool by which DNA damage is fixed into mutations.

A final feature of mitochondrial genetics that could impact mutagenesis is mtDNA transcription. MtDNA is transcribed in a multi-cistronic manner, meaning that multiple protein-coding genes are transcribed during a single transcription event, similar to bacterial operons[35]. Nascent mRNA molecules are ultimately cleaved at appropriate sites to generate mature mRNAs, which are then handed over to the translation machinery in an orchestrated manner. Although this mechanism is very efficient, it is likely to be very sensitive to DNA damage, because a single DNA lesion could block transcription of multiple genes that are downstream from the lesion. In the nucleus, this problem is solved by transcription coupled DNA repair (TCR), which identifies transcription complexes that are stalled in front of DNA lesions, and recruits the DNA repair machinery with the help of the CSA and CSB proteins[36]. Although CSA and CSB have been localized to mitochondria[37], it is still unclear if, and how transcription coupled repair takes place. For instance, UV lesions are important targets for TCR in the nucleus, where it recruits the nucleotide excision repair machinery to excise the damaged DNA. However, UV lesions are not repaired in mitochondria, and nucleotide excision repair itself does not take place. However, it is possible that, instead of recruiting nucleotide excision repair to stalled transcription complexes, CSA and CSB recruit BER instead[36]. This would be consistent with the idea that oxidative damage is the most likely lesion to occur on mtDNA.

Multi-cistronic transcription also raises the chances of collisions between the transcription bubble and a replication fork. How often such collisions occur, and how they are resolved is still an unexplored question. In the nucleus, these events may be important sources of DNA breaks, which warrants future investigation in studies on mitochondrial mutagenesis.

3. Measuring mutagenesis in mitochondrial DNA

One of the most important problems facing mtDNA researchers today is that there are only a few tools available to detect mutations. Several tools have been tried over the years, but it was often difficult to draw conclusions from the data because different labs observed different results[38]. The source of this confusion is probably the methodology itself, which is inherently error prone[39, 40]. For the field to grow, it will therefore be essential to develop new tools that match the sensitivity and versatility of those used to detect nDNA mutations.

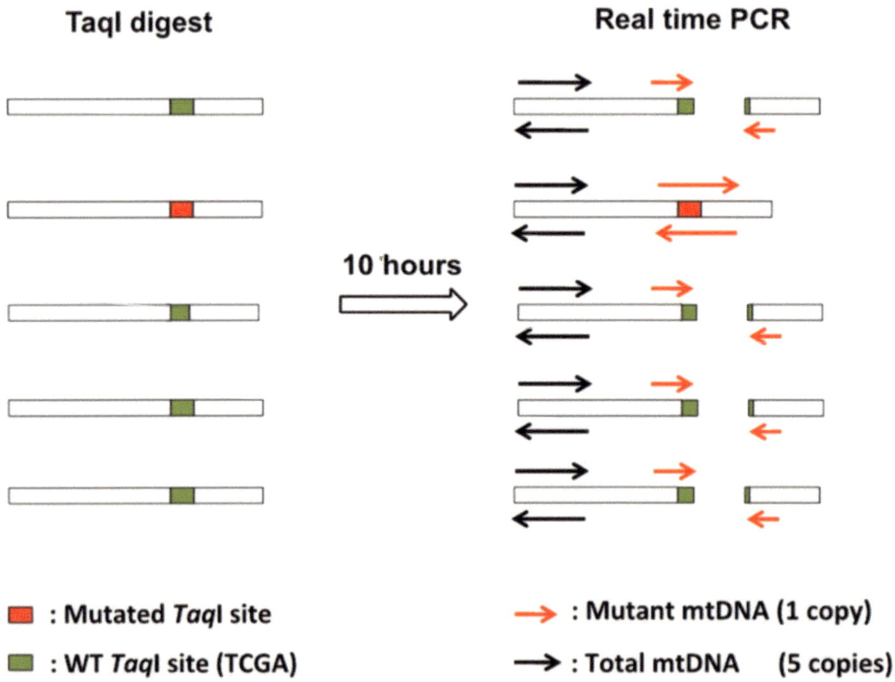

Figure 3. Random mutation capture assay. In this cartoon of the RMC-assay, 5 molecules of MtDNA are digested with TaqI. Four of these molecules contain a WT TaqI restriction site (green boxes) and one molecule contains a mutation in the TaqI site (red box). Using primers that flank the TaqI restriction site, real time PCR can then be used to count the number of copies of mtDNA that contain a mutation in the TaqI restriction site: WT molecules have been cleaved by TaqI, and thus do not provide a viable amplicon for PCR amplification. A second PCR reaction is then performed with primers that are near to, but not affected by the TaqI restriction site. This reaction counts the total number of molecules that were screened for mutations. The true power of this assay is revealed when it is used in a 96 well format, or droplet PCR, so that millions of molecules can be screened at once for mutated TaqI sites.

Most nDNA mutation assays rely heavily on transgenic technology[41], and custom-made mutation markers like fluorescent proteins[42] or the lacZ gene[43] are routinely inserted into a genome of interest. These markers can be genetically engineered to detect any type of mutation, including single base pair substitutions, insertions, deletions or gross rearrangements. Endogenous mutation markers such as the Hprt and Aprt locus can also be used for this purpose[44]. These assays have helped us identify genes that are involved in DNA repair[45], DNA sequences that are prone to mutagenesis[26], and identified environmental compounds that induce mutagenesis[26].

Because mammalian mtDNA cannot be transformed, it is not possible to repeat these experiments in a mitochondrial context. And the small size of mtDNA excludes the possibility of finding large, endogenous genes, such the Hprt or Aprt locus to detect mutations in. However, there are several mutations in the 16S rRNA gene that confer resistance to the drug chloramphenicol[46], which can be used in a mutation screen[47]. Only a handful of bases confer resistance though, which limits the scope of the screen. Another limitation of this assay is that, in order to acquire chloramphenicol resistance, most of the mtDNA molecules inside a cell need to carry the resistance mutation. Thus, mutations that are present in only one, or a few molecules will be missed, including most *de novo* mutations. This assay therefore detects primarily pre-existing mutations that have clonally expanded inside a cell. This extra requirement of clonal expansion introduces an unknown parameter into the assay that could confound the results.

It is important to understand that because of these technical limitations our knowledge of mitochondrial mutagenesis is still fairly limited. This is especially true when it is compared to our knowledge of nDNA, where we know of countless molecules that can induce mutagenesis[45]. We have detailed information about the lesions that are generated during exposure[45], identified the DNA repair proteins that repair these lesions[48], and we know the doses at which mutagenesis occurs. Moreover, we can predict the spectrum of mutations that will arise, and we know the time it takes for these mutations to be fixed. We even know how replication and transcription respond if they encounter these lesions, which signaling pathways are activated as a result, and what the ultimate physiological response of the cell is to these insults[49]. In contrast, we only know of a few chemicals that induce mtDNA mutations, and we have very little information about the proteins that repair these lesions, or how the cell as a whole responds to mtDNA damage.

To fill in these gaps in our knowledge, we probably need to find a way to transform the mitochondrial genome first. Until that time comes though, we will have to rely on biochemical assays and sequencing techniques to detect mutations. These assays need to inform us about two important end points. One assay should help us detect mtDNA mutations in bulk samples. This assay will inform us about the overall mutation rate, frequency and spectrum of mtDNA of large amounts of tissue, so that the impact of diet, age, treatment or gene deletions on mitochondrial mutagenesis can be determined. The second assay should help us determine how rapidly mtDNA mutations are expanding within single cells under these conditions. MtDNA mutations must clonally expand before they affect a cell and this assay will therefore provide us with a direct read out of the physiological impact of mutations on an organism.

Currently, the most accurate assay to measure the mutation frequency in bulk samples is the "random mutation capture assay' (RMC-assay)[39, 50]. This tool measures point mutations, deletions or insertions within restriction sites, and can detect one mutation among $1x10^8$ WT bases (Figure 2). The RMC-assay is a broadly applicable tool, because it uses naked DNA as a template, so that any type of DNA can be interrogated. Moreover, it's a very economic tool, which can screen approximately 25 million bases for $50 in a 4-hour time span. The development of this tool was an important step forward for the field, because it was the first assay to accurately document the spontaneous mutation frequency and spectrum of mtDNA in multiple tissues.

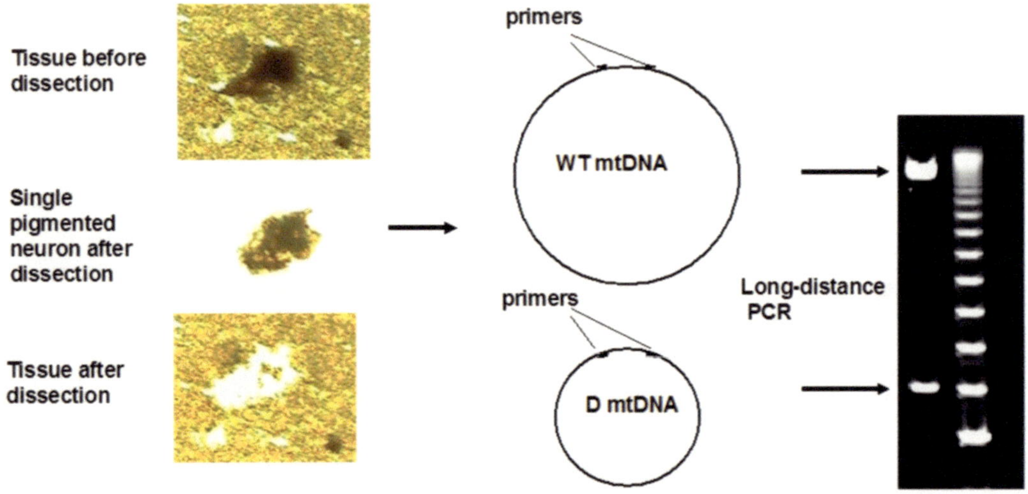

Figure 4. Detection of mtDNA with large deletions in single cells by long-distance PCR. To analyze mtDNA from single cells, a cell is first cut out of a tissue sections using a laser capture microdissection. MtDNA is extracted from the cell in a small scale lysis reaction, which is then used to in a PCR reaction. Primers that encompass the entire genome are used to amplify mtDNA, which can then be analyzed on a gel, or sequenced directly. In the example above, both WT molecuies, and deleted mtDNA molecules were present in a single isolated neuron.

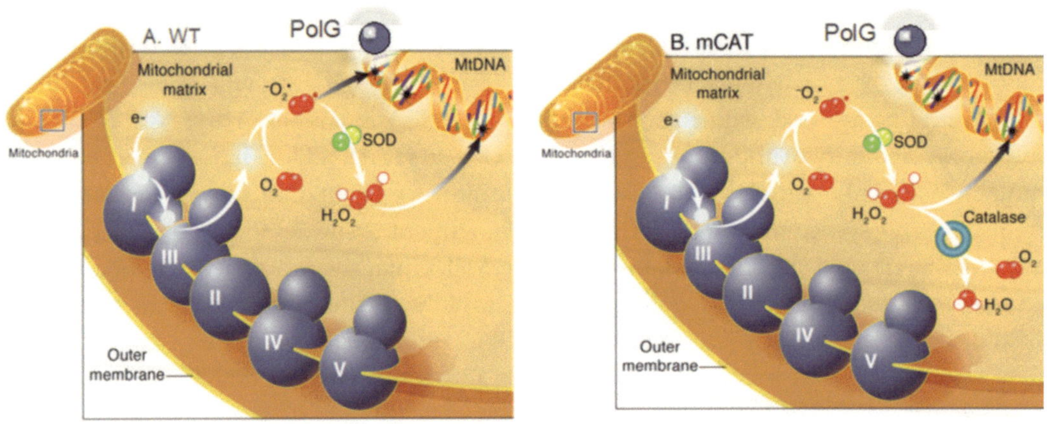

Figure 5. MtDNA in mitochondria, with or without a human catalase. MtDNA is located near the ETC, which exposes it to oxidative damage in the form of the superoxide anion $O_2\bullet$. Superoxide can be broken down into the less reactive H_2O_2 by SOD1 or SOD2. Ultimately, this can react to form a hydroxyl radical, OH• (not shown). Catalase targeted to mitochondria breaks H_2O_2 down into water and oxygen, which decreases the amount of oxidative damage in mitochondria and lowers the mutation rate.

The most precise assay to measure the expansion rate of mtDNA mutations is single cell sequencing of mtDNA[51]. Multiple labs have now established single cell, and even single molecule techniques to do this. In these assays, single cells are collected from tissue slides using laser capture micro-dissection. These cells are lysed in a micro-reaction, after which the mitochondrial genome is PCR amplified and sequenced[51]. If the DNA is diluted to a single molecule level before PCR amplification, it is even possible to sequence single molecules, so that the number of WT and mutant molecules can be determined to estimate the level of heteroplasmy in a cell. This tool is especially powerful when it is used in conjunction with a histo-chemical staining technique to identify cells that have lost mitochondrial function[52].

Each assay does have its drawbacks though. For instance, an important limitation of the random mutation capture assay is that it only detects mutations in *TaqI* restriction sites. This makes it harder to extrapolate the results to the rest of the genome. Single cell sequencing on the other hand, is very costly and laborious. It is expected though, that modern sequencing techniques will greatly improve the cost-effectiveness of this assay in the near future.

4. Mechanisms of mitochondrial mutagenesis

Accurate tools to measure mitochondrial mutagenesis in mammalian cells have only recently been developed. Thus, the mechanisms that cause mtDNA mutations in mammals are still being elucidated. However, several trends are already clear. First, mtDNA is anchored to the inner mitochondrial membrane, a structure that also harbors the electron transport chain. Along this chain, energy is produced in the form of ATP along 5 major complexes (complex I-V). Along this chain, electrons are shuttled from complex to complex, which stores up energy in the form of an electrochemical gradient that drives ATP synthesis. During this process, electrons can escape from the ETC, and react with oxygen to create a highly toxic form of reactive oxygen species. For instance, if the ETC becomes highly reduced, excess electrons from complex I or complex III can react directly with oxygen (O_2) to generate the short-lived superoxide anion $O_2^{\bullet}$. This molecule can react directly with mtDNA, or be converted into another form of reactive oxygen species, hydrogen peroxide (H_2O_2), by manganese superoxide dismutase (Sod2) or cupper/zinc super oxide dismutase[53] (Sod1). Hydrogen peroxide is a less reactive than $O_2^{\bullet}$, which makes it safer, but it is also longer-lived, and can diffuse far enough into the cell to reach the cytosol or the nucleus. H_2O_2 can be further reduced into a hydroxyl radical ($OH^{\bullet}$), the most potent oxidizing ROS, when it encounters a reduced metal or $O2^{\bullet}$. Under normal physiological conditions though, ROS production is kept at a low level. However, if ROS production increases (for instance if the respiratory chain is inhibited and electrons accumulate on the ETC carriers), it can exceed the antioxidant defenses of the mitochondria. If this happens, ROS can damage mtDNA, forming lesions on the genome that can be fixed into mutations. Thus, ROS is bound to be an important source of mitochondrial mutagenesis, and multiple experiments support this hypothesis. For instance, direct oxidative damage and drugs that induce oxidative damage cause homoplasmic mtDNA mutations in mammalian mtDNA[55].

And the reverse was also shown: reducing ROS by expressing a human catalase (an enzyme that breaks hydrogen peroxide down into oxygen and water) that is targeted to mitochondria lowers the mutation rate (figure 3). Mice expressing mitochondrial catalase were shown to display a 2.5-lower mutation burden than normal mice[40, 56]. These, and many other experiments show not only that ROS *can* cause mutations during periods of oxidative stress, but also that ROS drive the endogenous mutation rate in the absence of stress. This hypothesis is further supported by the mutation spectrum of WT mtDNA (figure 4), in which the predominant mutation is a GC::AT transition, which is the most common mutation caused by reactive oxygen species[23, 57]. These GC::AT transitions are generated by cytosine deamination through cytosine glycol and uracil glycol intermediates. Ultimately, uracil will pair with adenine during replication, causing the observed GC::AT transitions[23, 57]. Interestingly, GC::AT transitions are also the predominant mutations found in phylogenetic analyses[30] and many congenital mtDNA diseases[58], suggesting that these mutations may also be the result of oxidative damage.

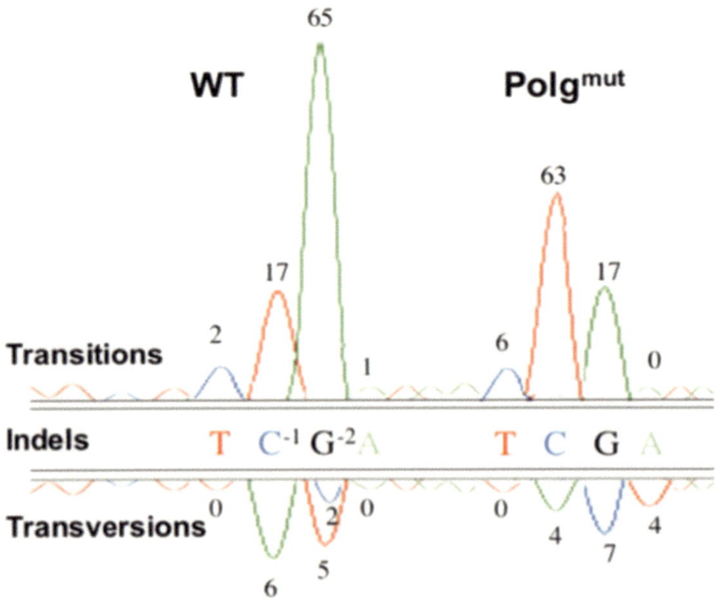

Figure 6. Mutation spectrum of WT mice and mitochondrial mutator mice. The mutation spectrum across 3 TaqI sites is shown in the form of electrophoretograms. The numbers above the peaks indicate the percentage of mutations found for that substitution. For instance, 65% of the mutations that occur at the TaqI sites in WT mice occur at the 3rd base pair (G), and are G to A transitions (green peak, corresponding to an adenine base). The mutator mice primarily display GC to AT transitions as well, but at a different base.

Reactive oxygen species are an unavoidable by-product of ATP synthesis[59]. Thus, it can be expected that the rate of ATP synthesis itself will affect mtDNA mutagenesis. Indeed, in a recent study in human cells, a strong correlation was found between mitochondrial respiration and mitochondrial mutagenesis. Cells that relied on mitochondrial respiration displayed higher mutation frequencies than cells that used glycolysis for ATP production.

This included cancer cells, which are known to prefer glycolysis for ATP production over respiration, a phenomenon referred to as the Warburg effect[60]. This observation is so far just a correlation though, and more mechanistic experiments will need to substantiate this finding.

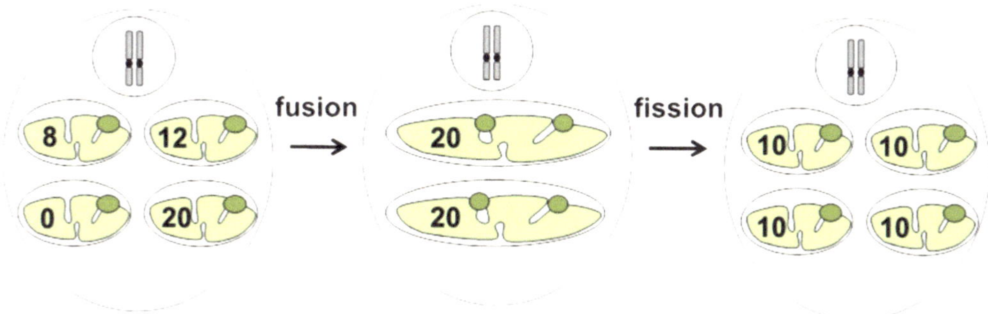

Figure 7. Mitochondrial fusion and fission homogenize protein content. Three cartoons of cells are depicted. Each celll contains one nucleus (circle with two chromosomes), and 4 mitochondria, which carry a variable number of DNA repair proteins (numbers inside matrix). After a cycle of fusion and fission, these proteins are homogenized.

Taken together, these experiments suggest that any mechanism controlling the production of oxygen radicals will impact mtDNA mutagenesis in some way. Such mechanisms could include homoplasmic mtDNA mutations[61]), proteins that control how tightly the electron transport chain is coupled, such as UCP1-3[62], or associated proteins that control major biological processes inside mitochondria such as ANT1[63].

Another potential source of mtDNA mutations is DNA polymerase gamma, the enzyme that replicates the mitochondrial genome[64]. DNA polymerase gamma is the enzyme that fixes DNA damage into mutations during DNA replication, but it is also possible that PolgA makes spontaneous errors on undamaged templates, which could contribute to the mutation rate as well. How large this contribution is, is difficult to say though, because DNA polymerase gamma has such a low error rate that it cannot be determined with current *in vitro* assays[65]. Thus, creating a mutation spectrum *in vitro*, and searching for traces of that spectrum *in vivo* is impossible. What is possible though, is to increase the error rate of DNA polymerase gamma by removing its proofreading domain. This proofreading domain has a 3′-5′ exonuclease activity[64] which corrects errors that are made during DNA synthesis. If this activity is disabled with a point mutation, the spontaneous error spectrum of the enzyme can be exposed[65]. Interestingly, a proofreading deficient DNA polymerase gamma induces predominantly GC::AT transitions *in vivo[66]*, similar the spectrum found in WT cells. However, a more careful comparison of these spectra at specific sites shows that the errors made by a proofreading deficient DNA polymerase gamma occur at different bases than those found in WT cells[40]. This suggests that spontaneous errors by DNA polymerase gamma on undamaged templates is not a major contributor to the mutation rate of WT cells. Its contribution may be greater under certain conditions though, for instance when rapid expansion of mtDNA is required, as is the case during development.

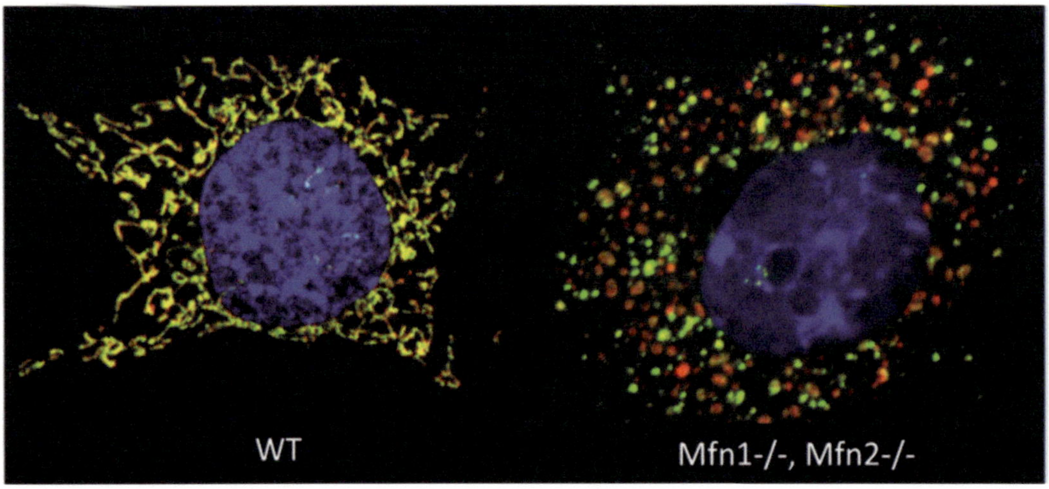

Figure 8. Mitochondrial fusion and fission homogenize protein content. One WT cell, and one fusion deficient cell line is depicted. In each cell, the nucleus is labeled with DAPI (blue) and immunocytochemistry has been used to label cytochrome C green and hsp60 red. Mitochondria in the WT cell contain approximately the same amount of cytochrome C and hsp60, resulting in a yellow coloring. Mitochondria in fusion deficient cells on the other hand carry highly variable amounts of these proteins.

A third source of mitochondrial mutagenesis is protein heterogeneity. Mitochondria typically contain 700-1400 proteins[67], which support a wide variety of functions. MtDNA only encodes 13 of these, all of which contribute to energy production. The proteins that safeguard the integrity of mitochondrial genome are therefore all encoded by the nucleus and imported into mitochondria from of the cytoplasm. These proteins include DNA polymerase gamma, the mitochondrial helicase Twinkle, and a wide variety of DNA repair proteins. Ultimately these proteins must be delivered in the proper stoichiometry in each mitochondrion to enable DNA repair[68]. As far as we know though, there is no way for the nucleus to orchestrate the distribution of these proteins over hundreds of mitochondria in such a precise manner. As a result, the distribution of DNA repair proteins across the mitochondrial population is likely to be uneven. This type of protein heterogeneity could put mitochondria at risk for mutagenesis. Ultimately, this problem is resolved by mitochondrial dynamics. While mitochondria travel along the cytoskeleton, they frequently collide in an end-to-end fashion. When they do, they fuse their membranes together to form a single, continuous organelle. This fusion process is mediated by 3 GTP-ases. Mitofusin 1 (Mfn1) and mitofusin 2 (Mfn2) help to fuse the outer membranes together, while OPA1 fuses the inner membranes together[13]. Fusion of the inner membrane creates one uninterrupted matrix, which contains all the molecules of the original fusion partners, including any DNA repair proteins. This newly formed mitochondrion will eventually be split back into two smaller organelles during a fission event, which is mediated by at least 3 proteins, Drp1[69], Fis1[69] and Mff[70]. The emerging organelles receive a random distribution of DNA, RNA and proteins, so that each mitochondrion contains an "average" of all the molecules that the initial fusion partners shared, including any DNA repair proteins. Because mitochondria undergo constant cycles of fusion and fission, DNA repair proteins are continuously homogenized over the mitochondrial population, which results in a more equal distribution of protein content (see fig 8,9). This idea is supported by mouse models, as well as cell lines, in which mitochondrial fusion has been disabled by deletion of either Mfn1 or Mfn2. In these cases, proteins are no longer homogenized, resulting in greater protein heterogeneity, which increases mtDNA mutations in tissues and cultured cells[15].

However, whether protein heterogeneity also contributes to spontaneous mutagenesis in WT cells remains unclear. If it does though, its contribution is likely to differ between cell types and conditions. For instance, in neurons, mitochondria are located far away from the cell body where fewer fusion partners may exist, which could result in increased protein heterogeneity. This problem may be exacerbated in older cells, where mitochondrial motility is further compromised due to excessive traffic jams inside axons[71]. Another sensitive cell type may be muscle fibers, which carry tens of thousands of mitochondria, and countless genomes. Some of the most active mitochondria in muscle fibers are placed in very rigid positions, in a pair wise pattern along the z-axis of the fiber. Their constant, regular shape and precise placement suggest that they undergo limited fusion and fission, and are restricted in their movement. As a result, these mitochondria may rely primarily on the import of DNA repair proteins to maintain mtDNA stability. This holds the inherent risk of increased protein heterogeneity.

5. DNA repair in mitochondria

Until recently, it was very difficult to measure mtDNA mutations in mammalian cells. For this reason, it remains unclear which *in vivo* DNA repair mechanisms suppress mutations in mtDNA. In contrast, mitochondrial mutation assays are easier to perform in yeast, and as a result our knowledge about mtDNA repair in mitochondria comes largely from yeast experiments. We now know that base excision repair (BER), mismatch repair (MMR) and recombination repair all occur in yeast mitochondria. However, nucleotide excision repair (NER) is conspicuously absent. Of these DNA repair pathways, BER understandably the most active pathway in mitochondria, given the proximity of mtDNA to the ETC. Accordingly, loss of either Ogg1p[72] (which repairs 8-oxo-guanine), Ung1p[73] (which excises uracil from mtDNA), or Ntg1p[74] (which excises oxidized pyrimidines) results in a 2-10 fold increase in mtDNA mutations in yeast. BER activity in mitochondria is further supported by Apn1[75] (an AP endonuclease), Pif1[74] (a helicase), MTH1[76] (an 8-oxo-dGTPase and 8-oxodATPase), and MYH1[77] (which removes adenine opposite 8-oxodG). These experiments in yeast have guided similar efforts to explore DNA repair in mammalian mitochondria, where a similar set of BER proteins has been detected. These experiments demonstrated that both the short and long-patch version of BER is present in mammalian mitochondria, with Dna2[78] and Fen1 removing the intermediate flaps[79]. Remarkably though, simultaneous loss of OGG1 and MYH (which removes adenine incorporated opposite 8-oxo-guanine) did not result in an increased mutagenesis in mammalian cells[80], in contrast to yeast. The reason for this discrepancy is unclear, although it is possible that extensive back-up mechanisms are present in mammalian mitochondria in the form of NEIL1 and NEIL2[80, 81], two glycosylases whose functions overlap with OGG1. A second possibility is that 8-oxo-guanine is not a common a lesion in mammalian mitochondria. Historically, 8-oxo-guanine has been the most studied oxidative lesion in the DNA repair field because it is practically the only lesion that is easily detected. However, there are other oxidative lesions that are undoubtedly more mutagenic than 8-oxo guanine, and some of these may even occur more frequently, but because they are almost impossible to detect, they remain unstudied. Thus, our emphasis on this lesion may be the result of a bias in our studies. The fact that multiple DNA repair enzymes exist to remove this lesion clearly argues otherwise though. Regardless, it will be essential for our understanding of DNA repair in mitochondria to screen all available mouse models with deleted DNA repair genes for increases in mitochondrial mutagenesis using modern tools. This is the only way we will get a clear picture of the DNA repair mechanisms that are active inside mitochondria.

Although BER is now clearly defined in mammalian mitochondria, there is only limited evidence for other DNA repair pathways in mammalian mitochondria, including mismatch repair and recombination repair, two pathways that are active in yeast mitochondria. Nucleotide excision repair is absent in mammalian mitochondria though, as it is in yeast, since UV lesions are not repaired in mammalian mitochondria[82].

In the nucleus, mismatch repair is an extremely important DNA repair pathway, which corrects errors that are generated during DNA replication[83]. The mismatch repair (MMR) proteins are composed of heterodimeric polypeptides termed MutS and MutL, which differ in composition to match the type of lesion they repair (either single base mismatches or small insertion-deletion loops). These proteins orchestrate the DNA repair process by first detecting the mismatch, and then discriminating between the two DNA strands to identify the template strand (which is correct), and the newly synthesized strand (which contains the mistake). It is still unclear how strand discrimination occurs in mammalian cells though.

Mitochondrial MMR was first identified in yeast[84, 85]. And since then, numerous experiments have tried to detect MMR activity in mammalian mitochondria. Some of these studies have found that the lysate of isolated mitochondria can repair mismatched templates [86]. However, it is possible for these lysates to be contaminated with nuclear proteins, which could confound the results. Thus, numerous researchers have also tried to image cells, in order to detect fluorescently tagged MMR proteins in mitochondria. Most of these experiments failed to detect the MutS or MutL complexes in mitochondria. However, they did identify a new protein, YB-1, which seems to aid mitochondrial MMR[47]. This raises the surprising possibility that a different set of proteins governs mismatch repair in mitochondria compared to the nucleus. In contrast, BER in mitochondria is performed by the same set of proteins that are present in the nucleus. However, if it is true that mtDNA is replicated by a mechanism that is different from nuclear DNA, it may actually be expected that a different type of enzyme is required for MMR. For now though, it will be important to validate this observation further with mechanistic insight. It will be important to elucidate the function of YB-1 is in mitochondrial MMR, determine its activity, and demonstrate how it aids in strand-discrimination. And finally, since loss of the proofreading activity of DNA polymerase gamma results in a >1000-fold increase in mutations, it will be important to understand how this increase can be so large in the presence of MMR.

Besides single base lesions, oxidative damage can also causes single strand and double strand breaks in DNA, as well as DNA cross-links. For this reason, it would make sense for recombination repair to take place in mammalian mitochondria alongside BER. The most direct evidence for mitochondrial recombination comes from sequence analysis of mtDNA molecules. For instance, multiple labs have found evidence for intra-, as well as inter-strand recombination events[87-91]. These experiments suggest that micro-homology between mtDNA sequences is used for repair activity. Although the mechanisms of homologous recombination in mitochondria are still unclear, it seems as though the proofreading domain of DNA polymerase gamma plays an important role in this process. When the proofreading domain is deactivated, mtDNA molecules seem to recombine without the need for homology, with breakpoints that are reminiscent of NHEJ events[88]. Thus, one possibility is that PolgA aids homology searching during DNA repair processes. Besides PolgA, it is unclear which molecules function in double strand break repair. For instance, RAD51 and RAD52 have not been found in mammalian mitochondria[79]. A third protein, Mre11, which is involved in nuclear DSB repair, does aid recombination repair in yeast mitochondria though, and may be present in mammalian mitochondria as well[92, 93].

The relatively sparse evidence for DNA repair pathways in mitochondria has even led some to suggest that rather than being repaired, damaged mtDNA may be targeted for destruction. Destroying an entire genome instead making a few repairs seems wasteful though, and more data is needed to build the case for this hypothesis. For instance, the nuclease that would destroy damaged mitochondrial genomes has not yet been identified[79]. Others have suggested that mitochondria containing damaged DNA or excessive mutations may be targeted for mitophagy. If this hypothesis is true though, one would expect that mutated genomes are removed from a cell until only the WT genome persists. Clearly this is not the case though, because congenital mtDNA mutations persist in mitochondrial patients, and are not selectively removed.

6. The effect of mtDNA mutations on human health

MtDNA encodes 13 proteins, all of which are essential to mitochondrial energy production. Accordingly, mtDNA mutations have the strongest impact on cells that have persistent, and high energy demands, including neurons, muscle fibers and the endocrine system. Despite the fact that only a limited set of cells is strongly affected by mtDNA mutations, it is very difficult to predict the effect a mutation will have on a patient due to the complexity of mitochondrial genetics. This is especially true in the case of inherited mtDNA mutations.

First, it is possible for mutations in different genes to have nearly identical consequences. This is a direct result of the extensive functional overlap between different mtDNA genes.

One example of a mitochondrial disease that is caused by these types of mutations is Lebers hereditary optical neuropathy[94] (LHON), a disease that causes a sudden onset of blindness due to mutations in the mitochondrial ND1, ND2, ND3, ND4, ND5, ND6[95-97], COII, COIII[98], CYB[94], or ATP6 gene[58, 94] (see also figure 1).

Surprisingly though, there is also a second class of mutations, which occur in the same gene, but have completely different consequences. For example, missense mutations in the ND6 gene can either cause Leigh syndrome[99-101], generalized dystonia and deafness[101, 102], mitochondrial encephalomyopathy with lactic acidosis and stroke-like episodes[103] (MELAS), or Leber hereditary optic neuropathy[94].

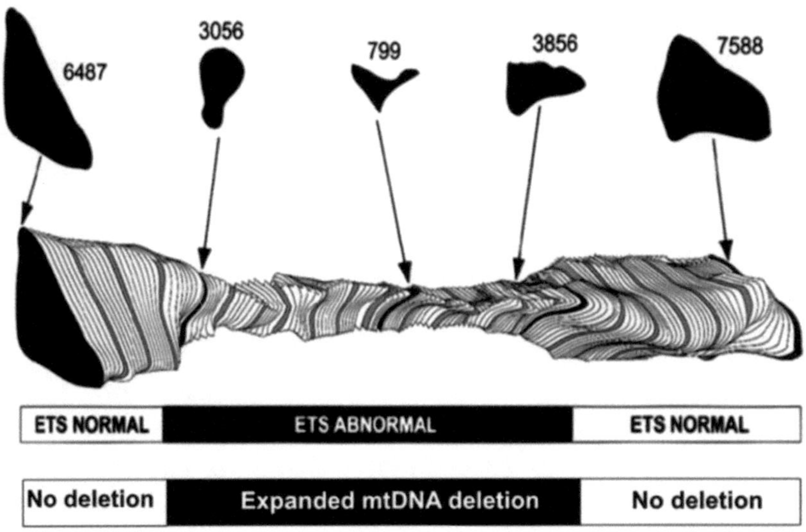

Figure 9. Data from Wanagat et al., FEBS 2000. A single muscle fiber is depicted in the cartoon above, which has been reconstructed from tens of sections that were taken across its length. Each slice represents one of these sections. Healthy sections of the fiber (at the edges) are larger than the sections that display mitochondrial dysfunction (center). The unhealthy sections contain excessive amounts of mtDNA deletions, whereas the healthy sections do not.

Finally, a third class exists, in which identical mutations in the same gene cause different phenotypes depending on the level of heteroplasmy at which that mutation is present in a patient's tissues. For instance, some of the most frequent mtDNA disorders are caused by mutations in the mitochondrial tRNA genes. The most common of these is an (A > G) mutation in the tRNALeu(UUR) gene[104]. When present at low levels of heteroplasmy (10%–30%) a patient may present with type II diabetes, with or without deafness[105]. This is actually the most common inherited cause of type II diabetes, accounting for approximately 1% of all type II diabetes in the world[59]. By contrast, when this mutation is present in >70% of the patients mtDNA molecules, it does not cause diabetes, but presents itself with more severe symptoms[104], including short stature, cardiomyopathy, CPEO (a mitochondrial myopathy that is frequently associated with ophthalmoplegia and ptosis, referred to as chronic progressive external ophthalmoplegia), and the MELAS syndrome.

The diagnosis of these diseases is further complicated by the fact that the symptoms of different mtDNA diseases can overlap with each other. For instance, some mtDNA rearrangements cause CPEO, which is characterized by a slow, progressive paralysis of the extraocular muscles[59]. These features can be completely mimicked by Kearns–Sayre syndrome (KSS) though, which can present itself just like CPEO[106]. However, KSS is a much more devastating myopathy, which progressses to a multisystemic disorder manifested through cardiac dysfunction, mental retardation, and various endocrine disorders. Another example of clinical phenotypes melding into each other is Pearson's pancreatic syndrome, which is caused by large mtDNA rearrangements that affect the bone marrow and frequently result in childhood pancytopenia. Pearson's syndrome can progress to KSS if the pancytopenia is treated successfully[59].

Like congenital mutations, somatic mtDNA mutations tend to impact muscles fibers and neurons as well. However, they do so in a very different way. If a congenital mutation is present in the zygote, it is passed from cell to cell during development, so that a single mutation is dispersed across a patient's tissues. Ultimately, the severity of the resulting disease is dictated by the amount of mutated mtDNA that ends up in a patient's muscle fibers and neurons. However, each cell carries the same mutation.

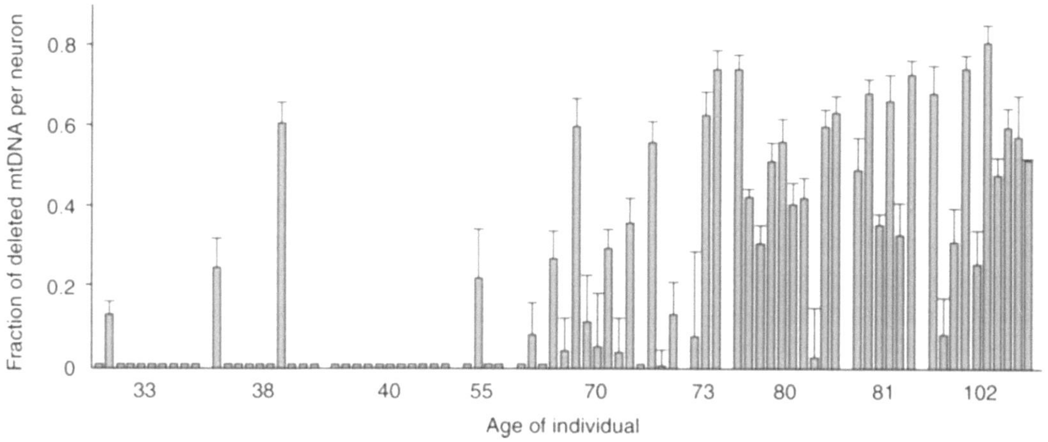

Figure 10. Data from Kraytsberg, Nature Genetics 2006. MtDNA was collected from single neurons, derived from nine individuals aged 33-102. The mtDNA of these individuals was then PCR amplified and sequenced. The fraction of the mtDNA molecules in those cells that contained an mtDNA deletion is represented. Each peak corresponds to the analysis of one neuron.

Somatic mutations on the other hand, arise in neurons and muscle fibers that are already differentiated, post-mitotic cells. Thus, each mutation only affects the cell it arises in, and will not be passed on from cell to cell. For a tissue to be affected then, countless mutations must occur, resulting in a mosaic pattern of cells that either do, or do not carry a mutation. Moreover, since each mutation results from a unique event, each cell carries a unique mtDNA mutation. This causes an important problem, because if each mutation only occurs once, they are almost impossible to detect in bulk samples, where millions of cells are monitored simultaneously. To find these mutations, a tissue must be screened one cell at a time. When single neurons, cardiomyocytes or muscle fibers with mitochondrial abnormalities are micro-dissected, and mtDNA is isolated from these single cells, it is indeed found that each cell carries a unique clonally expanded mtDNA mutation. These types of screens are very labor intensive, but they are the only way to measure the impact of somatic mtDNA mutations on a tissue.

These ground breaking single cell screens were first performed on muscle fibers[52, 107]. As we grow older, our muscle fibers tend to undergo age-related atrophy. This process is accompanied by segmental mitochondrial dysfunction, which can ultimately cause disruption of single fibers. If a dysfunctional muscle fiber is dissected, and mtDNA is isolated from different sections along the fiber, it is invariably found that dysfunctional sections carry a single mtDNA mutation that has clonally expanded to greater than 80-90% of all mtDNA molecules in the dysfunctional section of the fiber (figure 5). Healthy sections of the fiber on the other hand do not carry the mutation.

Similar experiments have been performed on neurons. As we grow older, our neurons tend to lose mitochondrial function, and the most prominent group of neurons to do so, are the dopaminergic neurons found in the substantia nigra[108, 109]. Like muscle fibers, these neurons can be recovered using laser capture micro-dissection, and their mtDNA can be sequenced. When the mtDNA is analyzed of patients with increasing age, it is indeed found that the substantia nigra of older patients carry more mtDNA deletions inside their cells than younger patients[109] (figure 6). Since the substantia nigra is the primary tissue affected by Parkinson's disease, it is thought that these mtDNA deletions may also play a key role in the etiology of this disease[110].

Besides post-mitotic cells, mtDNA mutations can also affect dividing cells. For instance, mtDNA mutations are frequently found in aging colonic stem cells[111]. Most dividing cells do not require enormous amounts of ATP production though, so that the pathology caused by mtDNA mutations is relatively mild in these cell types. Tumor cells may be an important exception to this rule[112]. In recent years, countless tumors have been tested for mutations in their mitochondrial genome. These screens revealed that human tumors frequently carry clonally expanded mtDNA mutations. Interestingly, these mutations seem to be present in almost every cell of the tumor, suggesting that they either originated in the original cancer stem cell, or were acquired during one of the genetic bottlenecks that define cancer progression. Since then, an important debate has erupted on the role of mtDNA mutations in human cancers. This debate centers around a single question: do mtDNA mutations provide a selective advantage to cancer cells?[112]

Opponents of this idea pointed out that, if it is true that mtDNA mutations induce carcinogenesis, then the offspring of a mouse that carries such mutations should develop tumors as well[113, 114]. However, no bias toward maternal inheritance of carcinogenesis has been reported. The foundation of this hypothesis was further shaken when it was shown that several studies that initially reported the presence of mtDNA mutations in human cancers[115-118] contained analytic or technical errors[119]. As a result, most researchers started to doubt whether mtDNA mutations had anything to do with carcinogenesis.

However, an unexplored possibility is that mtDNA mutations do not affect the initial oncogenic transformation of a tumor cell, but a later stage of cancer progression. One recent study tested this hypothesis by investigating whether mtDNA mutations control metastasis {!!!!Ishikawa, Hayashi, Science, 2008 please insert reference properly!!!}, the final, and most deadly stage of cancer progression. To do this, the authors of this study replaced the mtDNA of a cell line with low metastatic potential, with the mtDNA of a cell line that has high metastatic potential. They discovered that the metastatic potential of the malignant cell line was transferred to the benign cell line, not with its nDNA, but with its mtDNA. Conversely, if the mtDNA from a benign cell line was placed in the cytoplasm of the malignant cell line, the malignant cell line lost its metastatic potential. Additional analysis then showed that the mtDNA of the malignant cell line contained an mtDNA mutation in the ND6 gene, which increased ROS production. The increased ROS production must have been responsible for the metastatic potential of the cells, because if the malignant cell line was treated with ROS scavengers prior to implantation into mice to test its metastatic potential, the malignant cell line displayed a benign phenotype. This experiment was extremely important for the field for three reasons. First, it explained how mtDNA mutations can be involved in cancer progression without causing oncogenic transformation. Second, it demonstrated how proper mechanistic studies can be conducted in order to probe the role of mtDNA mutations in cancer progression. And third, it provided the first true mechanistic link between mtDNA mutations and metastasis.

7. Mouse models of mitochondrial mutagenesis

Because it is currently impossible to transform the mitochondrial genome of mice, it has been difficult to generate mouse models that mimic mitochondrial disease. Despite this frustrating technical limitation, there has actually been tremendous progress in this area over the last decade. During this time, three classes of mouse models have been developed: one class of mouse models was generated to study hereditary mtDNA diseases, a second to study somatic mtDNA diseases, and a third class was generated to study alleles of genes that cause disease in humans.

Of these three mouse models, the most difficult class to develop was a model to study inherited mtDNA mutations. Such a model requires a change to be made to the primary sequence of the mitochondrial genome. To do this, researchers have started using cybrid technology[120, 121]. Cybrid technology allows researchers to generate cells that are cytoplasmic hybrids of each other, or cybrids for short. A typical cybrid is created by fusing an enucleated donor cell, which contains an interesting mtDNA mutation, to an embryonic stem cell that has previously been depleted of its own mitochondrial DNA with ethidium bromide treatment. If the fusion process is successful, the resulting cell line will contain the nDNA of the ES cell and the mtDNA of the donor cell. This cell line can then be injected into blastocysts and placed in a foster mother to generate mice with an inherited mtDNA mutation.

Although this is technically challenging, the true task is to find a cell line that carries an interesting mtDNA mutation. One way to find such a cell line, is to exposing cells to drugs that interfere with the ETC, and then select for cells that become resistant. For instance, treatments with chloramphenicol[120, 122], antimycin A[123], or rotenone[124] have yielded several cell lines that contain harmful homoplasmic mutations in their mtDNA. These mutations can then be moved into the germ line of mice according to the protocol described above. Mice that carry the chloramphenicol resistance mutation for instance, exhibit pathology that is reminiscent of the MELAS syndrome, attesting to the usefulness of this approach.

Other "mito-mice" that have been generated carry either a 4.7kb DNA deletion[125, 126], or point mutations in the ND6, COI, or 16s rRNA gene[18]. These mice exhibit a spectrum of phenotypes that mimic mtDNA diseases and have become valuable tools to understand the natural history of inherited mtDNA mutations. So far, they have helped us understand how mtDNA mutations are transmitted through the germ line, how they are dispersed throughout the organism during development, and how they ultimately cause pathology in mature mice[18, 125]. There were also some surprises though. For instance, patients with large mtDNA deletions typically present with muscle weakness, exercise intolerance, abnormal mitochondria in their skeletal muscle fibers, diabetes, pancreatic dysfunction and hearing loss. However, mice that carry a 4.7 kb deletion have a very different phenotype. They do exhibit respiratory dysfunction in their muscle cells, but suffer no diabetes, hearing loss, exercise intolerance or loss of pancreatic function. Instead, these mice die overwhelmingly of kidney failure, which is not a typical symptom of mitochondrial disease.

A second class of mouse models was developed to study the effect of somatic mtDNA mutations on human health. To do this, researchers increased the somatic mutation rate of mtDNA, by generating error prone versions of DNA polymerase gamma, the enzyme that replicates the mitochondrial genome. DNA polymerase gamma contains a 3'-5' exonuclease domain, which corrects errors that are made during DNA synthesis, similar to the major DNA polymerases in the nucleus[127]. This proofreading domain can be knocked out with a single point mutation[128, 129], which vastly increases the mitochondrial mutation rate[40]. Numerous mouse models have been generated that express this error prone allele either as a tissue specific transgene[130], or systematically using gene substitution techniques[128, 129]. The most informative model has been the mice that express the error prone polymerase gamma from its native locus. Interestingly, both the heterozygous and homozygous carriers of the error prone allele display enormous increases in mutations. Thus, the error prone allele is partially dominant, which is consistent with previous results in yeast[131, 132]. It was found that homozygous carriers of the error prone allele display a 1000 fold increase in point mutations, whereas heterozygous carriers display a 100 fold increase in mutations[40], which is again comparable to experiments performed in yeast[131, 132]. These mouse models represent an extremely important contribution to the field because they are the first to directly manipulate the fidelity of the mitochondrial genome in mammals. As a result, they are also the first models to show a true mechanistic link between mtDNA mutations and disease. Interestingly, it was found that the homozygous carriers of the error prone allele suffer from extensive mitochondrial disease, which manifests itself as a premature aging like syndrome. These animals exhibit symptoms of alopecia (loss of fur), kyphosis (arching of the spine), premature deafness, premature blindness, cardiomyopathy, loss of subcutaneous fat, anemia, osteoporosis, loss of fertility and numerous other problems that resemble human aging[128, 129]. Partially, these problems are caused by extensive apoptosis, especially in dividing tissues[128]. Whether mitochondrial mutations cause these symptoms normal mice is still being debated. In later experiments it was shown that the heterozygous carriers of the proofreading deficient PolgA allele (which also display increased amounts of point mutations compared to normal mice) do not display overt symptoms of disease, or features of premature aging[40]. However, one important distinction is that, in contrast to the homozygous mice, the heterozygous carriers do not display an increased amount of mtDNA deletions, or aborted replication intermediates[88, 133]. It will be important to further test whether these deletions or the replication intermediates contribute significantly to the phenotype of the homozygous mice with the appropriate tools.

Finally, a third class of mouse models has been developed that carry specific alleles of genes that cause disease in humans. These mutations interfere with the normal function of proteins that are part of the mitochondrial replication fork. This fork consists of four core proteins: the catalytic subunit of DNA polymerase gamma, two accessory subunits, and the mitochondrial helicase Twinkle. Mutations in all these genes are known to cause various neuromuscular diseases, including progressive external opthalmoplegia (PEO), Alpers disease and ataxia neuropathy. Some of the mutations in Twinkle that are known to cause PEO in humans have now been reconstituted in mice[136]. These mouse models have provided us with fantastic new insight into the etiology of adult onset PEO. First, the muscles of these animals faithfully replicate all of the key histological, genetic, and biochemical features of PEO patients. Secondly, these alleles recreate a mitochondrial mutator phenotype by increasing the amount of mtDNA deletions that occur in somatic cells. Importantly, these mtDNA deletions do not result in a premature aging like syndrome, which is powerful evidence against a role for mtDNA deletions in aging. On the other hand though, it should be noted that these mice over-express transgenic copies of Twinkle. As a result, the expression pattern is likely to be mosaic, which may affect the severity of mitochondrial disease in these tissues. Regardless, these mouse models are extremely important successes on the road to battling mitochondrial disease. They provide a new system to test treatment options in, and to study the natural history of diseases that are notoriously difficult to track. As a result, these mouse models are our best to chance to make a difference in the lives of afflicted individuals, and their families. They will eventually illuminate the great unknowns of mitochondrial genetics, and open the door to improved health care and a longer health span, courtesy of mitochondrial medicine.

Author details

Marc Vermulst
Department of Chemistry, University of North Carolina, Chapel Hill, USA

Konstantin Khrapko
Department of Medicine, Division of Gerontology, Beth Israel Deaconess Hospital, Harvard, Boston, USA

Jonathan Wanagat
Department of Medicine, Division of Geriatrics, University of California Los Angeles, Los Angeles, USA

Acknowledgement

Dr. Lawrence A. Loeb, Dr. Jason Bielas, Dr. David Chan, Dr. Dorothy A. Erie, Dr. George Martin, Dr. Peter R. Rabinovitch.

8. References

[1] Ephrussi, B., Jakob, H., and Grandchamp, S. (1966). Etudes Sur La SuppressivitE Des Mutants a Deficience Respiratoire De La Levure. II. Etapes De La Mutation Grande En Petite Provoquee Par Le Facteur Suppressif. Genetics *54*, 1-29.

[2] Mounolou, J.C., Jakob, H., and Slonimski, P.P. (1966). Mitochondrial DNA from yeast "petite" mutants: specific changes in buoyant density corresponding to different cytoplasmic mutations. Biochem Biophys Res Commun *24*, 218-224.

[3] Schatz, G. (1963). The Isolation of Possible Mitochondrial Precursor Structures from Aerobically Grown Baker's Yeast. Biochem Biophys Res Commun *12*, 448-451.

[4] Tewari, K.K., Jayaraman, J., and Mahler, H.R. (1965). Separation and characterization of mitochondrial DNA from yeast. Biochem Biophys Res Commun *21*, 141-148.

[5] Ingman, M., Kaessmann, H., Paabo, S., and Gyllensten, U. (2000). Mitochondrial genome variation and the origin of modern humans. Nature *408*, 708-713.

[6] Cann, R.L., Stoneking, M., and Wilson, A.C. (1987). Mitochondrial DNA and human evolution. Nature *325*, 31-36.

[7] Wallace, D.C. Mitochondrial DNA mutations in disease and aging. Environ Mol Mutagen *51*, 440-450.

[8] McBride, H.M., Neuspiel, M., and Wasiak, S. (2006). Mitochondria: more than just a powerhouse. Curr Biol *16*, R551-560.

[9] Anderson, S., Bankier, A.T., Barrell, B.G., de Bruijn, M.H., Coulson, A.R., Drouin, J., Eperon, I.C., Nierlich, D.P., Roe, B.A., Sanger, F., et al. (1981). Sequence and organization of the human mitochondrial genome. Nature *290*, 457-465.

[10] Andrews, R.M., Kubacka, I., Chinnery, P.F., Lightowlers, R.N., Turnbull, D.M., and Howell, N. (1999). Reanalysis and revision of the Cambridge reference sequence for human mitochondrial DNA. Nat Genet *23*, 147.

[11] Frederick, R.L., and Shaw, J.M. (2007). Moving mitochondria: establishing distribution of an essential organelle. Traffic *8*, 1668-1675.

[12] Nakada, K., Inoue, K., Ono, T., Isobe, K., Ogura, A., Goto, Y.I., Nonaka, I., and Hayashi, J.I. (2001). Inter-mitochondrial complementation: Mitochondria-specific system preventing mice from expression of disease phenotypes by mutant mtDNA. Nat Med *7*, 934-940.

[13] Chan, D.C. (2006). Mitochondria: dynamic organelles in disease, aging, and development. Cell *125*, 1241-1252.

[14] Chomyn, A., Martinuzzi, A., Yoneda, M., Daga, A., Hurko, O., Johns, D., Lai, S.T., Nonaka, I., Angelini, C., and Attardi, G. (1992). MELAS mutation in mtDNA binding site for transcription termination factor causes defects in protein synthesis and in respiration but no change in levels of upstream and downstream mature transcripts. Proc Natl Acad Sci U S A *89*, 4221-4225.

[15] Chen, H., Vermulst, M., Wang, Y.E., Chomyn, A., Prolla, T.A., McCaffery, J.M., and Chan, D.C. (2010). Mitochondrial fusion is required for mtDNA stability in skeletal muscle and tolerance of mtDNA mutations. Cell *141*, 280-289.

[16] Case, J.T., and Wallace, D.C. (1981). Maternal inheritance of mitochondrial DNA polymorphisms in cultured human fibroblasts. Somatic Cell Genet *7*, 103-108.

[17] Stewart, J.B., Freyer, C., Elson, J.L., Wredenberg, A., Cansu, Z., Trifunovic, A., and Larsson, N.G. (2008). Strong purifying selection in transmission of mammalian mitochondrial DNA. PLoS Biol *6*, e10.

[18] Fan, W., Waymire, K.G., Narula, N., Li, P., Rocher, C., Coskun, P.E., Vannan, M.A., Narula, J., Macgregor, G.R., and Wallace, D.C. (2008). A mouse model of mitochondrial disease reveals germline selection against severe mtDNA mutations. Science *319*, 958-962.

[19] Sackton, T.B., Haney, R.A., and Rand, D.M. (2003). Cytonuclear coadaptation in Drosophila: disruption of cytochrome c oxidase activity in backcross genotypes. Evolution *57*, 2315-2325.

[20] Rand, D.M., Fry, A., and Sheldahl, L. (2006). Nuclear-mitochondrial epistasis and drosophila aging: introgression of Drosophila simulans mtDNA modifies longevity in D. melanogaster nuclear backgrounds. Genetics *172*, 329-341.

[21] Rand, D.M., Haney, R.A., and Fry, A.J. (2004). Cytonuclear coevolution: the genomics of cooperation. Trends Ecol Evol *19*, 645-653.

[22] Stumpf, J.D., and Copeland, W.C. Mitochondrial DNA replication and disease: insights from DNA polymerase gamma mutations. Cell Mol Life Sci *68*, 219-233.

[23] Kreutzer, D.A., and Essigmann, J.M. (1998). Oxidized, deaminated cytosines are a source of C --> T transitions in vivo. Proc Natl Acad Sci U S A *95*, 3578-3582.

[24] Frederico, L.A., Kunkel, T.A., and Shaw, B.R. (1990). A sensitive genetic assay for the detection of cytosine deamination: determination of rate constants and the activation energy. Biochemistry *29*, 2532-2537.

[25] Lindahl, T. (1993). Instability and decay of the primary structure of DNA. Nature *362*, 709-715.

[26] Barnes, D.E., Lindahl, T., and Sedgwick, B. (1993). DNA repair. Curr Opin Cell Biol *5*, 424-433.

[27] Yang, M.Y., Bowmaker, M., Reyes, A., Vergani, L., Angeli, P., Gringeri, E., Jacobs, H.T., and Holt, I.J. (2002). Biased incorporation of ribonucleotides on the mitochondrial L-strand accounts for apparent strand-asymmetric DNA replication. Cell *111*, 495-505.

[28] Holt, I.J., Lorimer, H.E., and Jacobs, H.T. (2000). Coupled leading- and lagging-strand synthesis of mammalian mitochondrial DNA. Cell *100*, 515-524.

[29] Pohjoismaki, J.L., Holmes, J.B., Wood, S.R., Yang, M.Y., Yasukawa, T., Reyes, A., Bailey, L.J., Cluett, T.J., Goffart, S., Willcox, S., et al. Mammalian mitochondrial DNA replication intermediates are essentially duplex but contain extensive tracts of RNA/DNA hybrid. J Mol Biol *397*, 1144-1155.

[30] Tanaka, M., and Ozawa, T. (1994). Strand asymmetry in human mitochondrial DNA mutations. Genomics *22*, 327-335.

[31] Holt, I.J., and Jacobs, H.T. (2003). Response: The mitochondrial DNA replication bubble has not burst. Trends Biochem Sci *28*, 355-356.

[32] Bogenhagen, D.F., and Clayton, D.A. (2003). Concluding remarks: The mitochondrial DNA replication bubble has not burst. Trends Biochem Sci *28*, 404-405.

[33] Bogenhagen, D.F., and Clayton, D.A. (2003). The mitochondrial DNA replication bubble has not burst. Trends Biochem Sci *28*, 357-360.

[34] Brown, T.A., Cecconi, C., Tkachuk, A.N., Bustamante, C., and Clayton, D.A. (2005). Replication of mitochondrial DNA occurs by strand displacement with alternative light-strand origins, not via a strand-coupled mechanism. Genes Dev *19*, 2466-2476.

[35] Asin-Cayuela, J., and Gustafsson, C.M. (2007). Mitochondrial transcription and its regulation in mammalian cells. Trends Biochem Sci *32*, 111-117.

[36] Hanawalt, P.C., and Spivak, G. (2008). Transcription-coupled DNA repair: two decades of progress and surprises. Nat Rev Mol Cell Biol *9*, 958-970.

[37] Kamenisch, Y., Fousteri, M., Knoch, J., von Thaler, A.K., Fehrenbacher, B., Kato, H., Becker, T., Dolle, M.E., Kuiper, R., Majora, M., et al. Proteins of nucleotide and base excision repair pathways interact in mitochondria to protect from loss of subcutaneous fat, a hallmark of aging. J Exp Med *207*, 379-390.

[38] Jacobs, H.T. (2003). The mitochondrial theory of aging: dead or alive? Aging Cell *2*, 11-17.

[39] Vermulst, M., Bielas, J.H., and Loeb, L.A. (2008). Quantification of random mutations in the mitochondrial genome. Methods *46*, 263-268.

[40] Vermulst, M., Bielas, J.H., Kujoth, G.C., Ladiges, W.C., Rabinovitch, P.S., Prolla, T.A., and Loeb, L.A. (2007). Mitochondrial point mutations do not limit the natural lifespan of mice. Nat Genet *39*, 540-543.
[41] Kuhn, R., and Wurst, W. (2009). Overview on mouse mutagenesis. Methods Mol Biol *530*, 1-12.
[42] Cariello, N.F., Narayanan, S., Kwanyuen, P., Muth, H., and Casey, W.M. (1998). A novel bacterial reversion and forward mutation assay based on green fluorescent protein. Mutat Res *414*, 95-105.
[43] Bielas, J.H. (2002). A more efficient Big Blue protocol improves transgene rescue and accuracy in a adduct and mutation measurement. Mutat Res *518*, 107-112.
[44] Skandalis, A., and Glickman, B.W. (1990). Endogenous gene systems for the study of mutational specificity in mammalian cells. Cancer Cells *2*, 79-83.
[45] Malling, H.V. (2004). History of the science of mutagenesis from a personal perspective. Environ Mol Mutagen *44*, 372-386.
[46] Kearsey, S.E., and Craig, I.W. (1981). Altered ribosomal RNA genes in mitochondria from mammalian cells with chloramphenicol resistance. Nature *290*, 607-608.
[47] de Souza-Pinto, N.C., Mason, P.A., Hashiguchi, K., Weissman, L., Tian, J., Guay, D., Lebel, M., Stevnsner, T.V., Rasmussen, L.J., and Bohr, V.A. (2009). Novel DNA mismatch-repair activity involving YB-1 in human mitochondria. DNA Repair (Amst) *8*, 704-719.
[48] Hoeijmakers, J.H. (2001). Genome maintenance mechanisms for preventing cancer. Nature *411*, 366-374.
[49] Friedberg EC , W.G., Siede W, Wood RD, Schultz R.A. Ellenberger T (2006). DNA repair and mutagenesis, 2 Edition, (Washington DC: American Society for Microbiology).
[50] Bielas, J.H., and Loeb, L.A. (2005). Quantification of random genomic mutations. Nat Methods *2*, 285-290.
[51] Kraytsberg, Y., Nicholas, A., Caro, P., and Khrapko, K. (2008). Single molecule PCR in mtDNA mutational analysis: Genuine mutations vs. damage bypass-derived artifacts. Methods *46*, 269-273.
[52] Wanagat, J., Cao, Z., Pathare, P., and Aiken, J.M. (2001). Mitochondrial DNA deletion mutations colocalize with segmental electron transport system abnormalities, muscle fiber atrophy, fiber splitting, and oxidative damage in sarcopenia. Faseb J *15*, 322-332.
[53] Zelko, I.N., Mariani, T.J., and Folz, R.J. (2002). Superoxide dismutase multigene family: a comparison of the CuZn-SOD (SOD1), Mn-SOD (SOD2), and EC-SOD (SOD3) gene structures, evolution, and expression. Free Radic Biol Med *33*, 337-349.
[54] Wallace, D.C., and Fan, W. (2009). The pathophysiology of mitochondrial disease as modeled in the mouse. Genes Dev *23*, 1714-1736.

[55] Mambo, E., Gao, X., Cohen, Y., Guo, Z., Talalay, P., and Sidransky, D. (2003). Electrophile and oxidant damage of mitochondrial DNA leading to rapid evolution of homoplasmic mutations. Proc Natl Acad Sci U S A *100*, 1838-1843.

[56] D Dai, L.S., M Vermulst, et al. (submitted). Overexpression of catalase targeted to mitochondria attenuates murine cardiac aging. Circulation.

[57] Wang, D., Kreutzer, D.A., and Essigmann, J.M. (1998). Mutagenicity and repair of oxidative DNA damage: insights from studies using defined lesions. Mutat Res *400*, 99-115.

[58] Wallace, D. (2012). Mitomap. (http://www.mitomap.org/).

[59] Wallace, D.C. (2005). A mitochondrial paradigm of metabolic and degenerative diseases, aging, and cancer: a dawn for evolutionary medicine. Annu Rev Genet *39*, 359-407.

[60] Bielas, N.E.M.K.M.V.K.S.J.O.S.J.J.S.J.H. (2012). Decreased Mitochondrial DNA Mutagenesis in Human Colorectal Cancer. PLOS Genetics.

[61] Ishikawa, K., Takenaga, K., Akimoto, M., Koshikawa, N., Yamaguchi, A., Imanishi, H., Nakada, K., Honma, Y., and Hayashi, J. (2008). ROS-generating mitochondrial DNA mutations can regulate tumor cell metastasis. Science *320*, 661-664.

[62] Krauss, S., Zhang, C.Y., and Lowell, B.B. (2005). The mitochondrial uncoupling-protein homologues. Nat Rev Mol Cell Biol *6*, 248-261.

[63] Graham, B.H., Waymire, K.G., Cottrell, B., Trounce, I.A., MacGregor, G.R., and Wallace, D.C. (1997). A mouse model for mitochondrial myopathy and cardiomyopathy resulting from a deficiency in the heart/muscle isoform of the adenine nucleotide translocator. Nat Genet *16*, 226-234.

[64] Copeland, W.C. (2008). Inherited mitochondrial diseases of DNA replication. Annu Rev Med *59*, 131-146.

[65] Longley, M.J., Nguyen, D., Kunkel, T.A., and Copeland, W.C. (2001). The fidelity of human DNA polymerase gamma with and without exonucleolytic proofreading and the p55 accessory subunit. J Biol Chem *276*, 38555-38562.

[66] Spelbrink, J.N., Toivonen, J.M., Hakkaart, G.A., Kurkela, J.M., Cooper, H.M., Lehtinen, S.K., Lecrenier, N., Back, J.W., Speijer, D., Foury, F., et al. (2000). In vivo functional analysis of the human mitochondrial DNA polymerase POLG expressed in cultured human cells. J Biol Chem *275*, 24818-24828.

[67] Mootha, V.K., Bunkenborg, J., Olsen, J.V., Hjerrild, M., Wisniewski, J.R., Stahl, E., Bolouri, M.S., Ray, H.N., Sihag, S., Kamal, M., et al. (2003). Integrated analysis of protein composition, tissue diversity, and gene regulation in mouse mitochondria. Cell *115*, 629-640.

[68] Fishel, M.L., Seo, Y.R., Smith, M.L., and Kelley, M.R. (2003). Imbalancing the DNA base excision repair pathway in the mitochondria; targeting and overexpressing N-methylpurine DNA glycosylase in mitochondria leads to enhanced cell killing. Cancer Res *63*, 608-615.

[69] Smirnova, E., Griparic, L., Shurland, D.L., and van der Bliek, A.M. (2001). Dynamin-related protein Drp1 is required for mitochondrial division in mammalian cells. Mol Biol Cell *12*, 2245-2256.

[70] Gandre-Babbe, S., and van der Bliek, A.M. (2008). The novel tail-anchored membrane protein Mff controls mitochondrial and peroxisomal fission in mammalian cells. Mol Biol Cell *19*, 2402-2412.

[71] Friedman, T., Dunlap, J.C., Goodwin, S.F. (2011). Advances in genetics, (Elsevier).

[72] Singh, K.K., Sigala, B., Sikder, H.A., and Schwimmer, C. (2001). Inactivation of Saccharomyces cerevisiae OGG1 DNA repair gene leads to an increased frequency of mitochondrial mutants. Nucleic Acids Res *29*, 1381-1388.

[73] Chatterjee, A., and Singh, K.K. (2001). Uracil-DNA glycosylase-deficient yeast exhibit a mitochondrial mutator phenotype. Nucleic Acids Res *29*, 4935-4940.

[74] O'Rourke, T.W., Doudican, N.A., Mackereth, M.D., Doetsch, P.W., and Shadel, G.S. (2002). Mitochondrial dysfunction due to oxidative mitochondrial DNA damage is reduced through cooperative actions of diverse proteins. Mol Cell Biol *22*, 4086-4093.

[75] Vongsamphanh, R., Fortier, P.K., and Ramotar, D. (2001). Pir1p mediates translocation of the yeast Apn1p endonuclease into the mitochondria to maintain genomic stability. Mol Cell Biol *21*, 1647-1655.

[76] Kang, D., Nishida, J., Iyama, A., Nakabeppu, Y., Furuichi, M., Fujiwara, T., Sekiguchi, M., and Takeshige, K. (1995). Intracellular localization of 8-oxo-dGTPase in human cells, with special reference to the role of the enzyme in mitochondria. J Biol Chem *270*, 14659-14665.

[77] Ohtsubo, T., Nishioka, K., Imaiso, Y., Iwai, S., Shimokawa, H., Oda, H., Fujiwara, T., and Nakabeppu, Y. (2000). Identification of human MutY homolog (hMYH) as a repair enzyme for 2-hydroxyadenine in DNA and detection of multiple forms of hMYH located in nuclei and mitochondria. Nucleic Acids Res *28*, 1355-1364.

[78] Duxin, J.P., Dao, B., Martinsson, P., Rajala, N., Guittat, L., Campbell, J.L., Spelbrink, J.N., and Stewart, S.A. (2009). Human Dna2 is a nuclear and mitochondrial DNA maintenance protein. Mol Cell Biol *29*, 4274-4282.

[79] Liu, P., and Demple, B. DNA repair in mammalian mitochondria: Much more than we thought? Environ Mol Mutagen *51*, 417-426.

[80] Halsne, R., Esbensen, Y., Wang, W., Scheffler, K., Suganthan, R., Bjoras, M., and Eide, L. Lack of the DNA glycosylases MYH and OGG1 in the cancer prone double mutant mouse does not increase mitochondrial DNA mutagenesis. DNA Repair (Amst) *11*, 278-285.

[81] Muftuoglu, M., de Souza-Pinto, N.C., Dogan, A., Aamann, M., Stevnsner, T., Rybanska, I., Kirkali, G., Dizdaroglu, M., and Bohr, V.A. (2009). Cockayne syndrome group B protein stimulates repair of formamidopyrimidines by NEIL1 DNA glycosylase. J Biol Chem *284*, 9270-9279.

[82] Clayton, D.A., Doda, J.N., and Friedberg, E.C. (1974). The absence of a pyrimidine dimer repair mechanism in mammalian mitochondria. Proc Natl Acad Sci U S A *71*, 2777-2781.

[83] Kunkel, T.A., and Erie, D.A. (2005). DNA mismatch repair. Annu Rev Biochem *74*, 681-710.

[84] Chi, N.W., and Kolodner, R.D. (1994). Purification and characterization of MSH1, a yeast mitochondrial protein that binds to DNA mismatches. J Biol Chem *269*, 29984-29992.

[85] Chi, N.W., and Kolodner, R.D. (1994). The effect of DNA mismatches on the ATPase activity of MSH1, a protein in yeast mitochondria that recognizes DNA mismatches. J Biol Chem *269*, 29993-29997.

[86] Mason, P.A., Matheson, E.C., Hall, A.G., and Lightowlers, R.N. (2003). Mismatch repair activity in mammalian mitochondria. Nucleic Acids Res *31*, 1052-1058.

[87] Bacman, S.R., Williams, S.L., and Moraes, C.T. (2009). Intra- and inter-molecular recombination of mitochondrial DNA after in vivo induction of multiple double-strand breaks. Nucleic Acids Res *37*, 4218-4226.

[88] Vermulst, M., Wanagat, J., Kujoth, G.C., Bielas, J.H., Rabinovitch, P.S., Prolla, T.A., and Loeb, L.A. (2008). DNA deletions and clonal mutations drive premature aging in mitochondrial mutator mice. Nat Genet *40*, 392-394.

[89] Zsurka, G., Kraytsberg, Y., Kudina, T., Kornblum, C., Elger, C.E., Khrapko, K., and Kunz, W.S. (2005). Recombination of mitochondrial DNA in skeletal muscle of individuals with multiple mitochondrial DNA heteroplasmy. Nat Genet *37*, 873-877.

[90] Kraytsberg, Y., Schwartz, M., Brown, T.A., Ebralidse, K., Kunz, W.S., Clayton, D.A., Vissing, J., and Khrapko, K. (2004). Recombination of human mitochondrial DNA. Science *304*, 981.

[91] Fukui, H., and Moraes, C.T. (2009). Mechanisms of formation and accumulation of mitochondrial DNA deletions in aging neurons. Hum Mol Genet *18*, 1028-1036.

[92] Dmitrieva, N.I., Malide, D., and Burg, M.B. Mre11 is expressed in mammalian mitochondria where it binds to mitochondrial DNA. Am J Physiol Regul Integr Comp Physiol *301*, R632-640.

[93] Larsen, N.B., Rasmussen, M., and Rasmussen, L.J. (2005). Nuclear and mitochondrial DNA repair: similar pathways? Mitochondrion *5*, 89-108.

[94] Abu-Amero, K.K., and Bosley, T.M. (2006). Mitochondrial abnormalities in patients with LHON-like optic neuropathies. Invest Ophthalmol Vis Sci *47*, 4211-4220.

[95] Brown, M.D., Trounce, I.A., Jun, A.S., Allen, J.C., and Wallace, D.C. (2000). Functional analysis of lymphoblast and cybrid mitochondria containing the 3460, 11778, or 14484 Leber's hereditary optic neuropathy mitochondrial DNA mutation. J Biol Chem *275*, 39831-39836.

[96] Brown, M.D., Zhadanov, S., Allen, J.C., Hosseini, S., Newman, N.J., Atamonov, V.V., Mikhailovskaya, I.E., Sukernik, R.I., and Wallace, D.C. (2001). Novel mtDNA mutations and oxidative phosphorylation dysfunction in Russian LHON families. Hum Genet *109*, 33-39.

[97] Wallace, D.C., Singh, G., Lott, M.T., Hodge, J.A., Schurr, T.G., Lezza, A.M., Elsas, L.J., 2nd, and Nikoskelainen, E.K. (1988). Mitochondrial DNA mutation associated with Leber's hereditary optic neuropathy. Science *242*, 1427-1430.

[98] Johns, D.R., Neufeld, M.J., and Hedges, T.R., 3rd (1994). Mitochondrial DNA mutations in Cuban optic and peripheral neuropathy. J Neuroophthalmol *14*, 135-140.

[99] Ugalde, C., Triepels, R.H., Coenen, M.J., van den Heuvel, L.P., Smeets, R., Uusimaa, J., Briones, P., Campistol, J., Majamaa, K., Smeitink, J.A., et al. (2003). Impaired complex I assembly in a Leigh syndrome patient with a novel missense mutation in the ND6 gene. Ann Neurol *54*, 665-669.

[100] Solano, A., Roig, M., Vives-Bauza, C., Hernandez-Pena, J., Garcia-Arumi, E., Playan, A., Lopez-Perez, M.J., Andreu, A.L., and Montoya, J. (2003). Bilateral striatal necrosis associated with a novel mutation in the mitochondrial ND6 gene. Ann Neurol *54*, 527-530.

[101] Jun, A.S., Brown, M.D., and Wallace, D.C. (1994). A mitochondrial DNA mutation at nucleotide pair 14459 of the NADH dehydrogenase subunit 6 gene associated with maternally inherited Leber hereditary optic neuropathy and dystonia. Proc Natl Acad Sci U S A *91*, 6206-6210.

[102] Leshinsky-Silver, E., Shuvalov, R., Inbar, S., Cohen, S., Lev, D., and Lerman-Sagie, T. Juvenile Leigh syndrome, optic atrophy, ataxia, dystonia, and epilepsy due to T14487C mutation in the mtDNA-ND6 gene: a mitochondrial syndrome presenting from birth to adolescence. J Child Neurol *26*, 476-481.

[103] Ravn, K., Wibrand, F., Hansen, F.J., Horn, N., Rosenberg, T., and Schwartz, M. (2001). An mtDNA mutation, 14453G-->A, in the NADH dehydrogenase subunit 6 associated with severe MELAS syndrome. Eur J Hum Genet *9*, 805-809.

[104] Goto, Y., Nonaka, I., and Horai, S. (1990). A mutation in the tRNA(Leu)(UUR) gene associated with the MELAS subgroup of mitochondrial encephalomyopathies. Nature *348*, 651-653.

[105] van den Ouweland, J.M., Lemkes, H.H., Trembath, R.C., Ross, R., Velho, G., Cohen, D., Froguel, P., and Maassen, J.A. (1994). Maternally inherited diabetes and deafness is a distinct subtype of diabetes and associates with a single point mutation in the mitochondrial tRNA(Leu(UUR)) gene. Diabetes *43*, 746-751.

[106] Finsterer, J. (2009). Mitochondrial ataxias. Can J Neurol Sci *36*, 543-553.

[107] Cao, Z., Wanagat, J., McKiernan, S.H., and Aiken, J.M. (2001). Mitochondrial DNA deletion mutations are concomitant with ragged red regions of individual, aged muscle fibers: analysis by laser-capture microdissection. Nucleic Acids Res *29*, 4502-4508.

[108] Bender, A., Krishnan, K.J., Morris, C.M., Taylor, G.A., Reeve, A.K., Perry, R.H., Jaros, E., Hersheson, J.S., Betts, J., Klopstock, T., et al. (2006). High levels of mitochondrial

DNA deletions in substantia nigra neurons in aging and Parkinson disease. Nat Genet *38*, 515-517.

[109] Kraytsberg, Y., Kudryavtseva, E., McKee, A.C., Geula, C., Kowall, N.W., and Khrapko, K. (2006). Mitochondrial DNA deletions are abundant and cause functional impairment in aged human substantia nigra neurons. Nat Genet *38*, 518-520.

[110] Banerjee, R., Starkov, A.A., Beal, M.F., and Thomas, B. (2009). Mitochondrial dysfunction in the limelight of Parkinson's disease pathogenesis. Biochim Biophys Acta *1792*, 651-663.

[111] Greaves, L.C., Preston, S.L., Tadrous, P.J., Taylor, R.W., Barron, M.J., Oukrif, D., Leedham, S.J., Deheragoda, M., Sasieni, P., Novelli, M.R., et al. (2006). Mitochondrial DNA mutations are established in human colonic stem cells, and mutated clones expand by crypt fission. Proc Natl Acad Sci U S A *103*, 714-719.

[112] Kulawiec, M., Salk, J.J., Ericson, N.G., Wanagat, J., and Bielas, J.H. Generation, function, and prognostic utility of somatic mitochondrial DNA mutations in cancer. Environ Mol Mutagen *51*, 427-439.

[113] Kaneda, H., Hayashi, J., Takahama, S., Taya, C., Lindahl, K.F., and Yonekawa, H. (1995). Elimination of paternal mitochondrial DNA in intraspecific crosses during early mouse embryogenesis. Proc Natl Acad Sci U S A *92*, 4542-4546.

[114] Shitara, H., Hayashi, J.I., Takahama, S., Kaneda, H., and Yonekawa, H. (1998). Maternal inheritance of mouse mtDNA in interspecific hybrids: segregation of the leaked paternal mtDNA followed by the prevention of subsequent paternal leakage. Genetics *148*, 851-857.

[115] Jeronimo, C., Nomoto, S., Caballero, O.L., Usadel, H., Henrique, R., Varzim, G., Oliveira, J., Lopes, C., Fliss, M.S., and Sidransky, D. (2001). Mitochondrial mutations in early stage prostate cancer and bodily fluids. Oncogene *20*, 5195-5198.

[116] Fliss, M.S., Usadel, H., Caballero, O.L., Wu, L., Buta, M.R., Eleff, S.M., Jen, J., and Sidransky, D. (2000). Facile detection of mitochondrial DNA mutations in tumors and bodily fluids. Science *287*, 2017-2019.

[117] Polyak, K., Li, Y., Zhu, H., Lengauer, C., Willson, J.K., Markowitz, S.D., Trush, M.A., Kinzler, K.W., and Vogelstein, B. (1998). Somatic mutations of the mitochondrial genome in human colorectal tumours. Nat Genet *20*, 291-293.

[118] Yeh, J.J., Lunetta, K.L., van Orsouw, N.J., Moore, F.D., Jr., Mutter, G.L., Vijg, J., Dahia, P.L., and Eng, C. (2000). Somatic mitochondrial DNA (mtDNA) mutations in papillary thyroid carcinomas and differential mtDNA sequence variants in cases with thyroid tumours. Oncogene *19*, 2060-2066.

[119] Salas, A., Yao, Y.G., Macaulay, V., Vega, A., Carracedo, A., and Bandelt, H.J. (2005). A critical reassessment of the role of mitochondria in tumorigenesis. PLoS Med *2*, e296.

[120] Bunn, C.L., Wallace, D.C., and Eisenstadt, J.M. (1974). Cytoplasmic inheritance of chloramphenicol resistance in mouse tissue culture cells. Proc Natl Acad Sci U S A *71*, 1681-1685.

[121] Wallace, D.C., Bunn, C.L., and Eisenstadt, J.M. (1975). Cytoplasmic transfer of chloramphenicol resistance in human tissue culture cells. J Cell Biol *67*, 174-188.

[122] Blanc, H., Wright, C.T., Bibb, M.J., Wallace, D.C., and Clayton, D.A. (1981). Mitochondrial DNA of chloramphenicol-resistant mouse cells contains a single nucleotide change in the region encoding the 3' end of the large ribosomal RNA. Proc Natl Acad Sci U S A *78*, 3789-3793.

[123] Howell, N., Appel, J., Cook, J.P., Howell, B., and Hauswirth, W.W. (1987). The molecular basis of inhibitor resistance in a mammalian mitochondrial cytochrome b mutant. J Biol Chem *262*, 2411-2414.

[124] Bai, Y., and Attardi, G. (1998). The mtDNA-encoded ND6 subunit of mitochondrial NADH dehydrogenase is essential for the assembly of the membrane arm and the respiratory function of the enzyme. EMBO J *17*, 4848-4858.

[125] Inoue, K., Nakada, K., Ogura, A., Isobe, K., Goto, Y., Nonaka, I., and Hayashi, J.I. (2000). Generation of mice with mitochondrial dysfunction by introducing mouse mtDNA carrying a deletion into zygotes. Nat Genet *26*, 176-181.

[126] Shoubridge, E.A. (2000). A debut for mito-mouse. Nat Genet *26*, 132-134.

[127] Graziewicz, M.A., Longley, M.J., and Copeland, W.C. (2006). DNA polymerase gamma in mitochondrial DNA replication and repair. Chem Rev *106*, 383-405.

[128] Kujoth, G.C., Hiona, A., Pugh, T.D., Someya, S., Panzer, K., Wohlgemuth, S.E., Hofer, T., Seo, A.Y., Sullivan, R., Jobling, W.A., et al. (2005). Mitochondrial DNA mutations, oxidative stress, and apoptosis in mammalian aging. Science *309*, 481-484.

[129] Trifunovic, A., Wredenberg, A., Falkenberg, M., Spelbrink, J.N., Rovio, A.T., Bruder, C.E., Bohlooly, Y.M., Gidlof, S., Oldfors, A., Wibom, R., et al. (2004). Premature ageing in mice expressing defective mitochondrial DNA polymerase. Nature *429*, 417-423.

[130] Zhang, D., Mott, J.L., Chang, S.W., Denniger, G., Feng, Z., and Zassenhaus, H.P. (2000). Construction of transgenic mice with tissue-specific acceleration of mitochondrial DNA mutagenesis. Genomics *69*, 151-161.

[131] Hu, J.P., Vanderstraeten, S., and Foury, F. (1995). Isolation and characterization of ten mutator alleles of the mitochondrial DNA polymerase-encoding MIP1 gene from Saccharomyces cerevisiae. Gene *160*, 105-110.

[132] Foury, F., and Vanderstraeten, S. (1992). Yeast mitochondrial DNA mutators with deficient proofreading exonucleolytic activity. Embo J *11*, 2717-2726.

[133] Bailey, L.J., Cluett, T.J., Reyes, A., Prolla, T.A., Poulton, J., Leeuwenburgh, C., and Holt, I.J. (2009). Mice expressing an error-prone DNA polymerase in mitochondria display elevated replication pausing and chromosomal breakage at fragile sites of mitochondrial DNA. Nucleic Acids Res.

[134] Edgar, D., Shabalina, I., Camara, Y., Wredenberg, A., Calvaruso, M.A., Nijtmans, L., Nedergaard, J., Cannon, B., Larsson, N.G., and Trifunovic, A. (2009). Random point

mutations with major effects on protein-coding genes are the driving force behind premature aging in mtDNA mutator mice. Cell Metab *10*, 131-138.

[135] Vermulst, M., Wanagat, J., and Loeb, L.A. (2009). On mitochondria, mutations, and methodology. Cell Metab *10*, 437.

[136] Tyynismaa, H., Mjosund, K.P., Wanrooij, S., Lappalainen, I., Ylikallio, E., Jalanko, A., Spelbrink, J.N., Paetau, A., and Suomalainen, A. (2005). Mutant mitochondrial helicase Twinkle causes multiple mtDNA deletions and a late-onset mitochondrial disease in mice. Proc Natl Acad Sci U S A *102*, 17687-17692.

Chapter 5

Molecular Mechanisms of Action of Antimutagens from Sage (*Salvia officinalis*) and Basil (*Ocimum basilicum*)

Biljana Nikolić, Dragana Mitić-Ćulafić,
Branka Vuković-Gačić and Jelena Knežević-Vukčević

Additional information is available at the end of the chapter

http://dx.doi.org/10.5772/50524

1. Introduction

DNA is a dynamic molecule that is constantly damaged and repaired. Major sources of DNA lesions are physical and chemical agents from the environment, intermediates of cellular metabolism, spontaneous chemical reactions of DNA, incorporation of foreign or damaged nucleotides, etc. [1,2]. As a response to DNA damage, essentially all organisms have developed elaborate DNA repair mechanisms to preserve the integrity of their genetic material: reversion, excision or tolerance of a lesion. These mechanisms are largely conserved among prokaryotes and eukaryotes, including human cells [3,4].

Unrepaired DNA lesions may block replication and transcription, potentially leading to cell death, or may give miscoding information, generating mutations. Mutations in germ cells can cause abnormal development of embryo, prenatal death or genetically defective offspring. Somatic mutations and rearrangements in DNA molecule can lead to development of many degenerative disorders including atherosclerosis, autoimmune diseases, Alzheimer's disease, certain types of diabetes, and aging [5-9]. Moreover, epidemiological studies indicate that many types of cancer are dependent on multiple mutational etiologies, as well as on inherited mutator phenotype [4,10-15]. With the increasing diversity and abundance of DNA damaging agents in the environment, it is very important for human health that active substances from medicinal and aromatic plants possess protective effects against genotoxic agents and under certain conditions could act as antimutagens.

2. Antimutagens

In order to protect human health, a relatively new area of research, designated as antimutagenesis and anticarcinogenesis, is continuously developing. The aim of antimutagenesis studies is to identify natural substances with antigenotoxic and antimutagenic potential and to determine the cellular and molecular mechanisms of their action. Possible application of plant antimutagens is in development of dietary and pharmaceutical supplements useful in primary prevention of mutation related diseases, including cancer.

Different prokaryotic and eukaryotic tests, routinely used to detect environmental mutagens and carcinogens, are suitably adapted for identifying agents with antigenotoxic, antimutagenic and anticarcinogenic potential, as well as for elucidating the mechanisms of their action. Due to rapidity and low costs, bacterial short-term tests are recommended to provide preliminary, but considerable information about cellular mechanisms of antimutagenesis. In combination with mammalian enzymes, they can provide information about the kind of metabolic activation or detoxification that an agent may undergo *in vivo*. However, for obvious reasons, bacterial short-term tests can not replace the antimutagenicity/antigenotoxicity studies in mammalian cells and *in vivo*, in order to identify mechanisms possibly relevant for human protection [16-19].

After several decades of research, antimutagenic effect of many naturally occurring compounds extracted from plants has been well established in bacteria and mammalian cells [20,21]. However, due to diversity of DNA lesions and the complexity of DNA repair pathways it is difficult to identify the processes involved in antimutagenesis. Antimutagens may be effective against single mutagen or a class of mutagens, may act by multiple, sometimes strictly interconnected or partially overlapping mechanisms, may be even mutagenic at certain concentrations or in certain test systems, which implies a discriminative approach in antimutagenesis studies, as well as careful interpretation of the results [22].

According to Kada et al. [23] antimutagens are placed in two major groups: desmutagens and bioantimutagens. Desmutagens are agents which prevent the formation of premutagenic lesions, while bioantimutagens prevent processing of premutagenic lesions into mutations by modulating DNA replication and repair. A revised and updated classification of antimutagens and anticarcinogenesis was given several times by different authors [18,19,24]. The classification took into consideration the multiple phases involved in the pathogenesis of cancer and other mutation related diseases. It analyzed first the inhibition of mutations and of cancer initiation, either extracellularly or inside the cells, and then the mechanisms interfering with promotion, progression, invasion and metastasis. A modified scheme incorporated possible points for intervention in primary, secondary and tertiary prevention.

Extensive search for natural compounds with antimutagenic effect often pointed at terpenes, a class of substances abundantly found in fruits, vegetables, and aromatic and medicinal plants. They are biosynthetically derived from isoprene units (C_5H_8) which may be linked to form monoterpenes (C_{10}), sesquiterpenes (C_{15}), diterpenes (C_{20}), triterpenes (C_{30}), tetraterpenes (C_{40}), and polyterpenes. Terpenes exist as hydrocarbons or have oxygen-containing substituents, such as hydroxyl, carbonyl, ketone, or aldehide groups; the latter usually are referred to as terpenoids. Both *in vitro* tests and epidemiological studies suggest that many dietary monoterpenes (including monoterpenoids) exert antimutagenic properties and could be helpful in the prevention and therapy of cancers [25-27].

The research efforts of our group have been focused on detection of antimutagenic properties of medicinal and aromatic plants of our region. In our initial search we screened crude extracts obtained from plants frequently used in our traditional medicine: sage (*Salvia officinalis* L.), lime-tree (*Tilia chordata* Mill.), mint (*Mentha piperita* L.), nettle (*Urtica dioica* L.), camomile (*Matricaria chammomilla* L.), aloe (*Aloe arborescens* L.), thyme (*Thymus serpyllum* L.), St. John's wort (*Hypericum perforatum* L.) and sweet basil (*Ocimum basilicum* L.). Analysis of the obtained data showed heterogeneous responses, depending on the extract, concentration applied, genetic background and end-point monitored. Comparison of obtained data promoted St. John's wort, mint, sweet basil and sage as potential source of antimutagens [28]. In further study, we focused our attention on antimutagenic effect of sage and sweet basil.

3. Medical properties of sage and basil

Salvia and *Ocimum* are genera of the family Lamiaceae consisting of about 900 and 35 species, respectively. *S. officinalis* and *O. basilicum* are employed as folklore remedy for a wide spectrum of ailments in many traditional medicines, including ours. Furthermore, the latter is irreplaceable spice of many national cuisines. Numerous biological activities of different extracts of *Salvia* and *Ocimum* species have been described, including antimicrobial, anti-inflammatory, antioxidative, antidiarrheal, blood-sugar lowering, immunomodulatory, a nervous system stimulatory, spasmolytic, and cholinergic binding [29-40]. Several reports also indicate antigenotoxic and chemopreventive activities of different extracts from *Salvia* and *Ocimum* species [41-45].

4. The strategy and assays for antimutagenesis study

In order to investigate the antimutagenic potential of plant extracts, we constructed and validated a new *Escherichia coli* K12 assay system, specially designed for detection of antimutagens and elucidation of molecular mechanisms of antimutagenesis [46,47]. We used this assay along with appropriatelly modified standard mutagenicity tests (*Salmonella*/microsome, *E. coli* WP2 and *S. cerevisiae* D7), to determine the antimutagenic potential, and applied comet assay for measuring the effect of antimutagen on mutagen induced DNA damage and repair. In all tests antimutagenic potential was determined in the range of non-toxic concentrations.

4.1. *E. coli* assay for bioantimutagens

The bacterial assay is composed of four tests measuring different end-points at the DNA level: spontaneous and induced mutagenesis in different genetic backgrounds, SOS induction and homologous recombination. To evaluate the effect on spontaneous and induced mutagenesis we first use reversion test on repair proficient strain SY252, constructed in our laboratory (Table 1). The strain contains an ochre mutation in the *argE3* gene, which can revert to prototrophy by base substitutions at the site of mutation or at specific suppressor loci [48]. We initially chose UV-irradiation (254 nm) to induce mutations for several reasons: (i) it mainly induces base substitutions [49] which can be detected in SY252; (ii) it shares cellular mechanisms of mutation avoidance (nucleotide excision and post-replication recombination repair) and mutation fixation (translesion error-prone replication mediated by SOS regulated UmuD'C complex) with many chemical mutagens and carcinogens [50,51] (iii) possible chemical interaction between mutagen and antimutagen is prevented, which is essential for detection of bioantimutagens. UV-mimetic mutagen 4-nitroquinoline-1-oxide (4NQO) [52,53] was recently used to provide comparison with the results obtained with UV. The possibility for chemical interaction between 4NQO and antimutagen was avoided in experimental procedure.

Since nucleotide excision repair (NER) is the major error-free pathway involved in repair of pyrimidine dimers and bulky DNA lesions such as 4NQO-DNA adducts [54], we also analyze potential of antimutagen to reduce mutagenesis in NER deficient *uvrA* counterpart of SY252. Comparison of results obtained in repair proficient and NER deficient strains indicates if observed antimutagenic effect involves increased capacity for NER.

Strain	Relevant marker	Reference
	E. coli K12	
SY252	*argE3*	[55]
IB101	SY252 *mutH471*::Tn*5*	[46]
IB103	SY252 *mutS215*::Tn*10*	[46]
IB105	SY252 *uvrA*::Tn*10*	[56]
IB106	SY252 *mutT*::Tn*5*	[57]
IB111	SY252 [λp(*sfiA::lacZ*)c*Iind1*] PHO^C	[56]
IB127	IB111 *uvrA*::Tn*10*	[58]
IB122	SY252/pAJ47	[57]
IB123	IB101/pAJ47	[57]
GY7066	*lacMS286* Φ80dII*lacBK1* *ΔrecA306* *srl*::Tn*10*	[59]
GY8281	GY7066/miniF*recA*$^+$	[59]
GY8252	GY7066/miniF*recA730*	[59]
	S. typhimurium	
TA98	*hisD3052* *rfa* Δ*uvrB*/pKM101	[60]

Strain	Relevant marker	Reference
TA100	*hisG46 rfa ΔuvrB*/pKM101	[60]
TA102	*hisG428 rfa*/pKM101 pAQ1	[60]
	E. coli B/r WP2	
IC185	*trpE65*	[61]
IC202	IC185 *ΔoxyR*/pKM101	[61]
	S. cerevisiae	
D7	ade2-40/119 trp5-12/27 ilv1-92/92	[62]
3A	a/α *gal1 leu2 ura3-52*	[63]

Table 1. Tester strains

To amplify the sensitivity of detection of spontaneous mutations, the isogenic mismatch repair (MMR) deficient strains, with increased frequency of spontaneous reversions were constructed and included in the assay. Due to deficiency in correcting replication errors, these strains can be used to detect agents affecting the fidelity of DNA replication.

To measure the level of SOS induction, which corresponds to the induction of mutagenic SOS repair [64], the repair proficient strain SY252 and NER deficient counterpart were lysogenized with non-inducible λ phage carrying *sfiA*::*lacZ* fusion. Since *sfiA* is under SOS regulation, the level of β-galactosidase in these strains reflects the level of SOS induction [65]. Both strains are constitutive for alkaline phosphatase, allowing simultaneous assessment of SOS induction and overall protein synthesis [56, 58].

To measure homologous recombination, we use the strains with two non-overlapping deletions in duplicated *lac* operon, in which intrachromosomal recombination results in the formation of Lac^{+} recombinants [66]. The strains carry different *recA* alleles and thus have different capacities for both recombination and SOS induction [59]. Strain GY8281 (*recA*$^{+}$) is recombination proficient, and an increased amount of activated RecA protein is formed only after DNA damaging treatments. On the contrary, strain GY8252 (*recA730*) is partially recombination deficient, but constitutive for SOS induction [67,68]. In this strain an increased level of activated RecA protein exists in the absence of DNA damaging treatments.

4.2. *E. coli* assay for desmutagens

A wide variety of compounds with antioxidative activity (vitamins, phenolic compounds, flavonoids, terpenes, etc.), have been shown to possess inhibitory or modulating effects on environmental mutagens and carcinogens. Natural antioxidants and their metabolites can modulate the mutagenesis and the initiation step in carcinogenesis by several desmutagenic mechanisms, such as scavenging of reactive oxygen species (ROS), inhibition of certain enzymes involved in the metabolic transformation, or inhibition of mutagen binding to DNA [69]. Antioxidants may also interfere with tumor promotion and progression by virtue of their multiple biological properties.

In order to identify antimutagens with antioxidative properties, we modified our *E. coli* K12 assay for bioantimutagens. In repair proficient strain SY252 mutations are induced by *t*-butyl hydroperoxide (*t*-BOOH), a latent donor of ROS, which promotes oxidative damage of DNA [70]. Since DNA damage induced by *t*-BOOH cause both transitions and transversions of AT base pairs, it can be used to increase *argE3* → Arg^+ reversions. The *mutT* strain was constructed for the assay in order to evaluate protective capacity of antioxidants against formation of oxidatively damaged bases in the cell pool [57]. Due to deficiency in removing 8-oxo-G, *mutT* strains have high frequency of A:8-oxo-G mispairs and show increased level of spontaneous AT→CG transversions [71]. During validation of the test we determined that the frequency of *argE3* → Arg^+ reversions is significantly increased in IB106 strain [47]. Since MMR is additionally involved in the repair of mispairs between normal and oxidized bases [72], MMR deficient strains from the assay for bioantimutagens are also included.

Considering that microsatellite instability (MSI) could be induced by oxidative DNA damage, by MMR deficiency or in many forms of cancer [73-75], we also designed the test for detection of MSI. The repair proficient and MMR deficient strains were transformed with the low copy number plasmid pAJ47 (Table 1). This plasmid contains dinucleotide repeats $(CA)_{11}$ placed out-of-frame within the coding region of β–lactamase gene. Cells harbouring plasmid are sensitive to β–lactam antibiotics, such as carbenicillin. Microsatellite sequence is a +2 frame construct and the mutation that restores the reading frame and provides resistance to carbenicillin is a 2 bp deletion. Repair-proficient strain is used for screening of *t*-BOOH-induced MSI, while *mutH* strain is used for monitoring of spontaneous MSI [57].

4.3. Other reversion tests

Preliminary screening of plant extracts included evaluation of possible mutagenic effects by standard *Salmonella*/microsome (Ames) test, recommended by OECD [76]. The mutagenicity was determined in strains TA98, TA100 and TA102 (Table 1) in plate incorporation assay [60]. For evaluation of antimutagenic effect, the tester strain and the mutagen were selected according to the mutational event monitored.

WP2 mutagenicity test, especially recommended for monitoring of oxidative mutagenesis [76], is used along with *E. coli* K12 assay to detect antimutagenic potential of plant extracts based on antioxidative properties. Test is performed on both OxyR proficient IC185 and OxyR deficient IC202 strains [61]. The OxyR protein is a redox-sensitive transcriptional activator of genes encoding antioxidative enzymes: catalase-hydroperoxidase I, alkyl hydroperoxide reductase and glutathione reductase [77]. Mutants in *oxyR* are deficient in inducible expression of antioxidant enzymes and thus very sensitive for detection of oxidative mutagens and antimutagenic effect of antioxidants. Comparison of results obtained in $OxyR^+$ and $OxyR^-$ strains indicates if observed antimutagenic effect is based on antioxidative properties.

To obtain preliminary information about mutagenic and antimutagenic potential of plant extracts in eukaryotic cells we used the *S. cerevisiae* diploid strain D7 [62], which permits simultaneous evaluation of point mutations (*ilv1-92*→Ilv^+), mitotic crossing over (*ade2*→Ade^+) and mitotic gene conversion (*trp5*→Trp^+).

4.4. Comet assay – direct monitoring of DNA damage

The alkaline comet assay was used in order to monitor the effect of plant extracts on formation and repair of DNA lesions induced by a mutagen. The comet assay or single-cell gel electrophoresis (SCGE) is a simple method for measuring DNA strand breaks, mostly in eukaryotic cells. It has become one of the standard methods for assessing DNA damage and found applications in different fields including genotoxicity/antigenotoxicity testing, human biomonitoring and molecular epidemiology, ecogenotoxicology, as well in fundamental research of DNA repair [78]. The assay was performed on repair proficient Vero cells, originated from the kidney of African green monkey (ECACC No: 88020401), and on two human cell lines: hepatoma HepG2 (ATCC HB-8065) and B lymphoid NC-NC cells (DSMZ ACC120). We also used the modified version of alkaline comet assay on *S. cerevisiae* 3A strain, designed by Miloshev et al. [63].

5. Antimutagenic potential of sage

Given the possibility to obtain large quantities of chemically characterized extracts from different varieties of sage, we focused our research on this plant. We screened the fractionated extracts of two varieties of sage. Wild sage originated from Pelješac, Croatia, while cultivated sage (variety D-70) was selected and grown at the Institute for Hop, Sorghum and Medicinal Plants, Bački Petrovac, Serbia. The most striking difference between the two plants is the composition of essential oils (EO). While both plants contain α+β Thujone (Thu), Camphor (Cam) is present only in traces in the wild sage, whereas it represents 1/5 of the monoterpenes in variety D-70 [79]. Extract 1 (E1) was prepared from the cultivated sage, collected during the flowering period, dried and subjected to ethanolic extraction as the whole herb. Extract 2 (E2) was prepared from the same herb as Extracts 1, but it was steam distilled prior to ethanolic extraction to remove EO. Extract 3 (E3) was prepared from the wild sage, treated in the same way as Extract 1. All extracts (E1-E3) were re-extracted by CO_2 at different pressure (200, 300, 400, 500 bar), resulting in the extracts (E/2-E/5) with high content of terpenes. The extracts obtained at low CO_2 pressure (200, 300 bar) contained mainly monoterpenes, while extraction at higher CO_2 pressure resulted in the increase of relative proportion of high molecular weight terpenes. Preliminary determination of antioxidative properties, performed with lipid peroxidation test, indicated significant antioxidative activity of the extracts obtained at high CO_2 pressure, which was attributed to diterpenes, such as 6-methyl-ether-γ-lactone carnosic acid and rosmanol-9-ethyl ether [79,80].

5.1. Desmutagenic potential of sage extracts

Protective effect of sage extracts against spontaneous and ethidium bromide (EtBr)-induced mutagenesis was monitored in *E. coli* K12 *mutT* and *S. typhimurium* TA98 strains, respectively [81]. The results showed that extracts of cultivated sage obtained at 500 bar (E1/5 and E2/5) exerted significant antimutagenic effect (Figure 1), indicating high molecular weight terpenes as active substances. The most effective extract E2/5, containing mainly rosmanol-9-ethyl ether (40%), was further investigated in order to elucidate the molecular mechanism of antimutagenicity. Different experimental procedures were applied: A – co-incubation of mutagen, extract and S9 fraction, followed by addition of bacteria and plating; B – pre-incubation of mutagen and S9, followed by addition of the extract, incubation, and final addition of bacteria and plating; C- pre-incubation of mutagen, S9 and bacteria, followed by removal of mutagen and S9, addition of the extract, and plating. The strongest inhibition was obtained when the mutagen and E2/5 were pre-incubated with S9 (procedure A), indicating that the main antimutagenic mechanism was inhibition of metabolic activation of EtBr. Extract E2/5 also moderately reduced spontaneous mutations (37%) in *mutT* strain.

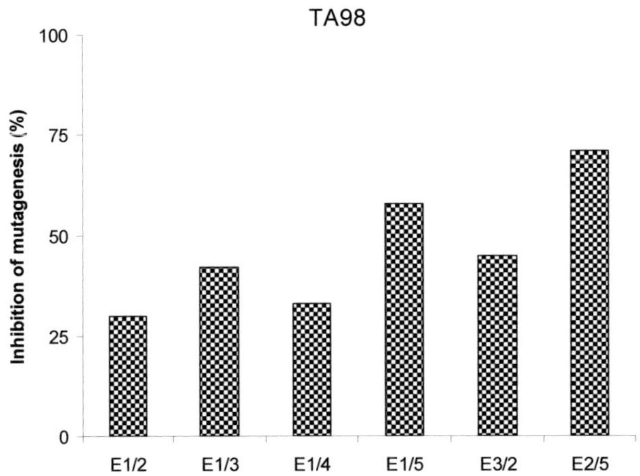

Figure 1. Effect of sage extracts against EtBr –induced mutagenesis in TA98 strain

5.2. Bioantimutagenic potential of sage

Bioantimutagenic potential of sage was evaluated in *E coli* K12 reversion assay by monitoring the effect of the extracts E1/3, E2/3 and E3/3 against UV-induced mutagenesis. The most interesting results were obtained with extract of cultivated sage E1/3 which strongly reduced UV-induced mutagenesis (60%) in repair proficient strain SY252, while no inhibition of mutagenesis was detected with extracts E2/3 and E3/3. Taking into consideration the extracts content, this result indicated the role of volatile terpenes from EO, especially Cam, in observed bioantimutagenic influence [82].

In the comparative study [22] monoterpenes-rich extracts of sage D-70 (EO, E1/2, E1/3) produced a significant antimutagenic response against UV-induced mutagenesis in repair proficient strain SY252 (Figure 2). The analysis of molecular mechanisms indicated no potential of the extracts to reduce spontaneous mutagenesis in IB103 (*mutS*) strain, as well as no inhibition of mutagenic SOS repair in IB111. However, EO and extracts stimulated UV-induced recombination in both GY8281 (*recA*$^{+}$) and GY8252 (*recA730*) strains. In addition, inhibition of UV-induced mutagenesis by E1/3 was significantly decreased by *uvrA* mutation, indicating the participation of both NER and recombination in the protection mechanisms.

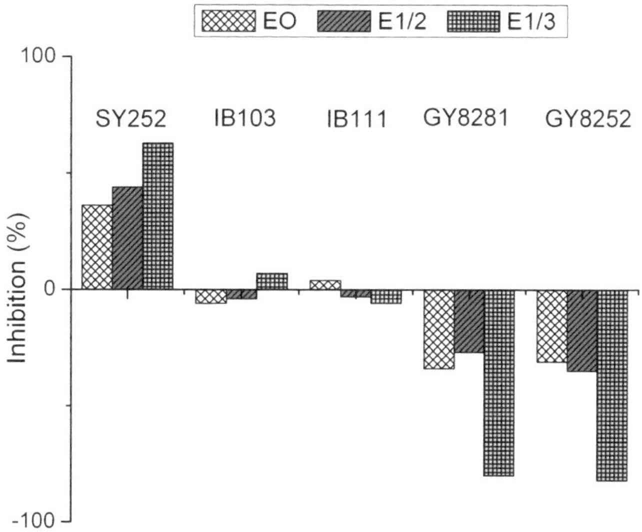

Figure 2. Bioantimutagenic effect of extracts of sage D-70

Additional evidence for bioantimutagenic effect of sage monoterpenes came from the study of EO of sage grown for industrial purposes by the Institute for Medicinal Plant Research "Dr. Josif Pančić", Belgrade, Serbia [83]. In contrast to wild sage and D-70 [79], this variety contains Eucalyptol (Euc, 1,8-cineole) in addition to Thu and Cam (Table 2). EO was fractionated by vacuum rectification to yield 5 fractions (F1-F5). The composition of EO and fractions was determined using analytical GC/FID and GC/MS techniques and Wiley/NBS library of mass spectra [84]. Fractions F1 and F2 contain exclusively monoterpenes, fractions F3 and F4 lack some of the monoterpenes and contain small proportion of sesquiterpenes, while fraction F5 contains about 40% of sesquiterpenes, the most abundant being α-humulene (Table 2).

Constituent	EO	F1	F2	F3	F4	F5
cis-Salven	0.518	0.134				
Tricyclen	0.123	0.146				
α-Thujene	0.178	0.100				
α-Pinene	5.059	5.194	0.620			
Camphene	3.683	6.017	1.361			
Sabinene	0.124	0.134				
β-Pinene	2.717	3.429	0.962			
Myrcene	0.874	0.295	0.042			
α-Felandren	0.062					
α-Terpinene	0.225					
p-Cymene	0.460	1.423	1.342	0.611	0.102	
Limonene	1.224	1.235	0.667	0.325		
Eucalyptol	**14.425**	31.661	21.864	4.853	0.475	
β-Ocimene	0.032	0.023	0.058	0.039		

Constituent	EO	F1	F2	F3	F4	F5
γ-Terpinene	0.391	0.101	0.144		0.236	
cis-Sabinene-hydrate	0.114			0.202	0.144	
cis-Linalol-oxide	0.069			0.123	0.135	
Terpinolene	0.262	0.095	0.135	0.125	0.924	
trans-Sabinene-hydrate	0.501	0.824	0.484	0.489	1.112	
α-Thujone	**37.516**	29.656	48.233	61.512	57.335	11.267
β -Thujone	**4.665**	3.002	4.781	7.439	7.895	2.150
Camphor	**13.777**	8.293	14.364	21.614	27.623	12.075
trans-Pinocamphon	0.461			0.364	0.545	
Borneol	0.753	0.903		0.509	1.200	4.227
cis-Pinocamphon	0.033			0.111	0.160	
Terpin-4-ol	0.351			0.155	0.337	0.997
p-Cimene-8-ol	0.025					
α-Terpinol	0.117			0.201	0.084	1.116
Mirtenal	0.208				0.236	
Bornylacetate	0.391	0.508		0.197	0.425	1.777
trans-Sabinilacetate	0.099				0.070	
α-Kubeben	0.029				0.048	
β -Burbonen	0.058				0.136	
Caryophyllene	1.824			0.185	0.454	
α-Humulene	4.994			0.239	0.586	29.852
allo-Aromadendren	0.085					
γ-Murolen	0.053					
Viridiflorene	0.109				0.054	
γ-Cadinene	0.031					
δ-Cadinene	0.066					
Caryophyllene-oxide	0.089					
Viridiflorol	1.371					8.745
Humulene-epoxide	0.340					2.683
Manool	0.277					1.892
Identified in total	**98.762**	**93.172**	**95.058**	**99.293**	**99.205**	**88.315**

Sage was cultivated by the Institute for Medicinal Plant Research "Dr. Josif Pančić", Belgrade, Serbia. EO was prepared according to Ph. Jug. IV and ISO 9909 and analyzed by GC/FID and GC/MS.

Table 2. Composition of essential oil of sage and its fractions (% m/m)

A comparative study of bioantimutagenic potential of sage EO and fractions was performed in *E. coli* K12, *Salmonella*/microsome and *S. cerevisiae* reversion assays [85]. The summarized effects against UV-induced mutagenesis in strains SY252, TA102 and D7 are presented in Figure 3. In all test organisms protective effect was obtained with EO and fractions containing only monoterpenes (F1 and F2). Fractions F3 and F4 were bioantimutagenic depending on the test organism, while fraction F5 was ineffective. The results confirmed bioantimutagenic potential of monoterpenes and pointed at Thu, Euc and Cam as candidates for further bioantimutagenesis study.

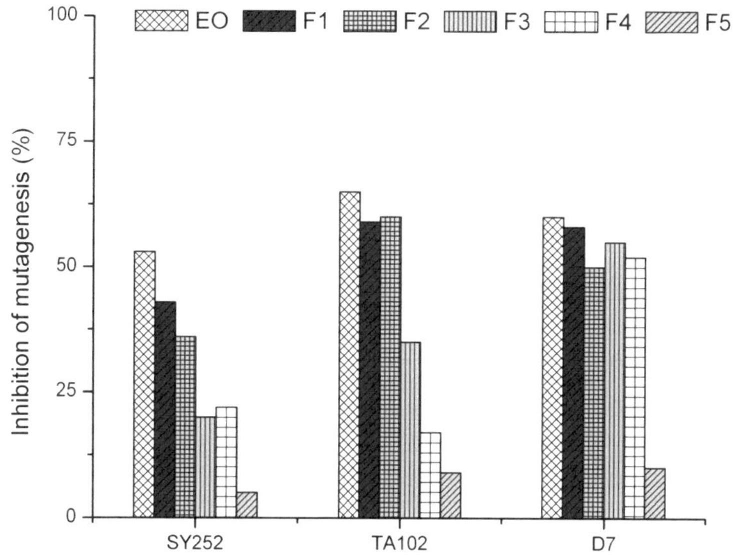

Figure 3. Antimutagenic effect of sage EO and fractions against UV-induced mutagenesis

6. Antimutagenic potential of basil

Sweet basil is well known for its antioxidative properties, but its antigenotoxic potential has not been extensively investigated. In order to monitor desmutagenic and bioantimutagenic potential of sweet basil, EO and its dominant component Linalool (Lin) were screened in *E. coli* reversion assays. The EO was prepared from the plant cultivated by the Institute for Medicinal Plant Research "Dr. Josif Pančić", Belgrade, Serbia. The GC/FID and GC/MS analyses of the oil confirmed the high content of Lin (69.2%, Table 3).

Constituent	% (m/m)	Constituent	% (m/m)
α-Terpinene	0.005	α-Murolene	0.090
Camphene	0.006	Naphthalene	0.270
α-Pinene	0.100	α-Copaen	0.400
β- Myrcene	0.300	α-Humulene	0.500
Limonene	0.900	β-Caryophyllene	0.560
p-Cimen-8-ol	0.025	Zingiberene	0.600
Terpinen-4-ol	0.040	β-Elemene	0.800
Carvone	0.060	α-Bergamotene	1.020
trans-β-Ocimene	0.100	β- Selinene	1.040
endo-Borneol	0.270	α-Guaiene	1.110
endo-Bornylacetate	0.300	δ-Cadinene	1.130
Camphor	0.300	α-Selinene	1.670
Nerol	0.400	δ-Guaiene	2.100
cis- β-Ocimene	0.400	γ-Cadinene	2.500

Constituent	% (m/m)	Constituent	% (m/m)
α-Terpinolene	0.400	Nerodiol	0.110
Thiogeraniol	0.560	cis-Farnesol	0.180
α-Terpineol	0.700	trans-Murolol	0.430
Eucalyptol	0.800	α-Cadinol	2.560
Geraniol	1.900		
Linalool	**69.200**		
Eugenol	1.400		
Estragole	2.400		
β-Burbonene	0.080	**Identified in total**	**97.716%**

Basil was cultivated by the Institute for Medicinal Plant Research "Dr. Josif Pančić" Belgrade, Serbia. EO was prepared according to Ph. Jug. IV and ISO 9909, and analyzed by GC/FID and GC/MS.

Table 3. Composition of essential oil of basil (*Ocimum basilicum* L.)

6.1. Desmutagenic potential of basil

The desmutagenic effect of EO of basil and Lin was monitored in *E. coli* K12 assay system with *t*-BOOH as a mutagen. Strong reduction was observed in repair proficient SY252 for both EO and Lin (Figure 4). Moreover, the spontaneous base substitutions in MMR deficient strain IB101 (*mutH*) were slightly decreased by EO, and moderately by Lin. Both basil derivatives also moderately decreased *t*-BOOH-induced MSI in repair proficient strain IB122 and spontaneous MSI in its MMR deficient counterpart IB123 (*mutH*). Antimutagenic potential determined in all tests was tentatively attributed to antioxidative properties and indicated Lin as principal active substance. The confirmation of proposed mechanism was obtained in *oxyR* deficient IC202 strain, where reduction of *t*-BOOH-induced mutagenesis was 72% and 70%, for EO and Lin, respectively [86].

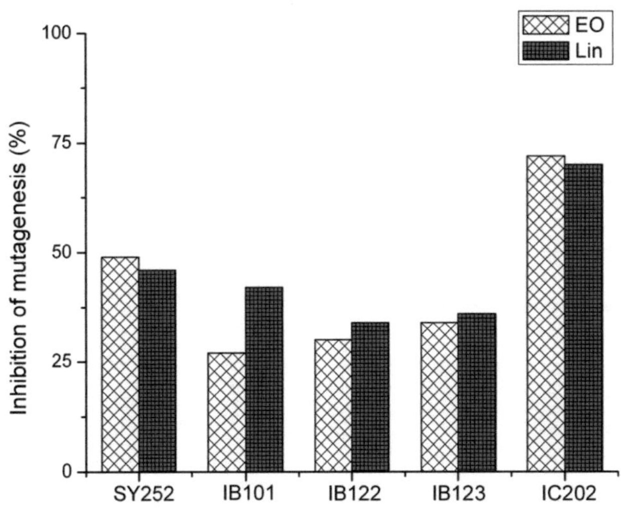

Figure 4. Antimutagenic effect of basil derivatives against *t*-BOOH

6.2. Bioantimutagenic potential of basil

The bioantimutagenic effect of EO and Lin against UV-induced mutagenesis was investigated in repair proficient and NER deficient strains of *E. coli* K12 assay [56]. Both basil derivatives reduced UV-induced mutagenesis only in repair proficient SY252, but not in NER deficient IB105 strain (Figure 5), suggesting potential of basil to modulate NER. No reduction of spontaneous mutagenesis was detected, even in the strain with increased sensitivity (*mutS*). However, Berić et al. [86] reported the inhibition of spontaneous mutagenesis in isogenic *mutH* strain. Similar discrepancy between responses obtained in different MMR mutants was already noted by Vuković-Gačić and Simić [28].

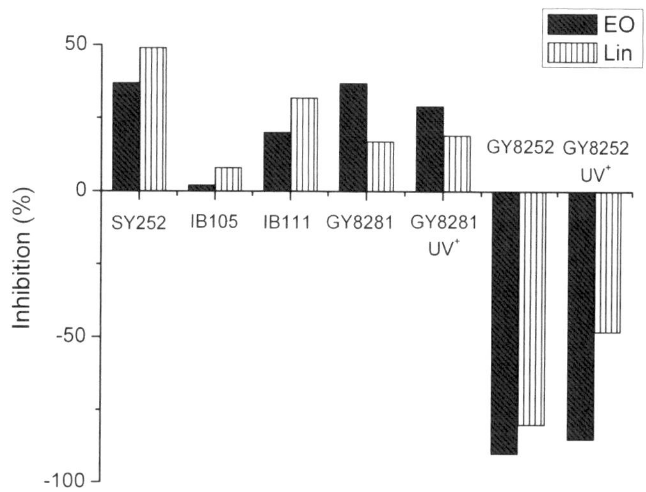

Figure 5. Bioantimutagenic effect of basil derivatives

In the further study EO and Lin showed inhibitory effect on SOS induction. Moreover, they stimulated spontaneous and UV-induced recombination only in strain GY8281 constitutively expressing RecA protein. Both effects were probably caused by inhibition of protein synthesis, as determined by comparing the inhibition of the levels of β-galactosidase and alkaline phosphatase in IB111 strain. Moreover, basil derivatives also decreased the growth rate. Based on all obtained results, we proposed that, by retaining bacterial growth and cell divisions, EO and Lin increased the time for error free repair of pyrimidine dimers by NER.

All obtained data directed our further study to investigation of antimutagenic and antigenotoxic potential of pure monoterpenes from sage and basil: Thu, Cam, Euc and Lin. An acyclic monoterpene Myrcene (Myr), widely distributed in many other medicinal and aromatic plants, was also included in the study.

7. Desmutagenic potential of linalool, myrcene and eucalyptol

Protective effect against oxidative DNA damage and mutagenesis was determined for Lin, Myr and Euc, since their antioxidative potential has been confirmed by TBA assay [87]. All tested monoterpenes slightly reduced *t*-BOOH-induced mutagenesis in strains SY252 and

IC185 (Figure 6). The obtained results indicated that the protective effect of antioxidant monoterpenes was low in the strains proficient in induction of antioxidative enzymes, presumably due to efficient antioxidative defense.

To increase the sensitivity of the assay, the strain deficient in the induction of antioxidative enzymes (IC202 *oxyR*) was included in the study. In this strain Lin and Myr strongly reduced *t*-BOOH-induced mutagenesis, while the effect of Euc was significantly lower. The suppression of mutagenesis by monoterpenes correlated with their antioxidative properties. Euc also decreased spontaneous mutations in IC202, indicating additional mechanisms of antimutagenesis [88].

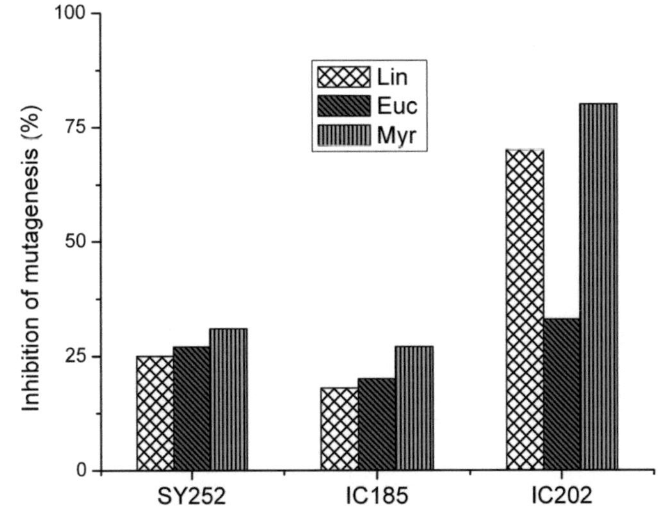

Figure 6. Antimutagenic effect of monoterpenes against *t*-BOOH-induced mutagenesis

Protective capacity of Lin, Myr and Euc against oxidative DNA damage was determined in human hepatoma HepG2 and human B lymphoid NC-NC cell lines by alkaline comet assay [87]. Experiments were performed in two experimental protocols, (i) co-treatment of cells with genotoxic agent (*t*-BOOH) and monoterpene, which tested the ability of the monoterpenes to directly scavenge ROS and (ii) a 20h pre-treatment with monoterpene followed by co-treatment (pre+co-treatment) which, in addition to direct scavenging activity, allows accumulation of monoterpenes in the cell and induction of enzymatic and non-enzymatic cellular antioxidants and detoxifying (Phase II) enzymes.

The results obtained in co-treatment experiments indicated that all three monoterpenes showed protective effect in NC-NC cells, while only Myr exerted weak protection in HepG2 cells (Figure 7). The different response obtained in two cell lines could be tentatively ascribed to the differences in absorption rates caused by different culturing conditions: while HepG2 cells were growing as monolayer, NC-NC cells were growing in suspension and were consequently more exposed to monoterpenes. In line with this presumption is the result obtained in preliminary testing of toxicity: all monoterpenes were more toxic to NC-NC cells. Lazarova et al. [89] reported that hepatocytes possess better antioxidative defense than lymphocytes. Therefore, we could conclude that stronger protective effect was obtained in cells with reduced antioxidative capacity, similarly as in bacteria.

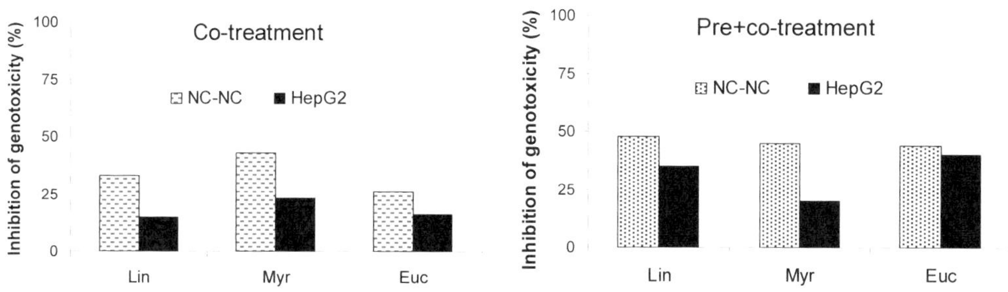

Figure 7. Antigenotoxic effect of monoterpenes against *t*-BOOH-induced genotoxicity

The results of pre+co-treatment experiments showed that Lin and Euc reduced *t*-BOOH-induced genotoxicity in both cell lines, while Myr was effective only in NC-NC cells. Since HepG2 cells retained the activities of many enzymes involved in metabolic transformation of xenobiotics, including monoterpenes [90-96], we proposed that metabolic transformation was responsible for the loss of the protective effect of Myr and reduced protective effect of Lin obtained in this cell line. Consistent with this explanation are our unpublished data that in bacteria treated with *t*-BOOH the presence of microsomal enzymes (S9) significantly reduced antimutagenicity of Lin and diminished antimutagenicity of Myr.

Protective effect of Lin in eukaryotic cells was also reported by Berić et al. [86]. In the comet assay performed on *S. cerevisiae* 3A diploid strain, Lin significantly reduced H_2O_2-induced DNA damage. Stronger protection was obtained in pre-treated (70% inhibition), than in co-treated cells (50% inhibition), indicating that in addition to direct scavenging of ROS, other protective mechanisms, such as accumulation of Lin in the cells, and/or elevation of cellular antioxidative defense, were also included.

8. Bioantimutagenic potential of camphor, eucalyptol and thujone

Preliminary screening of bioantimutagenic potential of Cam, Euc, Thu, Lin and Myr, performed in *E. coli* K12 repair proficient strain, indicated that UV-induced mutagenesis was strongly reduced with Cam, Euc and Thu [97], moderately reduced with Lin [56], while no protective effect of Myr was detected [88]. In order to elucidate the mechanisms involved in strong bioantimutagenic potential of Cam, Euc and Thu, we further analyzed their effect on DNA repair processes [58].

Using the *E. coli* K12 repair proficient strain, we showed that Cam, Euc and Thu, in addition to UV, significantly reduced mutagenesis induced by UV-mimetic 4NQO. Moreover, the extent of mutagenesis inhibition was similar as with UV (Figure 8). In IB105 (*uvrA*) strain protective effect of monoterpenes against both mutagens was diminished, indicating that NER proficiency is necessary for bioantimutagenic activity of Cam, Euc and Thu.

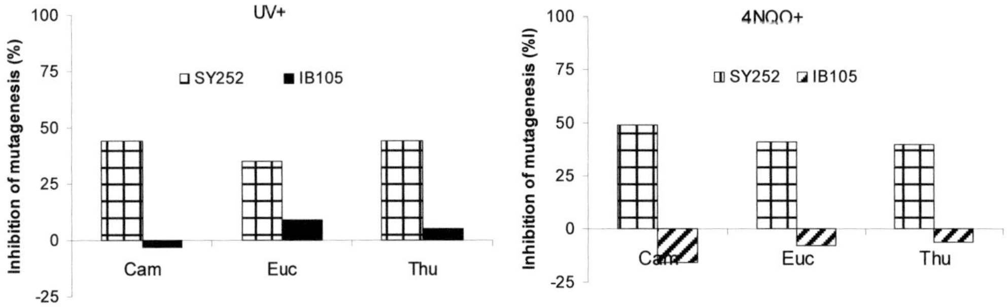

Figure 8. Antimutagenic effect of monoterpenes in repair proficient and NER deficient strain

Experiments in repair proficient IB111 and NER deficient IB127 strain showed that in both strains UV-induced levels of β-galactosidase were higher and persisted longer in cultures with Cam. Euc also maintained induced levels of β-galactosidase longer than in corresponding controls, but due to inhibition of protein synthesis they were slightly lower than in the control ones. Since SOS induction in *E. coli* increases the efficiency of NER [98], obtained results indicated that increased induction of NER could be involved in antimutagenic effect of Cam and Euc.

On the contrary, the kinetics of UV-induced SOS response was not affected by Thu, but it significantly decreased the levels of β-galactosidase and the growth rate of both strains. These effects were attributed to inhibition of protein synthesis [58]. According to obtained results we proposed that, by retaining bacterial growth and cell divisions, Thu increased the time for error free repair of pyrimidine dimers by NER, similarly as Lin. Consistent with the effect on protein synthesis, stimulation of homologous recombination by Thu and Euc was observed only in strain with constitutive expression of RecA (*recA730*) protein, while Cam was additionally effective in *recA*+ strain (Figure 9).

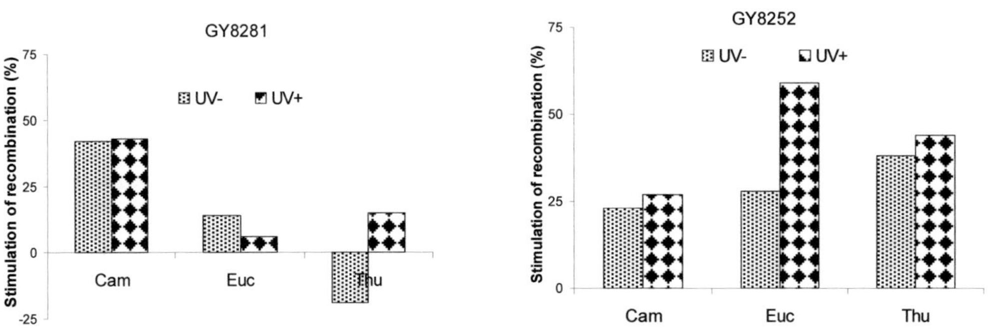

Figure 9. Effect of monoterpenes on homologous recombination

The effect of monoterpenes on the repair of 4NQO-induced DNA damage was also monitored with comet test on repair proficient Vero cell line. Obtained results showed that in cells pre-treated with 4NQO incubation with low doses of monoterpenes resulted in significant reduction of tail moment compared with control, indicating more efficient repair of 4NQO-induced DNA lesions (Figure 10A).

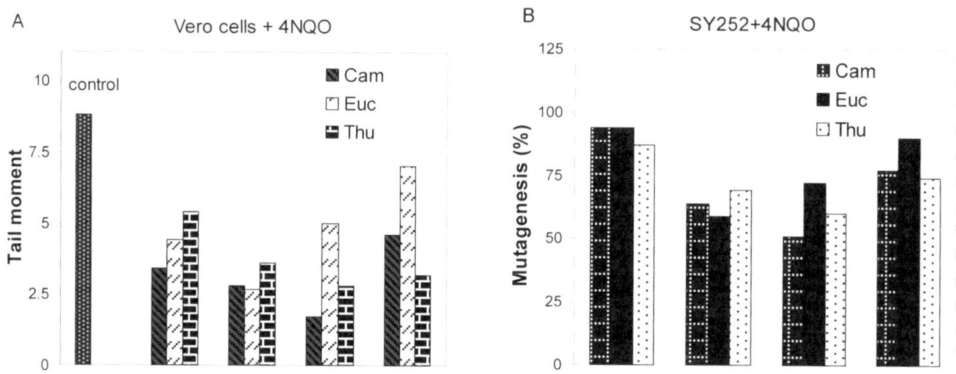

Figure 10. Antigenotoxic/antimutagenic effect of monoterpenes against 4NQO

A common feature of Cam, Euc and Thu was that the mutagenicity and genotoxicity were not reduced in a dose-dependent manner. On the contrary, U-shaped concentration-response curves were obtained (Figure 10). This type of response is usually interpreted as indication of mutagenicity/genotoxicity at higher concentrations of the agent. In bacteria we did not detect mutagenic effect of Cam, Euc and Thu in repair proficient strain in the range of tested concentrations. Moreover, none of the monoterpenes alone could induce SOS response. However, Thu slightly increased spontaneous mutagenesis in *uvrA* mutant. In addition, indication of co-mutagenic effect of all three monoterpenes was observed. Moreover, slower fading of SOS response obtained with Cam and Euc, and stimulation of homologous recombination by all three monoterpenes alone also indicated possible genotoxicity.

In order to determine if monoterpenes could induce DNA lesions, in further work we applied higher doses and evaluated their genotoxicity in the comet assay. Obtained results showed that applied doses induced DNA lesions, providing direct confirmation of genotoxicity of Cam, Euc and Thu (Figure 11). The genotoxicity of high concentrations of Lin was previously determined in comet assay on yeast cells [86].

Taken together, our results led us to propose that, by making a small amount of DNA lesions, low concentrations of monoterpenes stimulated error-free DNA repair (mainly NER), and therefore reduced genotoxicity induced by UV or 4NQO. The results fitted in hormesis phenomenon, defined as beneficial response to a low dose of a stressor agent [99]. Hormesis is now generally accepted as a real and reproducible biological phenomenon, being highly generalized and independent of biological model, end-point measured and chemical/physical stressor applied [100].

Hormesis hypothesis could successfully explain controversial literature and our data concerning genotoxicity/antigenotoxicity of monoterpenes. No mutagenicity of Cam and Euc was detected in the *Salmonella*/microsome assay [97,101,102]. No DNA damage by Euc was observed in cultured Chinese hamster ovary cells [103], and in human leukemic K562 cells [104]. Cam did not induce significant mutagenicity in bone marrow cells of pregnant rats [105]. However, in SMART test Cam was genotoxic [106] and Euc induced apoptosis in two human leukemia cell lines and inhibited DNA synthesis in plant cells [107,108]. On the other hand, Cam and Euc reduced aflatoxin B_1-induced mutagenesis in *S. typhimurium* TA100 [109], Cam reduced γ-radiation-induced increase in SCE frequency in mice bone marrow cells [110] and Euc reduced mutagenesis induced by several model and environmental mutagens in *Salmonella*/microsome reversion assay [102].

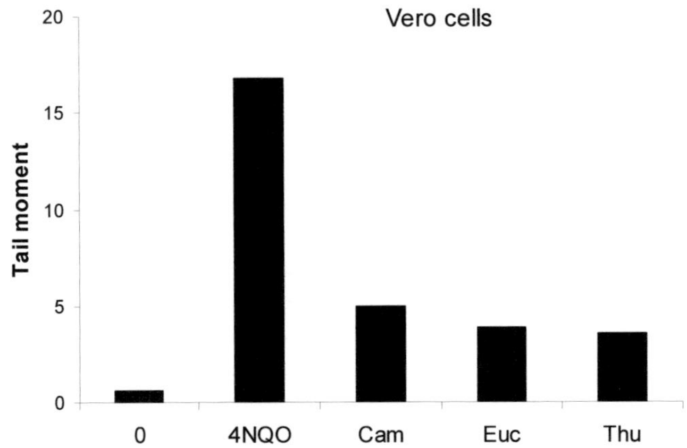

Figure 11. Genotoxic effect of Cam, Euc and Thu

Many literature data indicate no mutagenic or genotoxic effect of Lin in prokaryotic and eukariotic cells [102,111-116]. However, Lin was genotoxic in mouse lymphoma L5178Y TK+ cells [114] and in *B. subtilis* Rec-assay [113]. Evidence for antimutagenic effect was provided by Stajković et al. [102], who reported that Lin reduced UV- and 4NQO-induced mutagenesis in *Salmonella*/microsome reversion assay.

Although no genotoxicity of Thu was detected in SMART test [106], Kim et al. [109] reported co-mutagenic effect on aflatoxin B_1-induced mutagenesis in *S. typhimurium* TA100. Besides our results, no antigenotoxic effect of Thu was reported, but there is evidence about antigenotoxicity of plant extracts containing high proportion of Thu [45,117].

It is clear that antimutagenic and antigenotoxic features of tested monoterpenes depend on the cell type, genetic background, mutagen applied and other experimental conditions. Moreover, our results indicate the special importance of applied concentrations for antimutagenic response.

Considering implications of our hypothessis, our work in progress analyzes if pre-treatment with low doses of monoterpenes could induce DNA repair mechanisms and protect from subsequent exposure to genotoxic agent. Preliminary results indicate that pre-treatment with Cam, Euc and Thu reduces UV-induced mutagenesis in repair proficient strain. In NER deficient strain protective effect of Thu is diminished, while Cam and Euc are even co-mutagenic. Moreover, pre-treatment of repair proficient strain with low doses of 4NQO also provides protection against UV-induced mutagenesis. In our opinion, this strongly supports proposed mechanism of bioantimutagenicity.

9. Conclusions

The identification of natural substances with antigenotoxic/antimutagenic potential and estimation of molecular mechanisms involved are very important to establish their value for chemoprevention strategies [19]. Our comparative study of sage and basil extracts and pure monoterpenes showed that multiple mechanisms are involved in their antimutagenicity/antigenotoxicity. Desmutagenic mechanisms of antioxidants from sage and basil include radical scavenging activity of Lin, Myr and Euc, and the inhibition of metabolic activation of promutagen by high molecular weight terpenes (Figure 12). Bioantimutagenic mechanism involves increased efficiency of error-free DNA repair, mainly NER, by Cam, Euc, Thu and Lin (Figure 13).

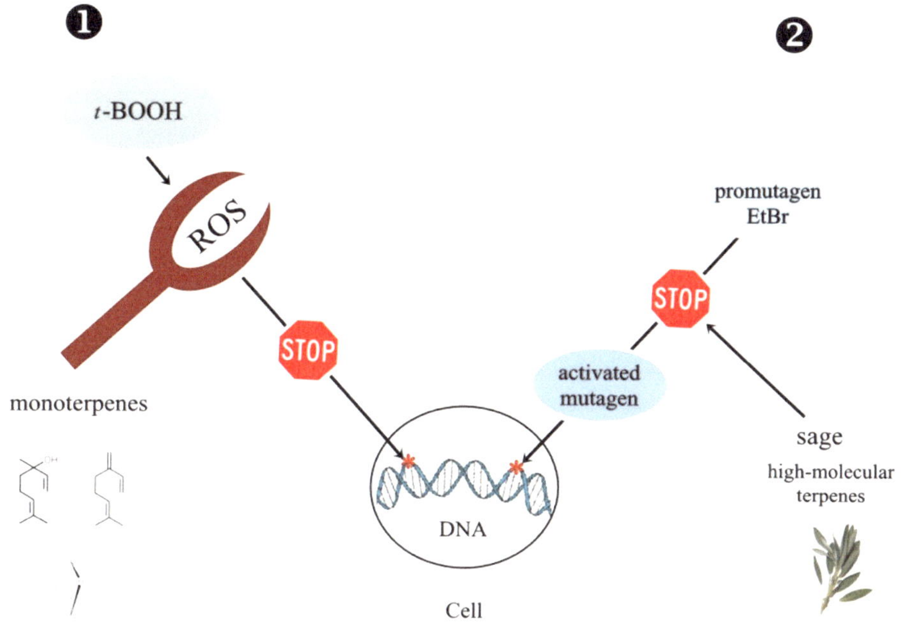

[1] Radical scavenging activity of monoterpenes Lin, Myr and Euc
[2] Inhibition of metabolic activation of promutagen EtBr

Figure 12. Desmutagenic effects of sage and basil

Dietary use of plant antimutagens has been seen by many authors as the most practical way of primary chemoprevention of cancer and many chronic degenerative diseases. Due to low cost and commercial availability of studied monoterpenes, they might be interesting candidates for further chemoprevention studies, but their genotoxicity must be taken in consideration and carefully analyzed.

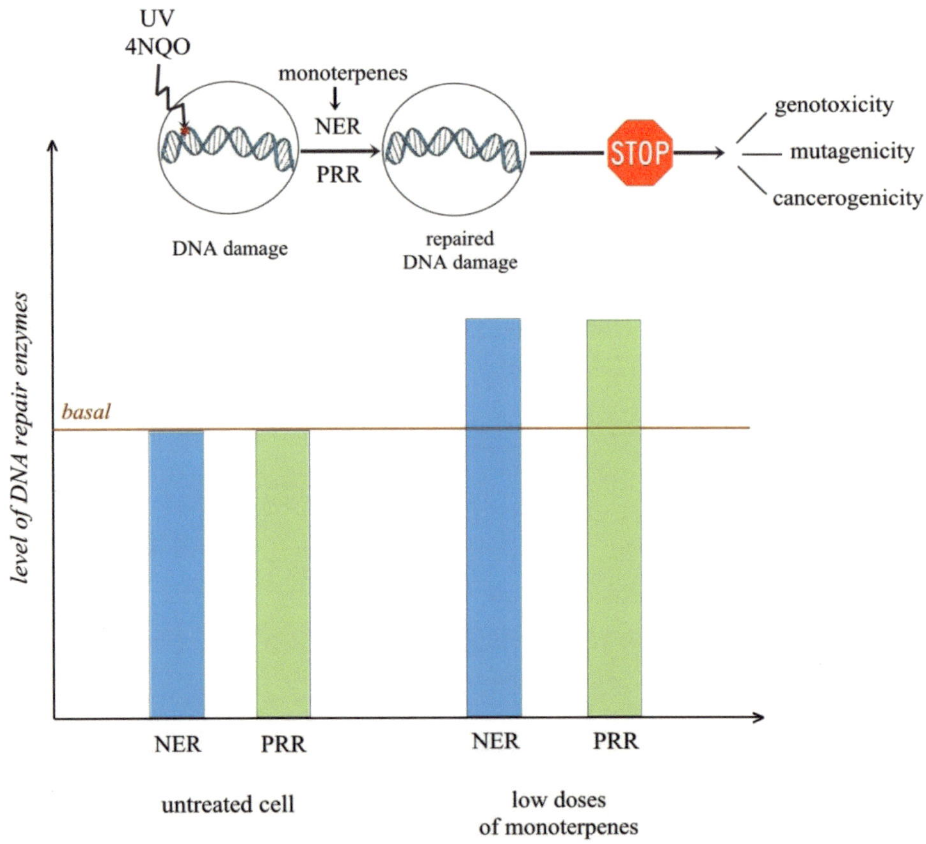

Figure 13. Bioantimutagenic effects of monoterpenes Cam, Euc, Thu and Lin

Author details

Biljana Nikolić*, Dragana Mitić-Ćulafić,
Branka Vuković-Gačić and Jelena Knežević-Vukčević
Chair of Microbiology, University of Belgrade, Faculty of Biology, Belgrade, Serbia

Acknowledgement

This study was supported by the Ministry of Education and Science of the Republic of Serbia, Project No. 172058. We thank PhD student Mina Mandić for creating artwork showing molecular mechanisms of antimutagenesis (Figures 12 and 13).

10. References

[1] Lindahl T, Wood R.D (1999) Quality Control by DNA Repair. Science 286: 1897-1905.

* Corresponding Author

[2] Friedberg E.C, Walker G.C, Siede W, Wood R.D, Schultz R.A, Elleberger T (2006) DNA Repair and Mutagenesis, 2nd Edition, Washington D.C: ASM Press. 9 p.
[3] Eisen J.A, Hanavalt P.C (1999) A phylogenomic study of DNA repair genes, proteins and processes. Mutat. Res. 435: 171-213.
[4] Hoeijmakers J.H.J (2001) Genome maintenance mechanisms for preventing cancer. Nature 411: 366-374.
[5] De Flora S, Izzotti A, Randerath K, Randerath E, Bartsch H, Nair J, Balansky R, van Schooten F, Degan P, Fronza G, Valsh D, Lewtas J (1996) DNA adducts and chronic degenerative diseases, Pathogenic relevanse and imlications in preventive medicine. Mutat. Res. 366: 197-238.
[6] Finkel T, Holbrook N.J (2000) Oxidants, oxidative stress and the biology of ageing. Nature 408: 239-247.
[7] Olinski R, Gackowski D, Foksinski M, Rozalski R, Roszkowski K, Jaruga P (2002) Oxidative DNA damage: assessment of the role in carcinogenesis, atherosclerosis, and acquired immunodeficiency syndrome. Free Radic. Biol. Med. 33: 192-200.
[8] Davydov V, Hansen L.A, Shackelford D.A (2003) Is DNA repair compromised in Alzheimer's disease? Neurobiol. Aging 5809: 1-16.
[9] Coppede F, Migliore L (2009) DNA damage and repair in Alzheimer's disease. Curr. Alzheimer Res. 6: 36–47.
[10] Loeb L.A (1991) Mutator phenotype may be required for multistage carcinogenesis. Cancer Res. 51: 3075–3079.
[11] Marnett L.J (2000) Oxyradicals and DNA damage. Carcinogenesis 21: 361-370.
[12] Loeb L.A, Loeb K.R, Anderson J.P (2003) Multiple mutations and cancer. Proc Natl Acad Sci USA 100: 776–781.
[13] Mills K.D, Ferguson D.O, Alt F.W (2003) The role of DNA breaks in genomic instability and tumorigenesis. Immunol Rev. 194: 77-95.
[14] Bielas J.H, Loeb K.R, Rubin B.P, True L.D, Loeb L.A (2006) Human cancers express a mutator phenotype. Proc Natl Acad Sci USA 103: 18238–18242.
[15] Hoffmann J-S, Cazaux C (2010) Aberrant expression of alternative DNA polymerases: A source of mutator phenotype as well as replicative stress in cancer. Semin.cancer biol. 20: 312-319.
[16] Kada T, Kaneko K, Matsuzaki T, Hara Y (1985) Detection and chemical identification of natural bioantimutagens: A case of green tea factor. Mutat. Res. 150: 127-132.
[17] Kuroda Y (1990) Antimutagenesis studies in Japan. In: Kuroda Y, Shankel D.M, Waters M.D, editors. Antimutagenesis and Anticancerogenesis Mechanisms II, New York and London: Plenum Press pp. 1-22.
[18] De Flora F, Izzotti A, D'Agostini F, Balansky R.M, Noonan D, Albini A (2001) Multiple points of intervention in the prevention of cancer and other mutation-related diseases. Mutat. Res. 480-481: 9-22.
[19] De Flora F, Ferguson R.L (2005) Overview of mechanisms of cancer chemopreventive agents. Mutat. Res. 591: 8-15.
[20] Ferguson L.R, Philpott M, Karunasinghe N (2004) Dietary cancer and prevention using antimutagens. Toxicology 198: 147–159.

[21] Ferguson L.R (2011) Antimutagenesis studies: Where have they been and where are they heading?. Genes Environ. 33: 71-78.
[22] Simić D, Vuković-Gačić B, Knežević-Vukčević J, Trninić S, Jankov R.M (1997) Antimutagenic effect of terpenoids from sage (*Salvia officinalis* L.). J. Environ. Pathol. Toxicol. Oncol. 16: 293-301.
[23] Kada T, Inoue T, Ohta T, Shirasu Y (1986) Antimutagens and their modes of action, In: Shankel D.M, Hartman P.H, Kada T, Hollaender A, editors. Antimutagenesis and Anticarcinogenesis Mechanisms. New York: Plenum Press. pp. 181-196.
[24] De Flora S (1998) Mechanisms of inhibitors of mutagenesis and carcinogenesis. Mutat. Res. 402: 151-158.
[25] Crowell P.L (1999) Prevention and therapy of cancer by dietary monoterpenes. J. Nutr. 129: 7755-7785.
[26] Paduch R, Kandefer-Szerszen M, Trytek M, Fiedurek J (2007) Terpenes: substances useful in human healthcare. Arch. Immunol. Ther. Exp. 55: 315-327.
[27] Bakkali F, Averbeck S, Averbeck D, Waomar M (2008) Biological effects of essential oils – a review. Food Chem. Toxicol. 46: 446-475.
[28] Vuković-Gačić B, Simić D (1993) Identification of natural antimutagens with modulating effects on DNA repair. In: G. Bronzzeti G, Hayatsu H, De Flora S, Waters M.D, Shankel D.M, editors. Antimutagenesis and Anticarcinogenesis Mechanisms III, New York: Plenum Press. pp. 269-277.
[29] Cuvelier M.E, Berset C, Richard H (1994) Antioxidant constituents in sage (*Salvia officinalis*). J. Agric. Food Chem. 42: 665-669.
[30] Chattopadhyay R.R (1999) A comparative evaluation of some blood sugar lowering agents of plant origin. J. Ethnopharmacol. 67: 367-372.
[31] Koga T, Hirota N, Takumi K (1999) Bactericidal activities of essential oils of Basil and Sage against a range of bacteria and the effect of these essential oils on *Vibrio parahaemolyticus*. Microbiol. Res. 154: 267-273.
[32] Offiah V.N, Chikwendu, U.A (1999) Antidiarrheal effects of *Ocimum gratissimum* leaf extract in experimental animals. J. Ethnopharmacol. 68: 327-330.
[33] Barićević D, Bartol T (2000) The biological/pharmacological activity of the *Salvia* genus V., Pharmacology, In: Kintzios S.E, editor. Sage, The Genus Salvia. Amsterdam: Haewood Academic Publishers. pp. 143-184.
[34] Klem M.A, Nair M.G, Sraassburg G.M, Dewitt D.L (2000) Antioxidant and cyclooxygenase inhibitory phenolic compounds from *Ocimum sanctum* Linn. Phytomedicine 7: 7-13.
[35] Zupko I, Hohmann J, Redei D, Falkay G, Janicsak G, Mathe I (2001) Antioxidant activity of leaves of *Salvia* species in enzyme-dependent and enzyme-independent system of lipid peroxidation and their phenolic constituents. Planta Med. 67: 366-368.
[36] Capasso R, Izzo A.I, Capasso F, Romussi G. Bisio A, Mascolo N (2004) A diterpenoid from *Salvia cinnabarina* inhibits mouse intestinal motility in vivo. Planta Med. 70: 375-377.
[37] Ren Y, Houghton P.J, Hider R.C, Howes M.J.R (2004) Novel diterpenoid acetylcholinesterase inhibitors from *Salvia miltiorhiz*. Planta Med. 70: 201-204.

[38] Mitić-Ćulafić D, Vuković-Gačić B, Knežević-Vukčević J, Stanković S, Simić D (2005) Comparative study on the antibacterial activity of volatiles from sage (*Salvia officinalis* L.). Arch. Biol. Sci. 57: 173-178.

[39] Šmidling D, Mitić-Ćulafić D, Vuković-Gačić B, Simić D, Knežević-Vukčević J (2008) Evaluation of antiviral activity of fractionated extracts of sage *Salvia officinalis* L. (Lamiaceae). Arch. Biol. Sci. 60: 421-429.

[40] Tsai K.D, Lin B.R, Perng D.S, Wei J.C, Yu Y.W, Cherng, J-M (2011) Immunomodulatory effects of aqueous extract of *Ocimum basilicum* (Linn.) and some of its constituents on human immune cells. J. Med. Plant Res. 5: 1873-1883.

[41] Gali-Muhtasib H.U, Affara N.I (2000) Chemopreventive effects of sage oil on skin papilomas in mice. Phytomedicine. 7: 129-136.

[42] Prakash J, Gupta S.K (2000) Chemopreventive activity of *Ocimum sanctum* seed oil. J. Ethnopharmacol. 72: 29-34.

[43] Vujošević M, Blagojević J (2004) Antimutagenic effects of extracts from sage (*Salvia officinalis*) in mammalian system *in vivo*. Acta Vet. Hung., 52: 439-443.

[44] Sidiqque Y.H, Ara G, Beg T, Afzal M (2007) Anti-genotoxic effect of *Ocimum sanctum* L. extract against cyproterone acetate induced genotoxic damage in cultured mammalian cells. Acta Biol. Hun. 58: 397-409.

[45] Patenković A, Stamenković-Radak M, Banjanac T, Anđelković, M (2009) Antimutagenic effect of sage tea in the wing spot test of *Drosophila melanogaster*. Food Chem Toxicol. 47: 180-183.

[46] Simić D, Vuković-Gačić B, Knežević-Vukčević J (1998) Detection of natural bioantimutagens and their mechanisms of action with bacterial assay-system. Mutat. Res. 402: 51-57.

[47] Vuković-Gačić B, Simić D, Knežević-Vukčević J (2006) *Escherichia coli* assay system for detection of plant antimutagens and their mechanisms of action. In: Verschaeve L, editor. Topical Issues in Applied Microbiology and Biotechnology, Kerala, India: Research Signpost. pp. 61-86.

[48] Tood P.A, Monti-Bragadini C, Glickman B.W (1979) MMS mutagenesis in strains of *E. coli* carrying the R46 mutagenic enhancing plasmid: phenotypic analysis of Arg^+ revertants. Mutat. Res. 62: 227-237.

[49] Young L.C, Hays J.B, Tron V.A, Andrew S.E (2003) DNA mismatch repair proteins: potential guardians against genomic instability and tumorigenesis induced by ultraviolet photoproducts. J Invest Dermatol. 121: 435-440.

[50] Walker G.C (1984) Mutagenesis and inducible responses to DNAdamage in *Escherichia coli*. Microbiol. Rev. 48: 60-93.

[51] Walker G.C (1985) Inducible DNA system. Annu.Rev. Bioch. 54: 425-457.

[52] Jones C.J, Edwards S.M, Waters R (1989) The repair of identified large DNA adducts induced by 4-nitroquinoline-1-oxide in normal or xeroderma pigmentosum group A human fibroblasts, and the role of DNA polymerase alpha or delta. Carcinogenesis. 10: 1197-1201.

[53] Hömme M, Jacobi H, Juhl-Strauss U, Witte I (2000) Synergistic DNA damaging effects of 4-nitroquinoline-1-oxide and non-effective concentrations of methyl methanesulfonate in human fibroblasts. Mutat. Res. 461: 211-219.

[54] Truglio J.J, Croteau D.L, van Houten B, Kisker C (2006) Prokaryotic nucleotide excision repair: The UvrABC System. Chem.Rev. 106: 233-252.

[55] Knežević J, Simić D (1982) Induction of phage lambda by bleomycin and nalidixic acid in the *lexA* mutant of *E. coli* K12. Genetika. 14: 77-91.

[56] Stanojević J, Berić T, Opačić B, Vuković-Gačić B, Simić D, Knežević-Vukčević J (2008) The effect of essential oil of basil (*Ocimum basilicum* L.) on UV-induced mutagenesis in *Escherichia coli* and *Saccharomyces cerevisiae*. Arch. Biol. Sci. 60: 93-102.

[57] Nikolić B, Stanojević J, Mitić D, Vuković-Gačić B, Knežević-Vukčević J, Simić D (2004) Comparative study of the antimutagenic potential of Vitamin E in different *E. coli* strains. Mutat. Res. 564: 31-38.

[58] Nikolić B, Mitić-Ćulafić D, Vuković-Gačić B, Knežević-Vukčević J (2011) Modulation of genotoxicity and DNA repair by plant monoterpenes Camphor, Eucalyptol and Thujone in *Escherichia coli* and mammalian cells. Food Chem. Tox. 49: 2035-2045.

[59] Dutreix M, Moreau P.L, Bailone A, Galibert A, Battista J.R, Walker G.C, Devoret R (1989) New *recA* mutations that dissociate the various RecA protein activities in *E. coli* provide evidence for an additional role for RecA protein in UV mutagenesis. J. Bacteriol. 171: 2415-2423.

[60] Maron D.M, Ames B.N (1983) Revised methods for the *Salmonella* mutagenicity test. Mutat. Res. 113: 173-215.

[61] Martinez A, Urios A, Blanco M (2000) Mutagenicity of 80 chemicals in *Escherichia coli* tester strains IC203, deficient in OxyR, and its *oxyR*$^{+}$ parent WP2 *uvrA*/pKM101: detection of 31 oxidative mutagens. Mutat. Res. 467: 41-53.

[62] Zimmermann F.K, Kern R. Rasenberger H.(1975).A yeast strain for simultaneous detection of induced mitotic crossing over, mitotic gene conversion and reverse mutation. Mutat. Res. 28: 381-388.

[63] Miloshev G, Mihaylov I, Anachkova B (2002) Application of the single cell gel electrophoresis on yeast cells. Mutat. Res. 513: 69-74.

[64] Radman M (1999) Enzymes of evolutionary change. Nature, 401: 866-869.

[65] Quillardet P, Hofnung M (1993) The SOS chromotest: a review. Mutat. Res. 297: 235-279.

[66] Konrad E.B (1977) Method for the isolation of *Escherichia coli* mutants with enhanced recombination between chromosomal duplications. J. Bacteriol. 130: 167-172.

[67] Lawery P.E, Kowalczykowski S.C (1992) Biochemical basis of the constitutive repressor cleavage activity of RecA730 protein. J. Biol. Chem. 267: 20648-20658.

[68] Ennis D.G, Levine A.S, Koch W.H, Woodgate R (1995) Analysis of *recA* mutants with altered SOS functions. Mutat. Res., 336: 39-48.

[69] Kohlimeir L, Simonsen M, Mottus K (1995) Dietary modifiers of carcinogenesis. Environ. Health Perspect. 103: 177-184.

[70] Urios A, Blanco M (1996) Specifity of spontaneous and *t*-butyl hydroperoxide-induced mutations in Δ*oxyR* strains of *Escherichia coli* differing with respect to the SOS mutagenesis proficiency and to the MutY and MutM functions. Mutat. Res. 354: 95-101.

[71] Fowler R.G, White S.J, Koyama C, Moore S.C, Dunn R.L, Schaaer R.M (2003) Interactions among the *Escherichia coli mutT, mutM,* and *mutY* damage prevention pathways. DNA Repair 2: 159-173.

[72] Mure K. Rossman,T.G (2001) Reduction of spontaneous mutagenesis in mismatch repair-deficient and proficient cells by dietary antioxidants. Mutat. Res. 480-481: 85-95.

[73] Jackson A.L, Loeb L.A.(1998) The mutation rate and cancer. Genetics 148: 1483-1490.

[74] Jackson A.L, Loeb L.A (2000) Microsatellite instability induced by hydrogen peroxide in *Escherichia coli*. Mutat. Res. 447: 187-198.

[75] Boyer J.C, Yamada N.A. Roques C.N. Hatch S.B. Riess K, Farber R.A (2002) Sequence dependent instability of mononucleotide microsatellites in cultured mismatch repair proficient and deficient mammalian cells. Hum. Mol. Genet. 11: 707-713.

[76] OECD (2007). ISSN : 2074-5788 (online) DOI : 10.1787/20745788.

[77] Blanco M, Urios A, Martinez A (1998) New *Escherichia coli* WP2 tester strains highly sensitive to reversion by oxidative mutagens. Mutat. Res. 413: 95-101.

[78] Collins A.R (2004) The Comet assay for DNA damage and repair. Mol. Biotech. 26: 249-261.

[79] Đarmati Z, Jankov R.M, Vujičić Z, Csanadi J, Đulinac B, Švirtlih E, Đorđević A, Švan K (1994) 12-Deoxocarnisol isolated from the wild type of sage from Dalmatia. J. Serb. Chem. Soc. 59: 291-299.

[80] Đarmati Z, Jankov R.M, Vujičić Z, Csanadi J, Švirtlih E, Đorđević A, Švan K (1993) Natural terpenoids isolated from grown variety of sage. J. Serb. Chem. Soc. 58: 515-523.

[81] Mitić D, Vuković-Gačić B, Knežević-Vukčević J, Berić T, Nikolić B, Stanković S, Simić D (2001) Natural antioxidants and their mechanisms in inhibition of mutagenesis, In: Kreft I, Škrabanja V, editors. Molecular and genetic interactions involving phytochemicals. Ljubljana, Slovenia: Univerza v Ljubljani in Slovenska akademija znanosti in umetnosti. pp. 67-74.

[82] Simić D, Vuković-Gačić B, Knežević-Vukčević J, Đarmati Z, Jankov R.M (1994) New assay system for detecting bioantimutagens in plant extracts. Arch. Biol. Sci. 46: 81-85.

[83] Brkić D, Stepanović B, Nastovski T, Brkić S (1999) Distilation of sage. In: Brkić D, editor. *Sage* (*S. officinalis L.*). Belgrade, Serbia: Institute for Medicinal Plant Research "Dr Josif Pančić". pp. 131-136.

[84] Marinković B, Marin P.D, Knežević-Vukčević J, Soković M.D, Brkić D (2002) Activity of essential oils of three *Micromeria* species (Lamiaceae) against micromycetes and bacteria. Phytoter. Res. 16: 336-339.

[85] Knežević-Vukčević J, Vuković-Gačić B, Stević T, Stanojević J, Nikolić B, Simić D (2005) Antimutagenic effect of essential oil of sage (*Salvia officinalis* L.) and its fractions against UV-induced mutations in bacterial and yeast cells. Arch. Biol. Sci. 57: 163-172.

[86] Berić T, Nikolić B, Stanojević J, Vuković-Gačić B, Knežević-Vukčević J (2008) Protective effect of basil (*Ocimum basilicum* L.) against oxidative DNA damage and mutagenesis. Food Chem. Toxicol. 46: 724-732.

[87] Mitić-Ćulafić D, Žegura B. Nikolić B. Vuković-Gačić B. Knežević-Vukčević J. Filipič M (2009) Protective effect of linalool, myrcene and eucalyptol against *t*-butyl

hydroperoxide induced genotoxicity in bacteria and cultured human cells. Food Chem. Tox. 47: 260-266.

[88] Nikolić B, Mitić-Ćulafić D, Vuković-Gačić B, Knežević-Vukčević J (2011) The antimutagenic effect of monoterpenes against UV-irradiation-, 4NQO- and *t*-BOOH-induced mutagenesis in *E. coli*. Arch. Biol. Sci. 63: 117-128.

[89] Lazarova M, Labaj J, Eckl P, Slamenova D (2006) Comparative evaluationof DNA damage by genotoxicants in primary rat cells applying the comet assay. Toxicol. Lett. 164: 54-62.

[90] Chadha A, Madyastha K.M (1984) Metabolism of geraniol and linalool in the rat and effects on liver and lung microsomal-enzymes. Xenobiotica 14: 365-374.

[91] Madyastha M.K, Srivatsan M (1987) Metabolism of β-myrcene *in vivo* and *in vitro*: its effects on rat-liver microsomal enzymes. Xenobiotica 17: 539-549.

[92] Knasmüller S, Parzefall W, Sanyal R, Ecker S, Schwab C, Uhl M, Mersch-Sundermann V, Williamson G, Hietsch G, Langer T, Darroudi F, Natarajan A.T (1998) Use of metabolically competent human hepatoma cells for the detection of mutagens and antimutagens, Mutat. Res. 402: 185-202.

[93] Miyazawa M, Shindo M, Shimada T (2001) Oxidation of 1,8-cineole, the monoterpene cyclic ether originated from *Eucalyptus polybractea,* by cytochrome P450 3A enzymes in rat and human liver microsomes. Drug Metab. Disposit. 29: 200-205.

[94] Kassie F, Mersch-Sundermann V, Edenharder R, Platt L.K, Darroudi F, Lhoste E, Humbolt C, Muckel E, Uhl M, Kundi M, Knasmuller S (2003) Development and application of test methods for the detection of dietary constituents which protect against heterocyclic aromatic amines. Mutat.Res. 523-524: 183-192.

[95] Ishida T (2005) Biotransformation of terpenoids by mammals, microorganisms and plant-cultured cells. Chem. Biodiver. 2: 569-590.

[96] Belsito D, Bickers D, Bruze M, Calow P, Greim H, Hanifin J.M, Rogers A.E, Saurat J.H, Sipes I.G, Tagami H (2008) A toxicologic and dermatologic assessment of cyclic and non-cyclic terpene alcohols when used as fragrance ingredients. Food Chem. Tox. 46: S1-S71.

[97] Vuković-Gačić B, Nikčević S, Berić-Bjedov T, Knežević-Vukčević J, Simić D (2006) Antimutagenic effect of essential oil of sage (*Salvia officinalis* L.) and its monoterpenes against UV-induced mutations in *Escherichia coli* and *Saccharomyces cerevisiae*. Food Chem. Toxicol. 44: 1730-1738.

[98] Crowley D.J, Hanawalt P.C, (1998) Induction of the SOS response increases the efficiency of global nucleotide excision repair of cyclobutane pyrimidine dimmers, but not 6-4 photoproducts, in UV-irradiated *Escherichia coli*. J. Bacteriol. 180: 3345-3352.

[99] Calabrese E.J, Baldwin L.A (2001) Hormesis: U-shaped dose responses and their centrality in toxicology. Trends Pharmacol. Sci. 22: 285-291.

[100] Calabrese E.J (2010) Hormesis is central to toxicology, pharmacology and risk assessment. Hum. Exp. Toxicol. 29: 249-261.

[101] Gomes-Carneiro M.R, Elzenszwalb I.F, Paumgartten F.J (1998) Mutagenicity testing (+/-) – camphor, 1,8-cineole, citral, citronellal, (-)-menthol and terpineol with *Salmonella*/ microsome assay. Mutat. Res. 416: 129-136.

[102] Stajković O, Berić-Bjedov T, Mitić-Ćulafić D, Stanković S, Vuković-Gačić B, Simić D, Knežević-Vukčević J (2007) Antimutagenic properties of basil (*Ocimum basilicum* L.) in *Salmonella typhimurium* TA100. Food Tech. Biotech. 45: 213-217.

[103] Ribeiro D.A, Matsumoto M.A, Marques M.E.A, Salvadori D.M.F (2007) Biocompatibility of gutta-percha solvents using in vitro mammalian test-system. Oral Surg. Oral Med. Oral Pathol. Oral Radiol. Endod. 103: e106-e109.

[104] Horvathova E, Turcaniova V, Slamenova D (2007) Comparative study of DNA-damaging and DNA-protective effects of selected components of essential plant oils in human leukamic cells K562. Neoplasma 54: 478-483.

[105] Alakilli S.Y.M (2009) Evaluation of camphor mutagenicity in somatic cells of pregnant rats. Asian J. Biotech. 1: 111-117.

[106] Pavlidou V, Karpouhtsis I, Franzios G, Zambetaki A, Scouras Z, Mavragani-Tsipidou P (2004) Insecticidal and genotoxic effects of essential oils of Greek sage, *Salvia fruticosa*, and mint, *Mentha pulegium*, on *Drosophila melanogaster* and *Bactrocera oleae* (Diptera: Tepritidae). J. Agr. Urban Entomol. 21: 39-49.

[107] Koitabashi R, Suzuki T, Kawazu T, Sakai A, Kuroiwa H, Kuroiwa T (1997) 1,8- Cineole inhibits root growth and DNA synthesis in the root apical meristem of *Brassica campestris* L. J. Plant Res. 110: 1-6.

[108] Moteki H, Hibasami H, Yamada Y, Katsuzaki H, Imai K, Komiya T (2002) Specific induction of apoptosis by 1,8-cineole in two human leukemia cell lines, but not in human stomach cancer cell line. Oncol. Rep. 9: 757-760.

[109] Kim J.O, Kim Y.S, Lee J.H, Kim M.N, Rhee S.H, Moon S.H, Park K.Y (1992) Antimutagenic effect of the major volatile compounds identified from mugworth (*Artemisiia asictica nakai*) leaves. J. Korean Soc. Food Nutr. 21: 308-313.

[110] Goel H.C, Singh S, Singh S.P (1989) Radiomodifying influence of camphor on sister-chromatid exchange induction in mouse bone marrow. Mutat. Res. 224: 157-160.

[111] Lutz D, Eder E, Neudecker T (1982) Structure-mutagenicity relationship in α,β-unsaturated carbonylic compounds and their corresponding allylic alcohols. Mutat. Res. 93: 305-315.

[112] Ishidate M Jr, Sofuni T, Yoshikawa K, Hayashi M, Nohmi T, Sawada M, Matsuoka A (1984) Primary mutagenicity screening of food additives currently used in Japan. Food Chem. Tox. 22: 623-636.

[113] Yoo Y.S (1986) Mutagenic and antimutagenic activities of flavoring agents used in foodstuffs. Osaka-shi Igakkai Zasshi [J Osaka City Medical Center] 34: 267-288.

[114] Heck J.D, Vollmuth T.A, Cifone M.A, Jagannath D.R, Myhr V, Curren R.D (1989) An evaluation of food flavoring ingredients in a genetic toxicity screening battery. Toxicologist 9: 257-264.

[115] Sasaki Y.F, Imanishi H, Ohta T, Shirasu Y (1989) Modifying effects of components of plant essence on the induction of sister chromatid exchanges in cultured Chinese hamster ovary cells. Mutat. Res. 226: 103-110.

[116] Di Sotto A, Mazzanti G, Carbone F, Hrelia P, Maffei F (2011) Genotoxicity of lavander oil, linalyl acetate, and linalool on human lymphocytes in vitro. Environ. Mol. Mutagen. 52: 69-71.

[117] Minnunni M, Wolleb U, Muellera O, Pfeifera A, Aeschbachera H.U (1992) Natural antioxidants as inhibitors of oxygen species induced mutagenicity. Mutat. Res. 269: 193-200.

Chapter 6

Chemical Modification of Oligonucleotides: A Novel Approach Towards Gene Targeting

Hidetaka Torigoe and Takeshi Imanishi

Additional information is available at the end of the chapter

http://dx.doi.org/10.5772/50393

1. Introduction

Artificial regulation of gene expression is quite important for basic study to analyze unknown biological functions of target genes. Comparison of phenotypes with and without knockdown of expression level of the target genes may be helpful to reveal unknown biological functions of the target genes. Artificial regulation of gene expression is also important for therapeutic applications to reduce expression level of mutated target genes. Knockdown of expression level of the mutated target genes may be useful to avoid undesirable effects produced by the mutated target genes. Antisense and antigene technologies are powerful tools to artificially regulate target gene expression. In antisense technology, a single-stranded oligonucleotide added from outside may bind with target mRNA to form oligonucleotide-RNA duplex[1,2]. The formed duplex may inhibit ribosome-mediated translation of target mRNA due to its steric hindrance, or RNaseH may cleave target mRNA in the formed duplex, which may result in reduction of expression level of target mRNA in both cases[1,2]. In antigene technology, a single-stranded homopyrimidine triplex-forming oligonucleotide (TFO) added from outside may bind with homopurine-homopyrimidine stretch in target duplex DNA by Hoogsteen hydrogen bonding to form pyrimidine motif triplex, where T•A:T and C^{+}•G:C base triplets are formed[3,4]. The formed triplex inhibits RNA polymerase and transcription factors-mediated transcription of target gene due to its steric hindrance, which may result in downregulation of expression level of target gene[3,4].

Serious difficulties, such as poor binding ability of added oligonucleotides with target mRNA or target duplex DNA[5-7], and low stability of added oligonucleotides against nuclease degradation[8], may limit practical applications of the antisense and antigene technologies *in vivo*. Many kinds of chemically modified oligonucleotides have been developed to overcome these serious difficulties[9]. In this context, we have developed a novel class of chemical modification of nucleic acids, 2′-*O*,4′-*C*-aminomethylene bridged nucleic acid (2′,4′-BNA^{NC}), in which 2′-O and 4′-C of the sugar moiety are bridged with the aminomethylene chain (Figure 1a)[10,11]. 2′,4′-BNA^{NC}-modified oligonucleotides showed higher binding affinity with target mRNA[11], stronger binding ability with target duplex DNA to form triplex[10-13], and higher stability against nuclease degradation than the corresponding unmodified oligonucleotides[11-13]. These excellent properties of 2′,4′-BNA^{NC}-modification of oligonucleotides may be favorable for their practical applications to the antisense and antigene technologies.

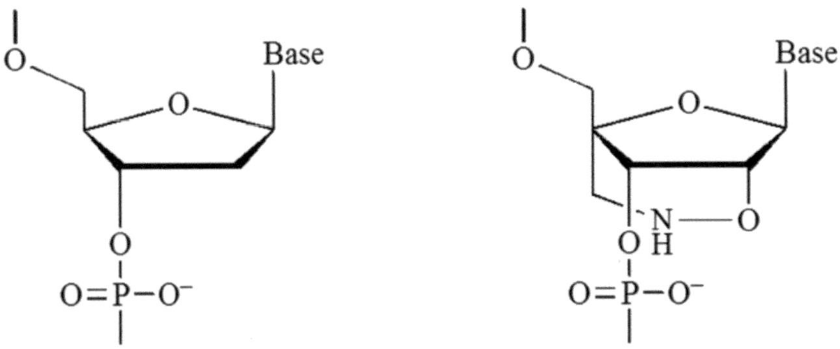

Phosphodiester (PO)

2’-*O*,4’-*C*-aminomethylene bridged nucleic acid (2’,4’-BNA^{NC})

(a)

Forward strand
F12 5′-d(GCGTTTTTTGCT)-3′
F12-1 5′-d(GCGTTTTTTGCT)-3′
F12-31 5′-d(GCGTTTTTTGCT)-3′
F12-32 5′-d(GCGTTTTTTGCT)-3′
F12-6 5′-d(GCGTTTTTTGCT)-3′
Complementary strand
R12R 3′-r(CGCAAAAAACGA)-5′
R12X 3′-r(CGCAAXAAACGA)-5′ (X=A, U, G, C)
R12D 3′-d(CGCAAAAAACGA)-3′
T: T with 2′-*O*,4′-*C*-aminomethylene bridge

(b)

Pyr15TM	5′-CTCTTCTTTTCTTTC-3′
Pyr15NC7-1	5′-CTCTTCTTTTCTTTC-3′
Pyr15NC7-2	5′-CTCTTCTTTTCTTTC-3′
Pyr15NC5-1	5′-CTCTTCTTTTCTTTC-3′
Pyr15NC5-2	5′-CTCTTCTTTTCTTTC-3′
Pur23A	5′-GCGCGAGAAGAAAAGAAAGCCGG-3′
Pyr23T	3′-CGCGCTCTTCTTTTCTTTCGGCC-5′
Pyr15NS2M	5′-TCTCTTCTTTTCTCT-3′

C: 5-methylcytosine

C: C with 2′-*O*,4′-*C*-aminomethylene bridge

T: T with 2′-*O*,4′-*C*-aminomethylene bridge

(c)

PCSK9-5-NC(T,C) 5′-d(GGTCCTCAGGGAACCAGGCC)-3′
(Target: 1287-1306 base of PCSK9 gene)

PCSK9-6-NC(T,C) 5′-d(GCCACCAGGTTGGGGTCAG)-3′
(Target: 1315-1334 base of PCSK9 gene)

PCSK9-7-NC(T,C) 5′-d(CTGGAGCAGCTCAGCAGCTC)-3′
(Target: 1453-1472 base of PCSK9 gene)

PCSK9-8-NC(T,C) 5′-d(TAGACACCCTCACCCCCAAA)-3′
(Target: 1552-1571 base of PCSK9 gene)

PCSK9-10-NC(T,C) 5′-d(GCCGGCTCCGGCAGCAGATG)-3′
(Target: 2037-2056 base of PCSK9 gene)

C: 5-methylcytosine

C: C with 2′-*O*,4′-*C*-aminomethylene bridge

T: T with 2′-*O*,4′-*C*-aminomethylene bridge

(d)

Figure 1. a) Structural formulas for phosphodiester (PO) and 2′-*O*,4′-*C*-aminomethylene bridged nucleic acid (2′,4′-BNA^{NC})-modified backbones. (b) Oligonucleotide sequences for the unmodified (F12) or 2′,4′-BNA^{NC}-modified (F12-1, F12-31, F12-32, F12-6) oligonucleotide, the complementary RNA (R12R), the complementary single-mismatched RNA (R12X) and the complementary DNA (R12D) (c) Oligonucleotide sequences for the target duplex (Pur23APyr23T), the specific TFOs (Pyr15TM, Pyr15NC7-1, Pyr15NC7-2, Pyr15NC5-1, and Pyr15NC5-2), and the nonspecific oligonucleotide (Pyr15NS2M). (d) Oligonucleotide sequences for the 2′,4′-BNA^{NC}-modified antisense oligonucleotides to target a certain region of PCSK9 gene.

In this chapter, we describe the excellent properties of 2′,4′-BNA^{NC}-modified oligonucleotides for higher ability to form duplex and triplex and for higher nuclease resistance[11-13]. We also show the biological application of 2′,4′-BNA^{NC}-modified oligonucleotides to reduce expression level of target mRNA in mammalian cells[14]. PCSK9 is a serine protease involved in the degradation of LDL receptor [15-17]. Suppression of PCSK9 by reducing expression level of PCSK9 mRNA may cause an increase in the amount of the LDL receptor, resulting in the reduction of serum LDL cholesterol level. Thus, the PCSK9 mRNA has the potential to be an antisense target for the treatment of hypercholesterolemia [15-17]. We present the excellent antisense effect of 2′,4′-BNA^{NC} modified antisense oligonucleotides to reduce the expression level of the PCSK9 mRNA.

2. Methods to prepare and characterize oligonucleotides

2.1. Preparation of oligonucleotides

We synthesized complementary oligonucleotides for duplex DNA and unmodified homopyrimidine TFO (Figure 1b, 1c) on a DNA synthesizer using the solid-phase cyanoethyl phosphoramidite method, and purified them with a reverse-phase high performance liquid chromatography on a Wakosil DNA or Waters X-Terra column. 2′,4′-BNA^{NC}-modified oligonucleotides (Figure 1b, 1c, 1d) were synthesized and purified as described previously[10,11]. 5′-Biotinylated oligonucleotides were prepared from biotin phosphoramidite for kinetic analyses by Biacore described below. The concentration of all oligonucleotides was determined by UV absorbance. The reported extinction coefficient for poly (dT) [ε_{265} = 8700 cm^{-1} (mol of base/liter) $^{-1}$][18] was used for unmodified and 2′,4′-BNA^{NC}-modified homopyrimidine TFO. Complementary strands for duplex DNA were annealed by heating at up to 90 °C, followed by a gradual cooling to room temperature. When the removal of unpaired single strands is necessary, the annealed sample was applied on a hydroxyapatite column (BioRad). The concentration of duplex DNA was determined by UV absorbance, considering DNA concentration ratio of 1 OD = 50 μg/ml.

2.2. UV melting

Heating of duplex results in monophasic strand dissociation based on the transition between the two states, duplex→2 single strands. Heating of triplex also leads to biphasic strand dissociation according to the transitions between the three states, triplex→ duplex + single strand→3 single strands. Base stacking interactions in the free strands are weaker than those in the bound strands, resulting in a hyperchromic increase in UV absorbance upon heating. UV melting monitors the process of duplex and triplex melting by the temperature dependent change in UV absorbance. First derivative plot of UV absorbance (dA/dT vs T) is calculated from the UV melting curve (A vs T). Peak temperatures in the first derivative plot correspond to the melting temperature, T_m, of the transition.

UV melting experiments for duplex and triplex study were carried out on a DU-650 and DU-640 spectrophotometer (Beckman Inc.), respectively, equipped with a Peltier type cell holder. The cell path length was 1 cm. UV melting profiles for duplex study were measured in 10 mM sodium phosphate buffer at pH 7.2 containing 100 mM NaCl. UV melting profiles for triplex study were measured in 10 mM sodium cacodylate-cacodylic acid at pH 6.8 containing 200 mM NaCl and 20 mM $MgCl_2$. UV melting profiles were recorded at a scan rate of 0.5 °C/min with detection at 260 nm. The first derivative was calculated from the UV melting profile. The peak temperatures in the derivative curve were designated as the melting temperature, T_m. The duplex and triplex nucleic acid concentration used was 4 μM and 1 μM, respectively.

2.3. CD spectroscopy

CD spectroscopy is sensitive to interactions of nearby bases vertically stacked in strands. Stacking interactions depend on the conformational details of nucleic acid structure. CD spectra provide a certain basis for suggesting the overall conformation of strands in duplex and triplex. The appearance of an intense negative band at the short wavelength range (210-220 nm) in CD spectra indicates the formation of triplex.

CD spectra of the triplex at 20 °C were recorded in 10 mM sodium cacodylate-cacodylic acid at pH 6.8 containing 200 mM NaCl and 20 mM $MgCl_2$ on a JASCO J-720 spectropolarimeter interfaced with a microcomputer. The cell path length was 1 cm. The triplex nucleic acid concentration used was 1 μM.

2.4. Electrophoretic mobility shift assay (EMSA)

The ^{32}P-radiolabelled band of triplex migrates slower than that of duplex in native polyacrylamide gel electrophoresis. The formation of triplex results in appearance of a novel radiolabelled band shifted to a new position corresponding to triplex. The percentage of the formed triplex was calculated using the following equation:

$$\%_{\text{triplex}} = [S_{\text{triplex}} / (S_{\text{triplex}} + S_{\text{duplex}})] \times 100$$

where S_{triplex} and S_{duplex} represent the radioactive signal for triplex and duplex bands, respectively. The dissociation constant, K_d, of triplex formation is determined from the concentration of the TFO, which causes half of target duplex to shift to triplex.

EMSA experiments for the triplex formation were performed essentially as described previously by a 15% native polyacrylamide gel electrophoresis[12,13,19-27]. In a 9 μl of reaction mixture, ^{32}P-labeled Pur23A•Pyr23T duplex (~1 nM) (Figure 1c) was mixed with increasing concentrations of the specific TFO (Pyr15TM, Pyr15NC7-1, Pyr15NC7-2, Pyr15NC5-1, or Pyr15NC5-2) (Figure 1c) and the nonspecific oligonucleotide (Pyr15NS2M) (Figure 1c) in buffer [50 mM Tris-acetate (pH 7.0), 100 mM NaCl, and 10 mM $MgCl_2$]. Pyr15NS2M was added to achieve equimolar concentrations of TFO in each lane as well as to minimize adhesion of the DNA (duplex and TFO) to plastic surfaces during incubation and subsequent losses during processing. After 6 h incubation at 37 °C, 2 μl of 50 % glycerol solution containing bromophenol blue was added without changing the pH and salt concentrations of the reaction mixtures. Samples were then directly loaded onto a 15 % native polyacrylamide gel prepared in buffer [50 mM Tris-acetate (pH 7.0) and 10 mM $MgCl_2$] and electrophoresis was performed at 8 V/cm for 16 h at 4 °C.

2.5. Thermodynamic analyses by isothermal titration calorimetry (ITC)

Isothermal titration calorimetry (ITC) relies upon the accurate measurement of heat changes caused by the interaction of molecules in solution and possesses the advantage of not requiring labeling or immobilization of the components[28]. ITC provides a great deal of thermodynamic information about the binding process from only a single experiment. This information includes the binding stoichiometry (n), the binding equilibrium constant (K_a), the enthalpy change (ΔH) of binding, the entropy change (ΔS) of binding and the Gibbs free energy change (ΔG) of the binding process[28]. A syringe containing a solution of one element (in the case of triplex formation, duplex) is incrementally titrated into a cell containing a solution of the second element (in the case of triplex formation, TFO). As the duplex is added to the TFO, heat is released upon the triplex formation. The heat for each injection is measured by the ITC instrument and is plotted as a function of time over the injection series. The heat signal from each injection is determined by the area underneath the injection peak. The heat is plotted against the molar ratio of the duplex added to the TFO added to the duplex in the cell. The titration plot provides the thermodynamic information of the triplex formation.

Isothermal titration experiments for the triplex formation were carried out on a VP ITC system (Microcal Inc., U.S.A.), essentially as described previously[12,13,19-22,25-27]. The TFO (Figure 1c) and Pur23A•Pyr23T duplex (Figure 1c) solutions were prepared by extensive dialysis against 10 mM sodium cacodylate-cacodylic acid at pH 6.1 or pH 6.8 containing 200 mM NaCl and 20 mM $MgCl_2$. The Pur23A•Pyr23T duplex solution in 10 mM sodium cacodylate-cacodylic acid at pH 6.1 or pH 6.8 containing 200 mM NaCl and 20 mM $MgCl_2$ was injected 20-times in 5-μl increments and 10-min intervals into the TFO solution without changing the reaction conditions. The heat for each injection was subtracted by the heat of dilution of the injectant, which was measured by injecting the Pur23A•Pyr23T duplex solution into the same buffer. Each corrected heat was divided by the moles of the Pur23A•Pyr23T duplex solution injected, and analyzed with Microcal Origin software supplied by the manufacturer.

2.6. Kinetic analyses by Biacore

Biacore is one example of a class of optical biosensors which can be used to determine the kinetic binding parameters of molecular interactions, such as association rate constant (k_{assoc}), dissociation rate constant (k_{dissoc}) and binding constant (K_a)[29,30]. Biacore measurements are usually performed with one partner immobilized (in the case of triplex formation, duplex) on a porous hydrogel to which the second component (in the case of triplex formation, TFO) then binds[29]. Changes in the mass loading at the sensor surface upon the triplex formation cause a shift in the resonance angle of light propagated through the wave-guiding structure immediately adjacent to the hydrogel. The time dependence of the change in resonance angle yields the kinetic information of the triplex formation.

Kinetic experiments for the triplex formation were performed on a BIACORE J instrument (GE Healthcare, U.S.A.), in which a real-time biomolecular interaction was measured with a laser biosensor, essentially as described previously[12,13,19-25,27]. The layer of a SA sensor tip with immobilized streptavidin was equilibrated with 10 mM sodium cacodylate-cacodylic acid at pH 6.8 containing 200 mM NaCl and 20 mM $MgCl_2$ at a flow rate of 30 μl/min. 40 μl of 50 mM NaOH containing 1 M NaCl was injected 3 times at a flow rate of 30 μl/min to reduce electrostatic repulsion from the surface. After equilibrating with 10 mM sodium cacodylate-cacodylic acid at pH 6.8 containing 200 mM NaCl and 20 mM $MgCl_2$, 160 μl of 0.2 μM Bt(biotinylated)-Pyr23T•Pur23A duplex (Figure 1c) solution was added at a flow rate of 30 μl/min to bind with the streptavidin on the surface. After extensive washing and equilibrating the Bt-Pyr23T•Pur23A-immobilized surface with 10 mM sodium cacodylate-cacodylic acid at pH 6.8 containing 200 mM NaCl and 20 mM $MgCl_2$, 70 μl of the TFO (Figure 1c) solution in 10 mM sodium cacodylate-cacodylic acid at pH 6.8 containing 200 mM NaCl and 20 mM $MgCl_2$ was injected over the immobilized Bt-Pyr23T•Pur23A duplex at a flow rate of 30 μl/min, and then the triplex formation was monitored for 2 min. This was followed by washing the sensor tip with 10 mM sodium cacodylate-cacodylic acid at pH 6.8 containing 200 mM NaCl and 20 mM $MgCl_2$, and the dissociation of the preformed triplex was monitored for an additional 2.5 min. Finally, 40 μl of 100 mM Tris-HCl (pH 8.0) for Pyr15TM (Figure 1c), or 40 μl of 10 mM NaOH (pH 12) for Pyr15NC7-1, Pyr15NC7-2, Pyr15NC5-1, and Pyr15NC5-2 (Figure 1c) was injected at a flow rate of 30 μl/min to completely break the Hoogsteen hydrogen bonding between the TFO and Pur23A, during which the Bt-Pyr23T•Pur23A duplex may be partially denatured. The Bt-Pyr23T•Pur23A duplex was regenerated by injecting 0.2 μM Pur23A. The resulting sensorgrams were analyzed with the BIA evaluation software supplied by the manufacturer to calculate the kinetic parameters.

2.7. Stability of oligonucleotides in human serum against nuclease degradation

Stability of oligonucleotides in human serum was examined by the following two procedures[12,13,26].

a. Analyses by native polyacrylamide gel electrophoresis

Oligonucleotide (Figure 1c) was 5′-end labeled with ^{32}P using [γ-^{32}P] ATP and T4 polynucleotide kinase by a standard procedure. 2 pmol ^{32}P-labeled oligonucleotide was incubated at 37 °C in 200 μl of human serum from human male AB plasma (Sigma-Aldrich Co., USA). Aliquots of 5 μl were removed after 10, 20, 40, 60, and 120 min of incubation, and mixed with 5 μl of stop solution (80 % formamide, 50 mM EDTA) to terminate the reactions. The samples were loaded on 15 % native polyacrylamide gels prepared in buffer [50 mM

Tris-acetate (pH 7.0), 100 mM $MgCl_2$], and electrophoresis was performed at 8 V/cm and 4 °C. The gels were scanned and analyzed by BAS system.

b. Analyses by anion-exchange HPLC

1 nmol oligonucleotide (Figure 1c) was incubated at 37 °C in 20 μl of 50 % human serum from human male AB plasma (Sigma-Aldrich Co., USA). After incubation for 20, 60 and 120 min, the samples were mixed with 13 μl of formamide to terminate the reactions, and stored at -80 °C until HPLC analyses. The samples were mixed with 400 μl of HPLC buffer [25 mM Tris-HCl (pH 7.0), 0.5 % CH_3CN], and analyzed by anion-exchange HPLC on JASCO LC-2000 Plus series with detection at 260 nm using a linear gradient of 0-0.5 M NH_4Cl in HPLC buffer over 45 min to resolve the products. The HPLC column used was TSK-GEL DNA-NPR (Tosoh, Japan). Under these conditions, peaks of all proteins from the human serum could be resolved from those of the intact and degraded TFO. Degradation data from the acquired chromatograms were processed using ChromNAV software as supplied by the manufacturer.

2.8. In vitro assay of PCSK9 gene expression in mouse hepatocyte cell line, NMuLi

Mouse hepatocyte cell line NMuLi (4.0 x 10^5 cells/ml) was cultivated in 6 well plates (2 ml/well) and incubated for 24 hr at 37 °C under 5 % CO_2. Antisense oligonucleotide to target a certain region of PCSK9 gene (Figure 1d), Lipofectamine 2000 (Invitrogen), and Opti-MEM (Invitrogen) were mixed. The final concentration of the antisense oligonucleotide in the mixture was adjusted to 1, 3, 10, 30 or 50 nM. After incubation of the mixture for 20 min at room temperature, the mixture was transfected into the cell line. Cell culture medium was exchanged into the new one at 4 hr after the transfection of the antisense oligonucleotide. Cells were collected at 20 hr after the exchange of the cell culture medium. The collected cells were homogenized by ISOGEN (Nippon Gene) to extract total RNA. Concentration of the extracted total RNA was measured by UV absorbance. Length of the extracted total RNA was analyzed by agarose gel electrophoresis. After adjusting the concentration of the total RNA to 4.0 μg/10 μl, we performed reverse transcription reaction using the total RNA to obtain 1st strand cDNA. Then, we carried out real-time PCR using the obtained cDNA to quantitate the expression levels of PCSK9 mRNA and control housekeeping GAPDH mRNA. We normalized the expression level of PCSK9 mRNA by that of control housekeeping GAPDH mRNA, because the antisense oligonucleotides did not affect the expression level of control housekeeping GAPDH mRNA. We examined the effect of the antisense oligonucleotides on the expression level of PCSK9 mRNA.

3. Stabilization of duplex by 2′,4′-BNA^{NC}-modification

Formation of stable duplexes with complementary single-stranded RNA (ssRNA) and single-stranded DNA (ssDNA) under physiological condition is essential for antisense and diagnostic applications. Thermal stability of duplexes formed between a 12-mer unmodified

(F12; Figure 1b) or 2′,4′-BNA^{NC}-modified (F12-1, F12-31, F12-32, F12-6; Figure 1b) oligonucleotide and each of its complementary 12-mer ssRNA (R12R; Figure 1b) and 12-mer ssDNA (R12D; Figure 1b) was examined at pH 7.2 by UV melting (Table 1). The T_m value of the duplex formed between the 12-mer oligonucleotide containing a single 2′,4′-BNA^{NC}-modification (F12-1) and the complementary 12-mer ssRNA (R12R) increased by 6 °C compared to that of the duplex formed between the 12-mer unmodified oligonucleotide (F12) and R12R. Further incremental increase in T_m was observed upon an increase in the number of 2′,4′-BNA^{NC}-modification (F12-31, F12-32, F12-6). The increase in T_m per 2′,4′-BNA^{NC}-modification (ΔT_m/modification) ranged from 5.3 to 6.3 °C. These results indicate that further 2′,4′-BNA^{NC}-modification of oligonucleotide should produce very stable duplex with complementary ssRNA. On the other hand, a single 2′,4′-BNA^{NC}-modification increased the T_m value of the duplex formed with complementary 12-mer ssDNA (R12D) by only 1 °C (F12:R12D vs. F12-1:R12D), indicating that stabilization of duplex by 2′,4′-BNA^{NC}-modification is highly complementary ssRNA-selective. Similar to the case of complementary ssRNA, T_m was further increased by increasing the number of 2′,4′-BNA^{NC}-modification (F12-31, F12-32, F12-6). ΔT_m/modification for complementary ssDNA (R12D) ranging from 1.0 to 3.8 °C was smaller than that for complementary ssRNA (R12R) ranging from 5.3 to 6.3 °C.

Oligonucleotide	R12R: 3′-r(CGCAAAAAACGA)-5′	R12D: 3′-d(CGCAAAAAACGA)-5′
F12: 5′-d(GCGTTTTTTGCT)-3′	45	50
F12-1: 5′-d(GCGTT$\underline{T}$TTTGCT)-3′[a]	51(+6.0)	51(+1.0)
F12-31: 5′-d(GCG$\underline{T}$T$\underline{T}$T$\underline{T}$TGCT)-3′[a]	64(+6.3)	55(+1.7)
F12-32: 5′-d(GCGTT$\underline{TTT}$TGCT)-3′[a]	61(+5.3)	57(+2.3)
F12-6: 5′-d(GCG$\underline{TTTTTT}$GCT)-3′[a]	83(+6.3)	73(+3.8)

[a]2′,4′-BNA^{NC}-modified positions are underlined.

Table 1. Melting temperatures of 4 μM duplexes formed between 12-base unmodified (F12) or 2′,4′-BNA^{NC}-modified (F12-1, F12-31, F12-32, F12-6) oligonucleotides and each of complementary single-stranded RNA (R12R) and single-stranded DNA (R12D) in 10 mM sodium phosphate buffer (pH 7.2) and 100 mM NaCl. The increase in melting temperature per 2′,4′-BNA^{NC}-modification (ΔT_m/modification) is shown in parentheses.

Because 2′,4′-BNA^{NC}-modified oligonucleotides (F12-1, F12-31, F12-32, F12-6) exhibited high binding affinity with complementary ssRNA (R12R) as described above, their ability to discriminate bases was evaluated using single-mismatched ssRNA strand (R12X; Figure 1b) (Table 2). Any mismatched base in the ssRNA strands (R12U, R12G, R12C) resulted in a substantial decrease in the T_m value of the duplexes formed with the 2′,4′-BNA^{NC}-modified oligonucleotide (F12-1). The T_m values of the duplexes formed with the 2′,4′-BNA^{NC}-modified oligonucleotide (F12-1:R12U, F12-1:R12G, F12-1:R12C) having a T-U, T-G, and T-C mismatched base pair were lower than those of the corresponding perfectly matched duplex (F12-1:R12A) by -14, -5, -17 °C, respectively. The mismatched base pair discrimination profile of the 2′,4′-BNA^{NC}-modified oligonucleotide (F12-1) was similar to that of the unmodified oligonucleotide (F12). These results indicate that 2′,4′-BNA^{NC}-modified oligonucleotide not only exhibits high-affinity RNA selective binding, but is also highly selective in recognizing bases.

Oligonucleotide	R12X: 3′-r(CGCAAXAAACGA)-5′			
	X = A (matched)	X = U	X = G	X = C
F12: 5′-d(GCGTTTTTTGCT)-3′	45	33	42	30
F12-1: 5′-d(GCGTT<u>T</u>TTTGCT)-3′[a]	51	37	46	34

[a]2′,4′-BNA^{NC}-modified position is underlined.

Table 2. Melting temperatures of 4 μM duplexes formed between 12-base unmodified (F12) or 2′,4′-BNA^{NC}-modified (F12-1) oligonucleotides and complementary single-stranded RNA containing a single-mismatched base (R12X) in 10 mM sodium phosphate buffer (pH 7.2) and 100 mM NaCl.

4. Stabilization of pyrimidine motif triplex at neutral pH by 2′,4′-BNA^{NC}-modification

Formation of stable triplex with TFO under physiological condition is essential for antigene application. Thermal stability of the pyrimidine motif triplexes formed between a 23-bp target duplex (Pur23A•Pyr23T; Figure 1c) and each of its specific 15-mer 2′,4′-BNA^{NC}-unmodified (Pyr15TM; Figure 1c) or 2′,4′-BNA^{NC}-modified (Pyr15NC7-1, Pyr15NC7-2, Pyr15NC5-1, or Pyr15NC5-2; Figure 1c) TFO was investigated at pH 6.8 by UV melting (Figure 2 and Table 3). UV melting curves in the both directions (heating and cooling) are almost superimposable in all cases, indicating that the dissociation and association processes are reversible. The triplex involving Pyr15TM showed two-step melting. Upon heating the first transition at lower temperature, T_{m1} (39.1 °C), was the melting of the triplex to a duplex and a TFO, and the second transition at higher temperature, T_{m2} (72.4 °C), was the melting of the duplex (Figure 2). On the other hand, the triplex involving each of the 2′,4′-BNA^{NC}-modified TFOs showed only one transition at higher temperature, T_m. As the magnitude in UV absorbance change at T_m for each of the 2′,4′-BNA^{NC}-modified TFOs was almost equal to the sum of those at T_{m1} and T_{m2} for Pyr15TM (Figure 2), the transition was identified as a direct melting of the triplex to its constituting single-strand DNAs upon heating. The 2′,4′-BNA^{NC} modification, therefore, increased the melting temperature of the triplex by more than 35 °C (Table 3), indicating that the 2′,4′-BNA^{NC} modification of TFO increased the thermal stability of the pyrimidine motif triplex at neutral pH.

<table>
<tr><th>TFO</th><th>T_{m1} (°C)</th><th>T_{m2} (°C)</th></tr>
<tr><td>Pyr15TM</td><td>39.1 ± 0.1</td><td>72.4 ± 0.4</td></tr>
<tr><td>Pyr15BNANC7-1</td><td colspan="2">79.4 ± 0.6</td></tr>
<tr><td>Pyr15BNANC7-2</td><td colspan="2">85.8 ± 0.1</td></tr>
<tr><td>Pyr15BNANC5-1</td><td colspan="2">75.6 ± 0.7</td></tr>
<tr><td>Pyr15BNANC5-2</td><td colspan="2">75.6 ± 0.7</td></tr>
</table>

Table 3. Melting temperatures of the triplexes between a 23-base pair target duplex (Pur23A•Pyr23T) and a 15-mer TFO (Pyr15TM, Pyr15BNANC7-1, Pyr15BNANC7-2, Pyr15BNANC5-1, or Pyr15BNANC5-2) in 10 mM sodium cacodylate-cacodylic acid (pH 6.8), 200 mM NaCl and 20 mM $MgCl_2$.

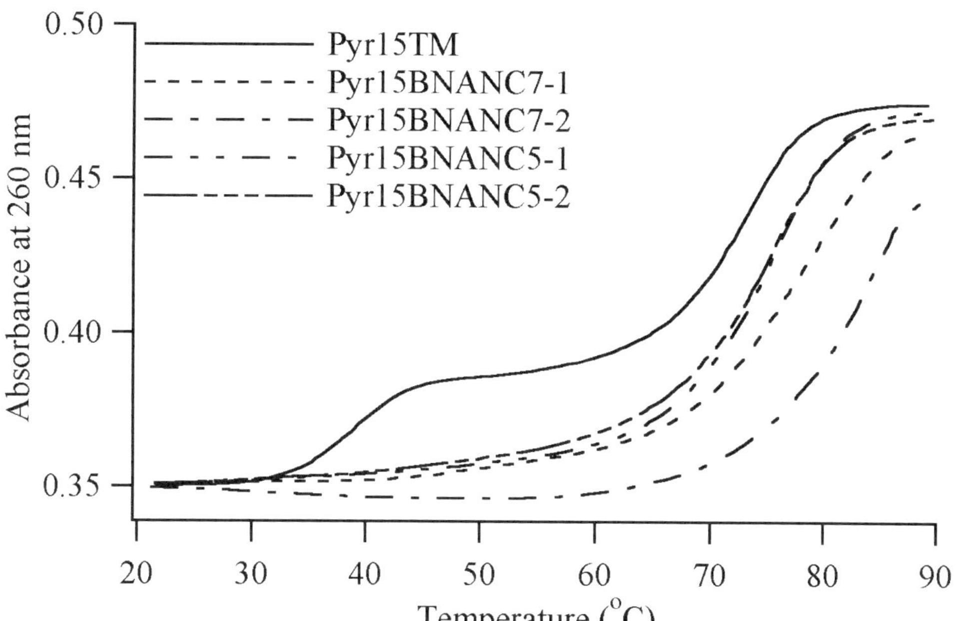

Figure 2. . UV melting profiles of the pyrimidine motif triplex with the specific TFO (Pyr15TM, Pyr15NC7-1, Pyr15NC7-2, Pyr15NC5-1, or Pyr15NC5-2) upon heating. The triplexes with Pyr15TM, Pyr15NC7-1, Pyr15NC7-2, Pyr15NC5-1, or Pyr15NC5-2 in 10 mM sodium cacodylate-cacodylic acid (pH 6.8), 200 mM NaCl and 20 mM $MgCl_2$ were melted at a scan rate of 0.5 °C/min with detection at 260 nm. The cell path length was 1 cm. The triplex nucleic acid concentration used was 1 μM.

5. No significant structural change of pyrimidine motif triplex at neutral pH by 2′,4′-BNA^{NC}-modification

Circular dichroism (CD) spectra of the pyrimidine motif triplexes between the target duplex (Pur23A•Pyr23T) and each of the 2′,4′-BNA^{NC}-unmodified (Pyr15TM) or 2′,4′-BNA^{NC}-modified (Pyr15NC7-1, Pyr15NC7-2, Pyr15NC5-1, or Pyr15NC5-2) TFO were measured at 20 °C and pH 6.8 (Figure 3). A significant negative band in the short-wavelength (210-220 nm) region was observed for all the profiles (Figure 3), confirming the triplex formation involving each TFO[31]. The overall shape of the CD spectra was quite similar among all the profiles (Figure 3), suggesting that no significant change may be induced in the higher order structure of the pyrimidine motif triplex by the 2′,4′-BNA^{NC} modification.

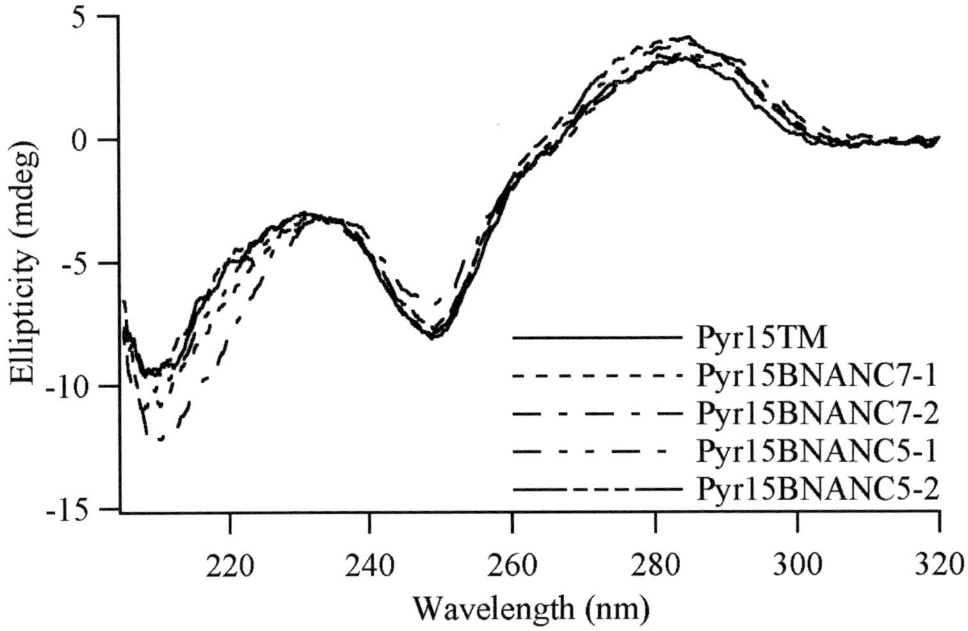

Figure 3. CD spectra of the pyrimidine motif triplex with the specific TFO (Pyr15TM, Pyr15NC7-1, Pyr15NC7-2, Pyr15NC5-1, or Pyr15NC5-2). The triplexes with Pyr15TM, Pyr15NC7-1, Pyr15NC7-2, Pyr15NC5-1, or Pyr15NC5-2 in 10 mM sodium cacodylate-cacodylic acid (pH 6.8), 200 mM NaCl and 20 mM $MgCl_2$ were measured at 20 °C in the wavelength range of 205-320 nm. The cell path length was 1 cm. The triplex nucleic acid concentration used was 1 μM.

6. Promotion of pyrimidine motif triplex formation at neutral pH by 2′,4′-BNA^{NC}-modification

The pyrimidine motif triplex formation between the target duplex (Pur23A•Pyr23T) and each of the 2′,4′-BNA^{NC}-unmodified (Pyr15TM) or 2′,4′-BNA^{NC}-modified (Pyr15NC7-1, Pyr15NC7-2, Pyr15NC5-1, or Pyr15NC5-2) TFO was examined at pH 7.0 by EMSA (Figure

4). Total oligonucleotide concentration ([specific TFO (Pyr15TM or 2′,4′-BNA^{NC}-modified TFO)] + [nonspecific oligonucleotide (Pyr15NS2M)]) was kept constant at 1000 nM to minimize loss of DNA during processing and to assess sequence specificity. While incubation with 1000 nM Pyr15NS2M alone did not cause a shift in electrophoretic migration of the target duplex (see *lane 1* for Pyr15TM), those with Pyr15TM or each of the 2′,4′-BNA^{NC}-modified TFOs at particular concentrations caused retardation of the duplex migration owing to triplex formation[32]. The dissociation constant, K_d, of triplex formation was determined from the concentration of the TFO, which caused half of the target duplex to shift to the triplex[32]. The K_d of the triplex with Pyr15TM was estimated to be between 250 and 1000 nM. In contrast, the K_d of the triplex with each of the 2′,4′-BNA^{NC}-modified TFOs was ~16 nM, indicating that the 2′,4′-BNA^{NC} modification of TFO increased the binding constant, K_a (= $1/K_d$), of the pyrimidine motif triplex formation at neutral pH by more than 10-fold. The increase in the K_a by the 2′,4′-BNA^{NC} modification of TFO was similar in magnitude among the four modified TFOs.

7. Thermodynamic analyses of pyrimidine motif triplex formation involving 2′,4′-BNA^{NC}-modified TFO at neutral pH

We examined the thermodynamic parameters of the pyrimidine motif triplex formation between the target duplex (Pur23A•Pyr23T) and each of the 2′,4′-BNA^{NC}-unmodified (Pyr15TM) or 2′,4′-BNA^{NC}-modified TFO at 25 °C and pH 6.8 by ITC. To investigate the pH dependence of the pyrimidine motif triplex formation, the thermodynamic parameters of the triplex formation between Pur23A•Pyr23T and Pyr15TM were also analyzed at 25 °C and pH 6.1 by ITC. Figure 5a shows a typical ITC profile for the triplex formation between Pyr15NC7-1 and Pur23A•Pyr23T at 25 °C and pH 6.8. An exothermic heat pulse was observed after each injection of Pur23A•Pyr23T into Pyr15NC7-1. The magnitude of each peak decreased gradually with each new injection, and a small peak was still observed at a molar ratio of [Pur23A•Pyr23T]/[Pyr15NC7-1]=2. The area of the small peak was equal to the heat of dilution measured in a separate experiment by injecting Pur23A•Pyr23T into the same buffer. The area under each peak was integrated, and the heat of dilution of Pur23A•Pyr23T was subtracted from the integrated values. The corrected heat was divided by the moles of injected solution, and the resulting values were plotted as a function of a molar ratio of [Pur23A•Pyr23T]/[Pyr15NC7-1], as shown in Figure 5b. The resultant titration plot was fitted to a sigmoidal curve by a nonlinear least-squares method. The binding constant, K_a, and the enthalpy change, ΔH, were obtained from the fitted curve[28]. The Gibbs free energy change, ΔG, and the entropy change, ΔS, were calculated from the equation, $\Delta G = -RT\ln K_a = \Delta H - T\Delta S$, where R is gas constant and T is temperature[28]. The titration plots for Pyr15TM at pH 6.8 and pH 6.1 are also shown in Figure 5b. The thermodynamic parameters for Pyr15TM at pH 6.8 and pH 6.1 were obtained from the titration plots in the same way. The ITC profiles and the titration plots for each of other 2′,4′-BNA^{NC}-modified TFOs at pH 6.8 were almost the same as those observed for Pyr15NC7-1 at pH 6.8. The thermodynamic parameters for each of other 2′,4′-BNA^{NC}-modified TFOs at pH 6.8 were obtained from the titration plots in the same way.

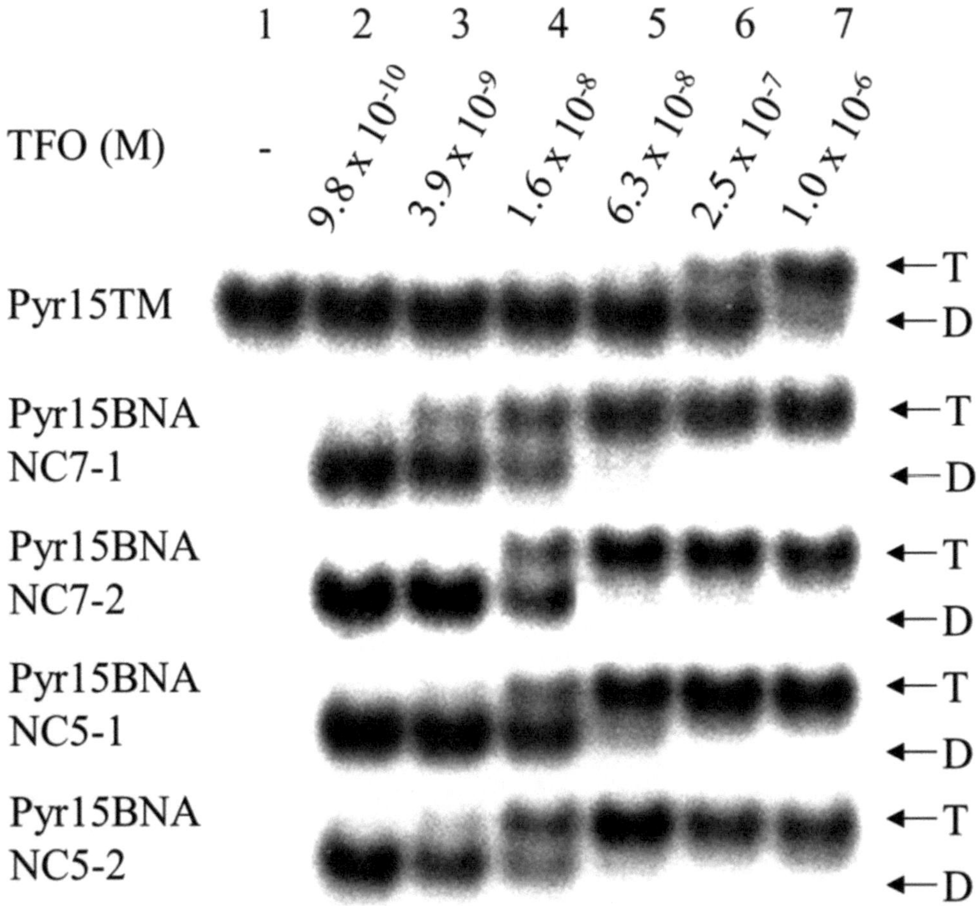

Figure 4. EMSA of the pyrimidine motif triplex formation with the specific TFO (Pyr15TM, Pyr15NC7-1, Pyr15NC7-2, Pyr15NC5-1, or Pyr15NC5-2) at neutral pH. Triplex formation was initiated by adding ^{32}P-labeled Pur23APyr23T duplex (~1 nM) with the indicated final concentrations of the specific TFO (Pyr15TM, Pyr15NC7-1, Pyr15NC7-2, Pyr15NC5-1, or Pyr15NC5-2). The nonspecific oligonucleotide (Pyr15NS2M) was added to adjust the equimolar concentrations (1000 nM) of TFO (Pyr15TM+Pyr15NS2M, Pyr15NC7-1+Pyr15NS2M, Pyr15NC7-2+Pyr15NS2M, Pyr15NC5-1+Pyr15NS2M, or Pyr15NC5-2+Pyr15NS2M) in each lane. Reaction mixtures involving Pyr15TM, Pyr15NC7-1, Pyr15NC7-2, Pyr15NC5-1, or Pyr15NC5-2 in 50 mM Tris-acetate (pH 7.0), 100 mM NaCl, and 10 mM $MgCl_2$ were incubated for 6 h at 37 °C, and then electrophoretically separated at 4 °C on a 15 % native polyacrylamide gel prepared in buffer [50 mM Tris-acetate (pH 7.0) and 10 mM $MgCl_2$]. Positions of the duplex (D) and triplex (T) are indicated.

Table 4 summarizes the thermodynamic parameters for the pyrimidine motif triplex formation with each of Pyr15TM and the 2′,4′-BNA^{NC}-modified TFOs at 25 °C and pH 6.8, and those with Pyr15TM at 25 °C and pH 6.1, obtained from ITC. The signs of both ΔH and ΔS were negative under each condition. Because an observed negative ΔS was unfavorable for the triplex formation, the triplex formation was driven by a large negative ΔH under each condition. The K_a for Pyr15TM at pH 6.1 was ~10-fold larger than that observed for Pyr15TM at pH 6.8, confirming, like others[5-7], that neutral pH is unfavorable for the pyrimidine motif triplex formation involving C^{+}•G:C triads. In addition, the K_a for each of the 2′,4′-BNA^{NC}-modified TFOs at pH 6.8 was ~10-fold larger than that observed for Pyr15TM at pH 6.8 (Table 4), indicating that the 2′,4′-BNA^{NC} modification of TFO increased the K_a for the pyrimidine motif triplex formation at neutral pH, which is consistent with the results of EMSA (Figure 4). The increase in the K_a by the 2′,4′-BNA^{NC} modification of TFO was similar in magnitude among the four 2′,4′-BNA^{NC}-modified TFOs. Also, although the K_a and ΔG for the triplex formation with each of the 2′,4′-BNA^{NC}-modified TFOs at pH 6.8 and those with Pyr15TM at pH 6.1 were quite similar, the constituents of ΔG, that is, ΔH and ΔS, were obviously different from each other (Table 4). The magnitudes of the negative ΔH and ΔS for each of the 2′,4′-BNA^{NC}-modified TFOs at pH 6.8 were significantly smaller than those observed for Pyr15TM at pH 6.1 (Table 4).

TFO	pH	K_a (M^{-1})	K_a (relative)	ΔG (kcal mol^{-1})	ΔH (kcal mol^{-1})	ΔS (cal mol^{-1} K^{-1})
Pyr15TM	6.1[a]	(5.81 ± 0.99) x 10^6	13.9	-9.23 ± 0.11	-92.0 ± 1.5	-278 ± 5
Pyr15TM	6.8[b]	(4.19 ± 2.0) x 10^5	1.0	-7.67 ± 0.38	-38.5 ± 7.5	-103 ± 26
Pyr15BNAN C7-1	6.8[b]	(5.30 ± 0.45) x 10^6	12.6	-9.17 ± 0.05	-56.1 ± 1.0	-157 ± 4
Pyr15BNAN C7-2	6.8[b]	(4.10 ± 0.69) x 10^6	9.8	-9.02 ± 0.11	-62.1 ± 2.4	-178 ± 8
Pyr15BNAN C5-1	6.8[b]	(3.73 ± 0.43) x 10^6	8.9	-8.96 ± 0.07	-54.3 ± 1.5	-152 ± 5
Pyr15BNAN C5-2	6.8[b]	(3.82 ± 0.29) x 10^6	9.1	-8.98 ± 0.05	-53.8 ± 1.1	-150 ± 4

[a]10 mM sodium cacodylate-cacodylic acid (pH 6.1), 200 mM NaCl and 20 mM $MgCl_2$. [b]10 mM sodium cacodylate-cacodylic acid (pH 6.8), 200 mM NaCl and 20 mM $MgCl_2$.

Table 4. Thermodynamic parameters for the triplex formation between a 23-base pair target duplex (Pur23A•Pyr23T) and a 15-mer TFO (Pyr15TM, Pyr15BNANC7-1, Pyr15BNANC7-2, Pyr15BNANC5-1, or Pyr15BNANC5-2) at 25 °C, obtained from ITC

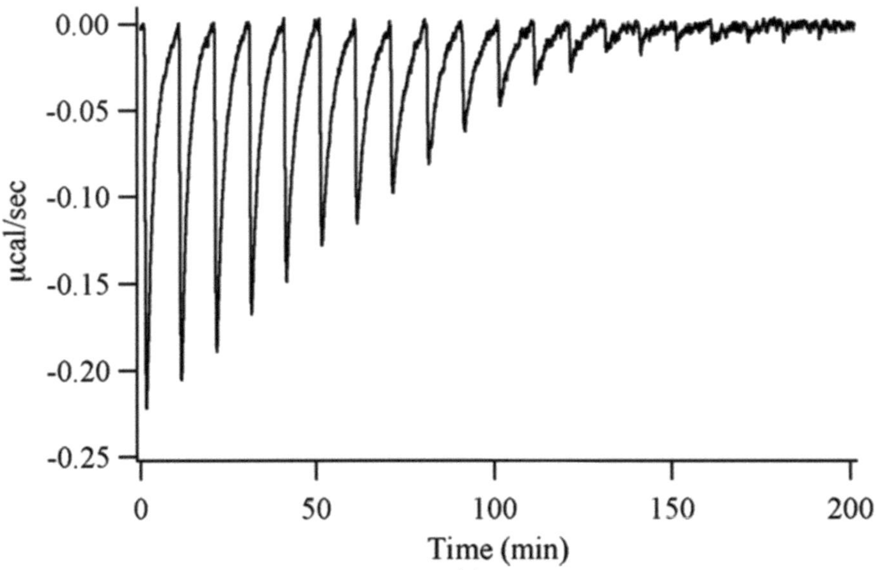

(a)

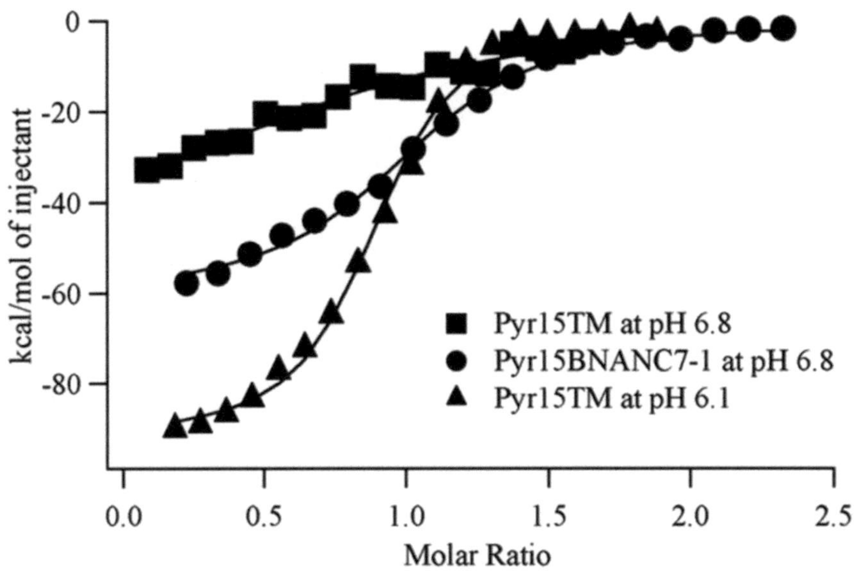

(b)

Figure 5. Thermodynamic analyses of the pyrimidine motif triplex formation with Pyr15TM or Pyr15NC7-1 at pH 6.8 and with Pyr15TM at pH 6.1 by ITC. (a) Typical ITC profiles for the triplex formation between Pyr15NC7-1 and Pur23APyr23T at 25 °C and pH 6.8. 114.7 µM Pur23APyr23T solution in 10 mM sodium cacodylate-cacodylic acid (pH 6.8), 200 mM NaCl and 20 mM $MgCl_2$ was injected 20 times in 5-µl increments into 3.58 µM Pyr15NC7-1 solution, which was dialyzed against the same buffer. Injections were occurred over 12 s at 10-min intervals. (b) The titration plots against the molar ratio of [Pur23APyr23T]/ [TFO]. The data were fitted by a nonlinear least-squares method.

8. Kinetic analyses of pyrimidine motif triplex formation involving 2′,4′-BNA^{NC}-modified TFO at neutral pH

To examine the putative mechanism involved in the increase in K_a of the pyrimidine motif triplex formation by the 2′,4′-BNA^{NC} modification of TFO (Figure 4 and Table 4), we assessed the kinetic parameters for the association and dissociation of TFO (Pyr15TM and the 2′,4′-BNA^{NC}-modified TFOs) with Pur23A•Pyr23T at 25 °C and pH 6.8 by BIACORE. Figure 6a shows the sensorgrams representing the triplex formation and dissociation involving the various concentrations of Pyr15NC7-1. The injection of Pyr15NC7-1 over the immobilized Bt(biotinylated)-Pyr23T•Pur23A caused an increase in response. As shown in Figure 6a, an increase in the concentration of Pyr15NC7-1 led to a gradual change in the response of the association curves. The on-rate constant (k_{on}) was obtained from the analysis of each association curve. Figure 6b shows a plot of k_{on} against the Pyr15NC7-1 concentrations. The resultant plot was fitted to a straight line by a linear least-squares method. The association rate constant (k_{assoc}) was determined from the slope of the fitted line[29,30]. The off-rate constant (k_{off}) was obtained from the analysis of each dissociation curve (Figure 6a). As k_{off} is usually independent of the concentration of the injected solution, the dissociation rate constant (k_{dissoc}) was determined by averaging k_{off} for several concentrations[29,30]. K_a was calculated from the equation, $K_a = k_{assoc}/k_{dissoc}$. The kinetic parameters for each of Pyr15TM and other 2′,4′-BNA^{NC}-modified TFOs were obtained in the same way.

TFO	k_{assoc} (M^{-1} s^{-1})	k_{assoc} (relative)	k_{dissoc} (s^{-1})	k_{dissoc} (relative)	K_a (M^{-1})	K_a (relative)
Pyr15TM	$(2.01 \pm 0.11) \times 10^2$	1.0	$(1.05 \pm 0.29) \times 10^{-3}$	1.0	$(1.91 \pm 0.88) \times 10^5$	1.0
Pyr15BNANC7-1	$(3.81 \pm 0.60) \times 10^2$	1.9	$(6.96 \pm 1.14) \times 10^{-5}$	0.066	$(5.47 \pm 2.10) \times 10^6$	28.6
Pyr15BNANC7-2	$(4.61 \pm 0.29) \times 10^2$	2.3	$(7.99 \pm 0.83) \times 10^{-5}$	0.076	$(5.77 \pm 1.07) \times 10^6$	30.2
Pyr15BNANC5-1	$(4.60 \pm 0.18) \times 10^2$	2.3	$(9.36 \pm 1.29) \times 10^{-5}$	0.089	$(4.91 \pm 1.01) \times 10^6$	25.7
Pyr15BNANC5-2	$(4.30 \pm 0.67) \times 10^2$	2.1	$(6.61 \pm 0.81) \times 10^{-5}$	0.063	$(6.51 \pm 2.06) \times 10^6$	34.1

Table 5. Kinetic parameters for the triplex formation between a 23-base pair target duplex (Pur23A•Pyr23T) and a 15-mer TFO (Pyr15TM, Pyr15BNANC7-1, Pyr15BNANC7-2, Pyr15BNANC5-1, or Pyr15BNANC5-2) at 25 °C and pH 6.8 in 10 mM sodium cacodylate-cacodylic acid, 200 mM NaCl and 20 mM $MgCl_2$, obtained from BIACORE interaction analysis system

Table 5 summarizes the kinetic parameters for the pyrimidine motif triplex formation with each of Pyr15TM and the 2′,4′-BNA^{NC}-modified TFOs at 25 °C and pH 6.8, obtained from BIACORE. The magnitudes of K_a calculated from the ratio of k_{assoc} to k_{dissoc} (Table 5) were consistent with those obtained from ITC (Table 4). The K_a for each of the 2′,4′-BNA^{NC}-modified TFOs at pH 6.8 was ~30-fold larger than that observed for Pyr15TM at pH 6.8, indicating that the 2′,4′-BNA^{NC} modification of TFO increased the K_a of the pyrimidine motif triplex formation at neutral pH, which supported the results of EMSA (Figure 4) and ITC (Table 4). The 2′,4′-BNA^{NC} modification of TFO decreased k_{dissoc} by ~15-fold, while it moderately increased k_{assoc} by ~2-fold. Thus, the much larger K_a by the 2′,4′-BNA^NC modification of TFO resulted mainly from the decrease in k_{dissoc} rather than the increase in k_{assoc}.

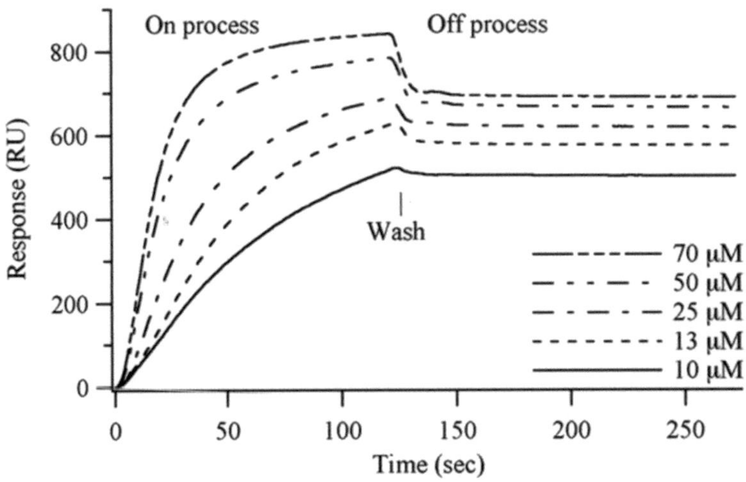

(a)

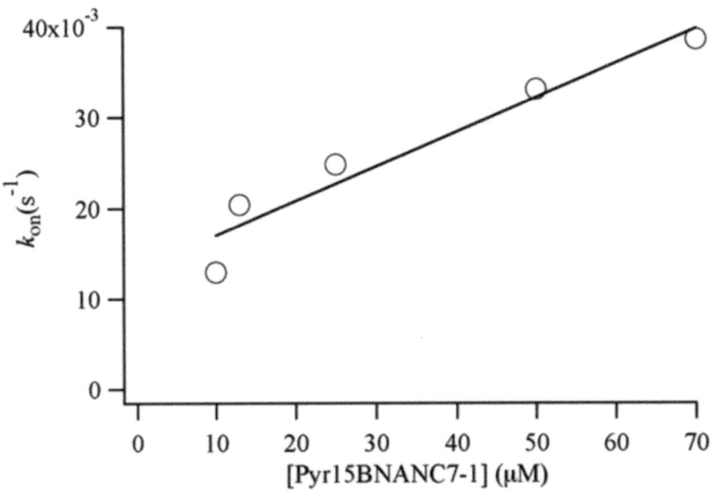

(b)

Figure 6. Kinetic analyses of the pyrimidine motif triplex formation with Pyr15NC7-1 in 10 mM sodium cacodylate-cacodylic acid (pH 6.8), 200 mM NaCl and 20 mM $MgCl_2$ by BIACORE interaction analysis system. (a) A series of sensorgrams for the triplex formation and the dissociation of the formed triplex between Pyr15NC7-1 and Pur23APyr23T at 25 °C and pH 6.8. The Pyr15NC7- 1 solutions, diluted in the buffer to achieve the indicated final concentrations, were injected into the Bt-Pyr23TPur23A-immobilized cuvette. The binding of Pyr15NC7-1 to Bt-Pyr23TPur23A and the dissociation of Pyr15NC7-1 from Bt-Pyr23TPur23A were monitored as the response against time. (b) Measured on-rate constants, k_{on}, of the triplex formation in (a) were plotted against the respective concentrations of Pyr15NC7-1. The plot was fitted to a straight line ($r^2 = 0.97$) by a linear least-squares method.

9. Increased stability of 2′,4′-BNA^{NC}-modified oligonucleotides in human serum against nuclease degradation

A major difficulty associated with the use of oligonucleotides as *in vivo* agents is the rapid degradation of oligonucleotides by nuclease *in vivo*[8]. To propose the possibility for the application of 2′,4′-BNA^{NC}-modified oligonucleotides to the various strategies *in vivo*, we examined the resistance of the 2′,4′-BNA^{NC}-unmodified (Pyr15TM) or 2′,4′-BNA^{NC}-modified oligonucleotides against nuclease degradation in human serum. The series of oligonucleotides 5′-end labeled with ^{32}P were incubated at 37 °C in human serum, and their degradation was assessed by 15 % native polyacrylamide gel electrophoresis (Figure 7). All of Pyr15TM was degraded and converted to shorter oligonucleotides within 20 min of incubation. In contrast, no significant degradation of Pyr15NC7-2 and Pyr15NC5-2 was observed even after 120 min of incubation. These results indicate that the 2′,4′-BNA^{NC} modification contributed to increase the stability of TFOs in human serum. Because Pyr15NC7-1 and Pyr15NC5-1 containing the 2′,4′-BNA^{NC} modification at the 5′-end were unable to be labeled with ^{32}P by T4 polynucleotide kinase, the stability of Pyr15NC7-1 and Pyr15NC5-1 in human serum was impossible to be examined. Thus, to investigate the resistance of all TFOs including Pyr15NC7-1 and Pyr15NC5-1 against nuclease degradation, their degradation was estimated by anion-exchange HPLC after incubating the TFOs at 37 °C in human serum. Figure 8 shows the percentage of the intact oligonucleotides as a function of the incubation time. Only 20 % of intact Pyr15TM was detected after 20 min of incubation with human serum, and Pyr15TM was completely degraded within 60 min. On the other hand, more than 50 % of the 2′,4′-BNA^{NC}-modified TFOs remained intact even after 120 min of incubation with human serum. These results indicate that the 2′,4′-BNA^{NC} modification significantly increased the nuclease resistance of TFOs in human serum. The results of anion-exchange HPLC are consistent with those of native polyacrylamide gel electrophoresis (Figure 7).

10. Excellent antisense effect of 2′,4′-BNA^{NC}-modified antisense oligonucleotides on the expression level of PCSK9 mRNA

We examined *in vitro* antisense effect of 2′,4′-BNA^{NC}-modified antisense oligonucleotides (Figure 1d) on the expression level of PCSK9 mRNA. A series of the concentrations of 2′,4′-BNA^{NC}-modified antisense oligonucleotides were transfected into mouse hepatocyte cell line, NMuLi, by lipofectamine 2000. Total RNA was extracted from the cells at 24 hr after the transfection. The expression level of PCSK9 mRNA was quantitated by real-time RT-PCR relative to that of control housekeeping GAPDH mRNA. In all cases of the 2′,4′-BNA^{NC}-modified antisense oligonucleotides, the expression level of PCSK9 mRNA was decreased upon increasing the concentration of the 2′,4′-BNA^{NC}-modified antisense oligonucleotides (Figure 9). The 2′,4′-BNA^{NC}-modified antisense oligonucleotides showed the excellent antisense effect to reduce the expression level of PCSK9 mRNA.

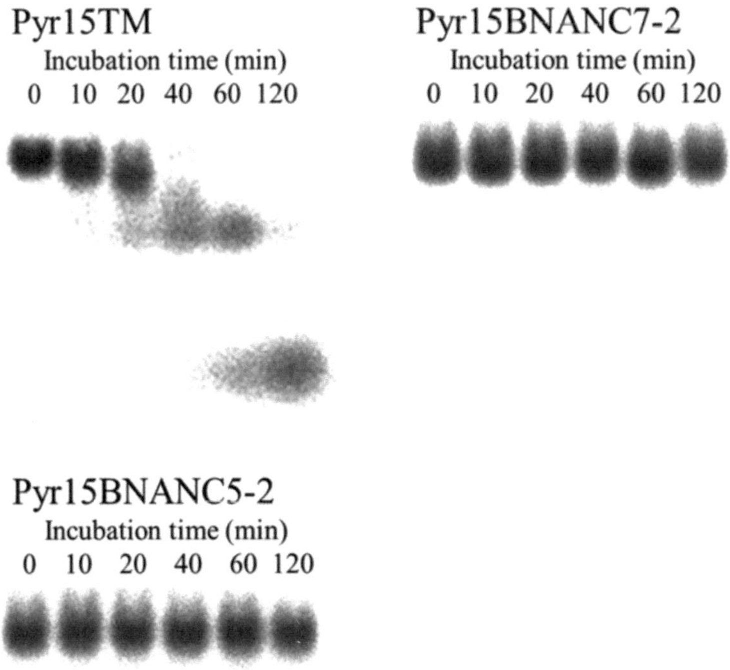

Figure 7. Stability of the specific TFOs (Pyr15TM, Pyr15BNANC7-2, and Pyr15BNANC5-2) in human serum. 2 pmol ^{32}P-labeled TFOs were incubated in human serum at 37°C, and aliquots were removed at the time points indicated and analyzed by 15 % native polyacrylamide gel electrophoresis.

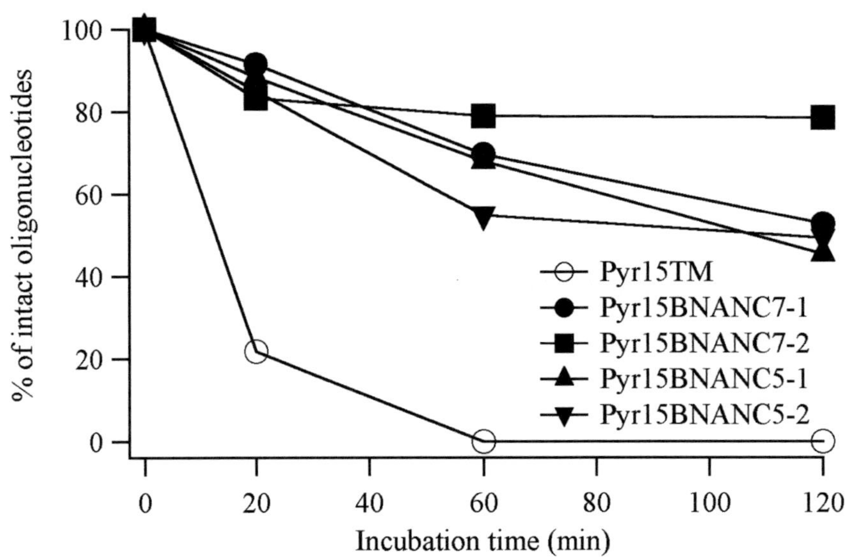

Figure 8. Stability of the specific TFOs (Pyr15TM, Pyr15NC7-1, Pyr15NC7-2, Pyr15NC5-1, and Pyr15NC5-2) in human serum. 1 nmol TFOs were incubated in human serum at 37 °C, and aliquots were removed at the time points indicated and analyzed by anion-exchange HPLC. The percentage of the intact oligonucleotides was determined and plotted as a function of the incubation time.

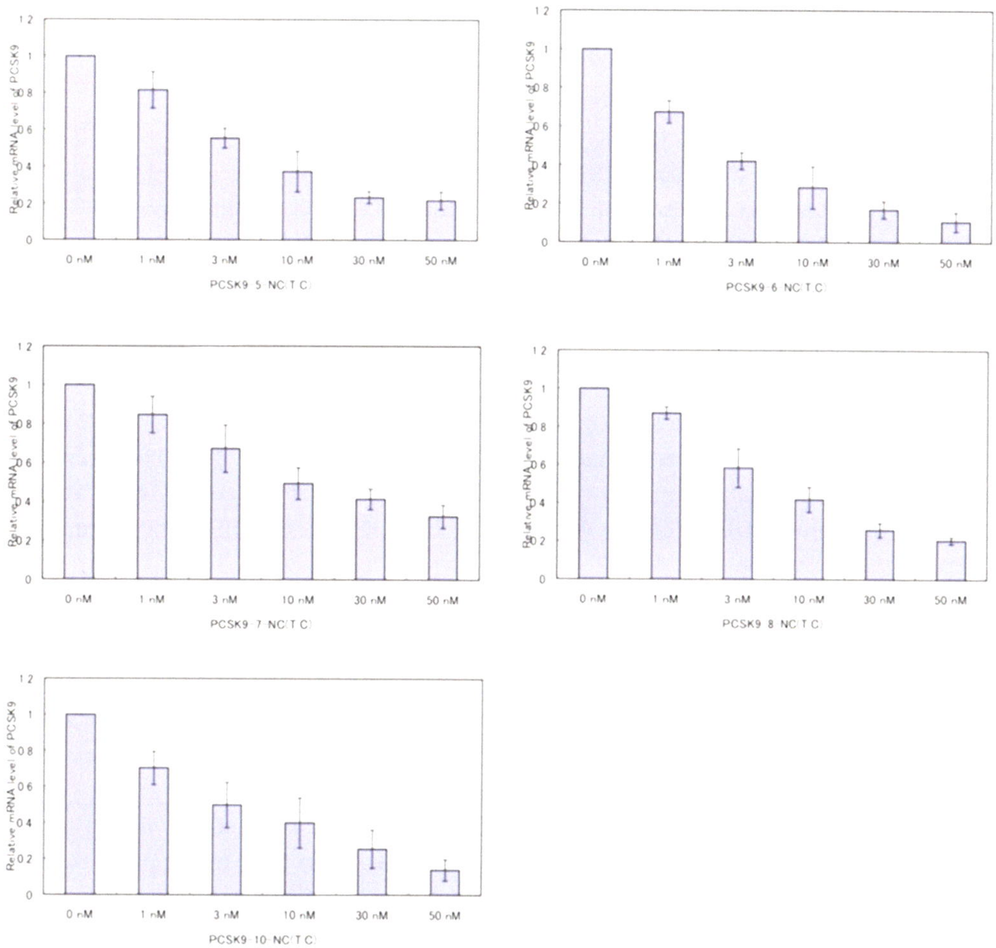

Figure 9. Relative expression level of PCSK9 mRNA as a function of the concentrations of 2′,4′-BNA^{NC}-modified antisense oligonucleotide. The expression level of PCSK9 mRNA was quantitated by real-time RT-PCR relative to that of control housekeeping GAPDH mRNA. The expression level of PCSK9 mRNA with the addition of 2′,4′-BNA^{NC}-modified antisense oligonucleotide was divided by that without the addition of the antisense oligonucleotide to obtain relative expression level of PCSK9 mRNA.

11. Discussion

The T_m value of the duplex formed between the 2′,4′-BNA^{NC}-modified oligonucleotide and the complementary ssRNA or ssDNA was significantly higher than that of the duplex formed between the corresponding unmodified oligonucleotide and the same complementary ssRNA or ssDNA (Table 1). The ΔT_m/modification value for the complementary ssRNA was significantly larger than that for the corresponding complementary ssDNA (Table 1), indicating that the 2′,4′-BNA^{NC}-modified oligonucleotide selectively bound to the complementary ssRNA with high affinity. In addition, the 2′,4′-BNA^{NC}-modified oligonucleotide recognized the base in the complementary strand with high selectivity (Table 2). These results indicate that the 2′,4′-BNA^{NC}-modification of oligonucleotide promotes the duplex formation involving the complementary ssRNA with high sequence selectivity at neutral pH.

The K_a of the pyrimidine motif triplex formation with Pyr15TM at pH 6.1 was ~10-fold larger than that observed with Pyr15TM at pH 6.8 (Table 4), which is consistent with the previously reported results that neutral pH is unfavorable for the pyrimidine motif triplex formation involving C^{+}·G:C triads[5-7]. On the other hand, the K_a of the pyrimidine motif triplex formation with each of the 2′,4′-BNA^{NC}-modified TFOs at pH 6.8 was ~10-fold larger than that observed with Pyr15TM at pH 6.8 (Table 4). The increase in K_a at neutral pH by the 2′,4′-BNA^{NC} modification of TFO was supported by the results of EMSA (Figure 4) and BIACORE (Table 5), although the magnitudes of K_d (=1/K_a) were different between EMSA (Figure 4) and each of ITC (Table 4) or BIACORE (Table 5) due to the difference in the experimental buffer conditions. In addition, the 2′,4′-BNA^{NC} modification of TFO increased the thermal stability of the pyrimidine motif triplex at neutral pH (Figure 2 and Table 3). These results indicate that the 2′,4′-BNA^{NC} modification of TFO promotes the pyrimidine motif triplex formation at neutral pH.

Because the formed triplex structure involving Pyr15TM at pH 6.1 and that involving Pyr15TM at pH 6.8 are the same, the magnitude of ΔH and ΔS upon the triplex formation measured by ITC could be the same between the two conditions. However, the magnitudes of ΔH and ΔS for Pyr15TM at pH 6.8 were significantly smaller than those observed for Pyr15TM at pH 6.1 (Table 4). When the ΔH and ΔS are calculated from the fitting procedure of ITC, the heat observed by ITC is divided not by the effective concentration really involved in the triplex formation, but by the apparent concentration added to the triplex formation[28]. The calculation does not take it into consideration how many percentage of the added concentration is really effectively involved in the triplex formation. Thus, if the triplex formation is substoichiometric under a certain condition, the magnitudes of ΔH and ΔS for the substoichiometric triplex formation estimated by ITC should be smaller than those observed for the more stoichiometric triplex formation under another condition. Therefore, the significantly smaller magnitudes of ΔH and ΔS for Pyr15TM at pH 6.8 relative to those for Pyr15TM at pH 6.1 (Table 4) suggest that the triplex formation with Pyr15TM at pH 6.8 was significantly more substoichiometric than that with Pyr15TM at pH 6.1, which was also supported by the significantly smaller magnitudes of K_a and ΔG for Pyr15TM at pH 6.8 (Table 4). In the substoichiometric triplex formation, the cytosine bases in the TFO may be protonated to a lesser extent, resulting in weaker hydrogen bonding interactions and possibly also weaker base stacking interactions. In contrast, the K_a and ΔG for Pyr15TM at pH 6.1 and those for the 2′,4′-BNA^{NC}-modified TFOs at pH 6.8 were quite similar (Table 4), suggesting that the triplex formations under these conditions were similarly quite stoichiometric. Thus, to discuss the promotion mechanism of the triplex formation by the 2′,4′-BNA^{NC} modification of TFO, the comparison of the ΔH and ΔS between Pyr15TM at pH 6.8 and the 2′,4′-BNA^{NC}-modified TFOs at pH 6.8 is not valid due to the significant substoichiometry for Pyr15TM at pH 6.8. The comparison of the ΔH and ΔS between Pyr15TM at pH 6.1 and the 2′,4′-BNA^{NC}-modified TFOs at pH 6.8 with similar stoichiometry will provide reasonable promotion mechanism of the triplex formation by the 2′,4′-BNA^{NC} modification of TFO, as discussed in the following.

Although the K_a and ΔG for Pyr15TM at pH 6.1 and those for the 2′,4′-BNA^{NC}-modified TFOs at pH 6.8 were quite similar (Table 4), the constituents of ΔG, that is, ΔH and ΔS, were obviously different. The magnitudes of the negative ΔH and ΔS for the 2′,4′-BNA^{NC}-modified TFOs at pH 6.8 were smaller than those observed for Pyr15TM at pH 6.1 (Table 4). The observed negative ΔH upon the triplex formation reflects major contributions from the hydrogen bonding and the base stacking involved in the triplex formation, the protonation of the cytosine bases upon the hydrogen bonding, and the accompanying deprotonation of the cacodylate buffer releasing the protons to bind with the cytosine bases[33-35]. The immobilization of electrostricted water molecules around polar atoms upon the triplex formation is also considered to be the major sources of the observed negative ΔH upon the triplex formation[33-35]. Because the degree of the protonation may be similar between the 2′,4′-BNA^{NC}-modified TFOs at pH 6.8 and Pyr15TM at pH 6.1 due to the similar stoichiometry discussed above and the protons to bind with the cytosine bases are released from the same cacodylate buffer in both cases, the ΔH derived from the protonation of the cytosine bases and the accompanying deprotonation of the cacodylate buffer should be similar between the two cases. Also, the CD spectra showed that the higher-order structure of the triplexes with each of the 2′,4′-BNA^{NC}-modified TFOs was quite similar to that with the corresponding unmodified TFO (Figure 3), suggesting no significant change in the hydrogen bonding and/or the base stacking of the triplex by the 2′,4′-BNA^{NC}-modification. Thus, the difference in ΔH between the 2′,4′-BNA^{NC}-modified TFOs at pH 6.8 and Pyr15TM at pH 6.1 (Table 4) should mainly result from the contribution that the 2′,4′-BNA^{NC}-modification may change the degree of the immobilization of water molecules around the polar atoms of the triplex. The polar nitrogen atom in the aminomethylene chain to bridge 2′-O and 4′-C of the sugar moiety may possibly achieve such change. On the other hand, the observed ΔS upon the triplex formation is mainly contributed by the two factors, a negative conformational entropy change due to the conformational restraint of TFO involved in the triplex formation, and a positive dehydration entropy change from the release of structured water molecules surrounding the TFO and the target duplex upon the triplex formation[33-35]. Therefore, one of the reason for the smaller magnitudes of the negative ΔS for the 2′,4′-BNA^{NC}-modified TFOs at pH 6.8 in comparison with that for Pyr15TM at pH 6.1 (Table 4) may be based on the negative conformational entropy change. The 2′,4′-BNA^{NC}-modified TFO in the free state may be more rigid than the corresponding unmodified TFO, because the 2′-O and 4′-C positions of the sugar moiety of the 2′,4′-BNA^{NC} are bridged with the aminomethylene chain. The increased rigidity in the free state may cause the smaller loss of the conformational entropy upon the triplex formation with the 2′,4′-BNA^{NC}-modified TFO. Another reason for the smaller magnitudes of the negative ΔS for the 2′,4′-BNA^{NC}-modified TFOs at pH 6.8 relative to that for Pyr15TM at pH 6.1 (Table 4) may be derived from the positive dehydration entropy change. The nitrogen atom in the aminomethylene chain may result in the increased degree of hydration of the 2′,4′-BNA^{NC}-modified TFO in the free state. The increased degree of hydration in the free state may cause the larger gain of the dehydration entropy upon the triplex formation with the 2′,4′-BNA^{NC}-modified TFO. We conclude that the smaller loss of the conformational entropy due to the increased rigidity in the free state and the larger gain of the dehydration entropy due to the increased degree of hydration in the free state may result in the smaller magnitudes of the negative ΔS for the 2′,4′-BNA^{NC}-modified TFOs at pH 6.8, which provides a favorable component to the ΔG and leads to the increase in the K_a of the triplex formation at neutral pH.

The increase in the K_a by the 2′,4′-BNA^{NC} modification was similar in magnitude among the four modified TFOs (Figure 4 and Tables 4 and 5), indicating that the number and position of the 2′,4′-BNA^{NC} modification did not significantly affect the magnitude of the increase in the K_a at neutral pH. The increased rigidity and the increased degree of hydration themselves of the 2′,4′-BNA^{NC}-modified TFO may be more important to achieve the increase in the K_a at neutral pH than the variation of the number and position of the 2′,4′-BNA^{NC} modification. Thus, other modification strategies to gain the increased rigidity of TFO and the increased degree of hydration of TFO may be also useful to increase the K_a at neutral pH.

Kinetic data have demonstrated that the 2′,4′-BNA^{NC} modification of TFO considerably decreased the k_{dissoc} of the pyrimidine motif triplex formation at neutral pH (Table 5). The decrease in the k_{dissoc} is a plausible kinetic reason to explain the remarkable gain in the K_a at neutral pH by the 2′,4′-BNA^{NC} modification (Figure 4 and Tables 4 and 5). Both our group[35] and others[36] have previously proposed a model that triplexes form along nucleation-elongation processes: in a nucleation step only a few base contacts of the Hoogsteen hydrogen bonds may be formed between TFO and the target duplex, and this may be followed by an elongation step in which Hoogsteen base pairings progress to complete triplex formation. Both groups[35,36] have also suggested that the observed K_a, which is the ratio of k_{assoc} to k_{dissoc}, may mostly reflect a rapid equilibrium of the nucleation step, which is probably the rate-limiting process of the triplex formation. In this sense, the 2′,4′-BNA^{NC} modification is considered to slow the collapse of the nucleation intermediate using the increased rigidity and the increased degree of hydration of 2′,4′-BNA^{NC}-modified TFO to increase the K_a of the pyrimidine motif triplex formation.

The nuclease resistance of the 2′,4′-BNA^{NC}-modified TFO in human serum was significantly higher than that of the unmodified TFO (Figures 7 and 8). The 2′,4′-BNA^{NC} modification increased the nuclease resistance of TFOs in human serum. Previously, the nuclease resistance of the phosphorothioate backbone, in which a nonbridging oxygen of a phosphodiester group was replaced by a sulfur atom, was known to be significantly higher than that of the unmodified backbone[37,38]. However, the K_a of the triplex formation with the phosphorothioate modified TFO was significantly smaller than that with the unmodified TFO[39,40]. Thus, the phosphorothioate modification increased the nuclease resistance of TFO, but it decreased the triplex forming ability. On the other hand, as discussed above, the 2′,4′-BNA^{NC} modification of TFO increased the triplex forming ability at neutral pH (Figure 4 and Tables 4 and 5). Therefore, the 2′,4′- BNA^{NC} modification enhanced both the nuclease resistance of TFO and the triplex forming ability at neutral pH. We conclude that due to these excellent properties the 2′,4′-BNA^{NC} modification may be more favorable than the phosphorothioate modification upon the application of TFO to the various triplex formation-based strategies *in vivo*.

The 2′,4′-BNA^{NC}-modified antisense oligonucleotides showed the excellent antisense effect to reduce the expression level of PCSK9 mRNA (Figure 9). As discussed above, the 2′,4′-BNA^{NC}-modified oligonucleotides exhibited significantly higher binding affinity with complementary ssRNA than unmodified oligonucleotides (Table 1). Also, the 2′,4′-BNA^{NC} modification significantly increased the nuclease resistance of oligonucleotides in human serum (Figures 7 and 8). The significantly higher binding affinity with complementary ssRNA and the significantly higher nuclease resistance of oligonucleotides in human serum may achieve the excellent antisense effect of the 2′,4′-BNA^{NC}-modified antisense oligonucleotides. We conclude that the 2′,4′-BNA^{NC}-modified antisense oligonucleotides may be useful to reduce the expression level of the target mRNA.

12. Conclusion

The present study has clearly indicated that the 2′,4′-BNA^{NC} modification increased the thermal stability of the duplex with complementary ssRNA and ssDNA at neutral pH. It has also clearly demonstrated that the 2′,4′-BNA^{NC} modification of TFO increased not only the thermal stability of the pyrimidine motif triplex but also the K_a of the pyrimidine motif triplex formation at neutral pH by more than 10-fold, which mainly resulted from the considerable decrease in the k_{dissoc}. It has also revealed that the 2′,4′-BNA^{NC} modification of TFO significantly increased the nuclease resistance of TFO in human serum. Our results certainly support the idea that the 2′,4′-BNA^{NC}-modified oligonucleotides may have a potential to be applied to the various duplex and triplex formation-based strategies *in vivo*, such as regulation of gene expression by antisense and antigene technology, mapping of genomic DNA, and gene-targeted mutagenesis. In fact, the 2′,4′-BNA^{NC}-modified antisense oligonucleotides showed the excellent antisense effect to reduce the expression level of the target mRNA. In addition, the present study has revealed that the increased rigidity and the increased degree of hydration of the 2′,4′-BNA^{NC}-modified TFO in the free state may enable the significant increase in the K_a for the pyrimidine motif triplex formation at neutral pH. We conclude that the design of oligonucleotides to bridge different positions of sugar moiety with polar atom-containing alkyl chain for the increased rigidity and the increased degree of hydration is certainly a promising strategy for the promotion of the duplex and triplex formation under physiological condition, and may eventually lead to progress in various duplex and triplex formation-based strategies *in vivo*.

Author details

Hidetaka Torigoe*
Department of Applied Chemistry, Faculty of Science, Tokyo University of Science, Shinjuku-ku, Tokyo, Japan

Takeshi Imanishi
Graduate School of Pharmaceutical Sciences, Osaka University, Suita, Osaka, Japan
BNA Inc., Ibaraki, Osaka, Japan

* Corresponding Author

Acknowledgement

We thank Dr. S. M. Abdur Rahman, Dr. Kiyomi Sasaki, Ms. Hiroko Takuma, Ms. Sayori Seki, and Mr. Haruhisa Yoshikawa for their technical assistance. We acknowledge Prof. Satoshi Obika and Prof. Kazuyuki Miyashita for their useful discussions. The present work was supported in part by Grant-in-Aid for Scientific Research on Innovative Areas (22113519), Grant-in-Aid for Exploratory Research (20655038), Grant-in-Aid for Scientific Research (B) (21350094) and Grant-in-Aid for JSPS Fellows (22-10383) from the Ministry of Education, Science, Sports, and Culture of Japan. This work was also supported partly by the Program for Promotion of Fundamental Studies in Health Sciences of the National Institute of Biomedical Innovation (NIBIO), the Program for Precursory Research for Embryonic Science and Technology (PRESTO), and the Creation and Support Program for Start-ups from the Universities of the Japan Science and Technology Agency (JST)

13. References

[1] Kole R, Krainer AR,Altman S (2012) Rna Therapeutics: Beyond Rna Interference and Antisense Oligonucleotides. Nat Rev Drug Discov. 11: 125-140.

[2] Watts JK,Corey DR (2012) Silencing Disease Genes in the Laboratory and the Clinic. J Pathol. 226: 365-379.

[3] Duca M, Vekhoff P, Oussedik K, Halby L,Arimondo PB (2008) The Triple Helix: 50 Years Later, the Outcome. Nucleic Acids Res. 36: 5123-5138.

[4] Jain A, Wang G,Vasquez KM (2008) DNA Triple Helices: Biological Consequences and Therapeutic Potential. Biochimie. 90: 1117-1130.

[5] Frank-Kamenetskii MD (1992) Protonated DNA Structures. Methods Enzymol. 211: 180-191.

[6] Singleton SF,Dervan PB (1992) Influence of Ph on the Equilibrium Association Constants for Oligodeoxyribonucleotide-Directed Triple Helix Formation at Single DNA Sites. Biochemistry. 31: 10995-11003.

[7] Shindo H, Torigoe H,Sarai A (1993) Thermodynamic and Kinetic Studies of DNA Triplex Formation of an Oligohomopyrimidine and a Matched Duplex by Filter Binding Assay. Biochemistry. 32: 8963-8969.

[8] Wickstrom E (1986) Oligodeoxynucleotide Stability in Subcellular Extracts and Culture Media. J Biochem Biophys Methods. 13: 97-102.

[9] Prakash TP (2011) An Overview of Sugar-Modified Oligonucleotides for Antisense Therapeutics. Chem Biodivers. 8: 1616-1641.

[10] Rahman SM, Seki S, Obika S, Haitani S, Miyashita K,Imanishi T (2007) Highly Stable Pyrimidine-Motif Triplex Formation at Physiological Ph Values by a Bridged Nucleic Acid Analogue. Angew Chem Int Ed Engl. 46: 4306-4309.

[11] Rahman SM, Seki S, Obika S, Yoshikawa H, Miyashita K,Imanishi T (2008) Design, Synthesis, and Properties of 2',4'-Bna(Nc): A Bridged Nucleic Acid Analogue. J Am Chem Soc. 130: 4886-4896.

[12] Torigoe H, Rahman SM, Takuma H, Sato N, Imanishi T, Obika S,Sasaki K (2011) 2'-O,4'-C-Aminomethylene-Bridged Nucleic Acid Modification with Enhancement of Nuclease Resistance Promotes Pyrimidine Motif Triplex Nucleic Acid Formation at Physiological Ph. Chemistry. 17: 2742-2751.

[13] Torigoe H, Rahman SM, Takuma H, Sato N, Imanishi T, Obika S,Sasaki K (2011) Interrupted 2'-O,4'-C-Aminomethylene Bridged Nucleic Acid Modification Enhances Pyrimidine Motif Triplex-Forming Ability and Nuclease Resistance under Physiological Condition. Nucleosides Nucleotides Nucleic Acids. 30: 63-81.

[14] Yamamoto T, Harada-Shiba M, Nakatani M, Wada S, Yasuhara H, Narukawa K, Sasaki K, Shibata M, Torigoe H, Yamaoka T, Imanishi T,Obika S (2012) Cholesterol-Lowering Action of Bna-Based Antisense Oligonucleotides Targeting Pcsk9 in Atherogenic Diet-Induced Hypercholesterolemic Mice. Molecular Therapy-Nucleic Acids. 1: e22.

[15] Mousavi SA, Berge KE,Leren TP (2009) The Unique Role of Proprotein Convertase Subtilisin/Kexin 9 in Cholesterol Homeostasis. J Intern Med. 266: 507-519.

[16] Abifadel M, Rabes JP, Devillers M, Munnich A, Erlich D, Junien C, Varret M,Boileau C (2009) Mutations and Polymorphisms in the Proprotein Convertase Subtilisin Kexin 9 (Pcsk9) Gene in Cholesterol Metabolism and Disease. Hum Mutat. 30: 520-529.

[17] Soutar AK (2010) Rare Genetic Causes of Autosomal Dominant or Recessive Hypercholesterolaemia. IUBMB Life. 62: 125-131.

[18] Riley M,Maling B (1966) Physical and Chemical Characterization of Two- and Three-Stranded Adenine-Thymine and Adenine-Uracil Homopolymer Complexes. J Mol Biol. 20: 359-389.

[19] Torigoe H, Ferdous A, Watanabe H, Akaike T,Maruyama A (1999) Poly(L-Lysine)-Graft-Dextran Copolymer Promotes Pyrimidine Motif Triplex DNA Formation at Physiological Ph. Thermodynamic and Kinetic Studies. J Biol Chem. 274: 6161-6167.

[20] Torigoe H, Hari Y, Sekiguchi M, Obika S,Imanishi T (2001) 2'-O,4'-C-Methylene Bridged Nucleic Acid Modification Promotes Pyrimidine Motif Triplex DNA Formation at Physiological Ph: Thermodynamic and Kinetic Studies. J Biol Chem. 276: 2354-2360.

[21] Torigoe H (2001) Thermodynamic and Kinetic Effects of N3'-->P5' Phosphoramidate Modification on Pyrimidine Motif Triplex DNA Formation. Biochemistry. 40: 1063-1069.

[22] Torigoe H,Maruyama A (2001) Promotion of Duplex and Triplex DNA Formation by Polycation Comb-Type Copolymers. Methods Mol Med. 65: 209-224.

[23] Torigoe H,Maruyama A (2005) Synergistic Stabilization of Nucleic Acid Assembly by Oligo-N3'-->P5' Phosphoramidate Modification and Additions of Comb-Type Cationic Copolymers. J Am Chem Soc. 127: 1705-1710.

[24] Torigoe H, Maruyama A, Obika S, Imanishi T,Katayama T (2009) Synergistic Stabilization of Nucleic Acid Assembly by 2'-O,4'-C-Methylene-Bridged Nucleic Acid Modification and Additions of Comb-Type Cationic Copolymers. Biochemistry. 48: 3545-3553.

[25] Torigoe H, Sasaki K,Katayama T (2009) Thermodynamic and Kinetic Effects of Morpholino Modification on Pyrimidine Motif Triplex Nucleic Acid Formation under Physiological Condition. J Biochem. 146: 173-183.

[26] Torigoe H, Nakagawa O, Imanishi T, Obika S,Sasaki K (2012) Chemical Modification of Triplex-Forming Oligonucleotide to Promote Pyrimidine Motif Triplex Formation at Physiological Ph. Biochimie. 94: 1032-1040.
[27] Torigoe H, Sato N,Nagasawa N (2012) 2'-O,4'-C-Ethylene Bridged Nucleic Acid Modification Enhances Pyrimidine Motif Triplex Forming Ability under Physiological Condition. J Biochem. in press.
[28] Wiseman T, Williston S, Brandts JF,Lin LN (1989) Rapid Measurement of Binding Constants and Heats of Binding Using a New Titration Calorimeter. Anal Biochem. 179: 131-137.
[29] Bates PJ, Dosanjh HS, Kumar S, Jenkins TC, Laughton CA,Neidle S (1995) Detection and Kinetic Studies of Triplex Formation by Oligodeoxynucleotides Using Real-Time Biomolecular Interaction Analysis (Bia). Nucleic Acids Res. 23: 3627-3632.
[30] Edwards PR, Gill A, Pollard-Knight DV, Hoare M, Buckle PE, Lowe PA,Leatherbarrow RJ (1995) Kinetics of Protein-Protein Interactions at the Surface of an Optical Biosensor. Anal Biochem. 231: 210-217.
[31] Manzini G, Xodo LE, Gasparotto D, Quadrifoglio F, van der Marel GA,van Boom JH (1990) Triple Helix Formation by Oligopurine-Oligopyrimidine DNA Fragments. Electrophoretic and Thermodynamic Behavior. J Mol Biol. 213: 833-843.
[32] Lyamichev VI, Mirkin SM, Frank-Kamenetskii MD,Cantor CR (1988) A Stable Complex between Homopyrimidine Oligomers and the Homologous Regions of Duplex Dnas. Nucleic Acids Res. 16: 2165-2178.
[33] Edelhoch H,Osborne JC, Jr. (1976) The Thermodynamic Basis of the Stability of Proteins, Nucleic Acids, and Membranes. Adv Protein Chem. 30: 183-250.
[34] Cheng YK,Pettitt BM (1992) Stabilities of Double- and Triple-Strand Helical Nucleic Acids. Prog Biophys Mol Biol. 58: 225-257.
[35] Kamiya M, Torigoe H, Shindo H,Sarai A (1996) Temperature Dependence and Sequence Specificity of DNA Triplex Formation: An Analysis Using Isothermal Titration Calorimetry. J Am Chem Soc. 118: 4532-4538.
[36] Rougee M, Faucon B, Mergny JL, Barcelo F, Giovannangeli C, Garestier T,Helene C (1992) Kinetics and Thermodynamics of Triple-Helix Formation: Effects of Ionic Strength and Mismatches. Biochemistry. 31: 9269-9278.
[37] Zon G,Geiser TG (1991) Phosphorothioate Oligonucleotides: Chemistry, Purification, Analysis, Scale-up and Future Directions. Anticancer Drug Des. 6: 539-568.
[38] Stein CA, Tonkinson JL,Yakubov L (1991) Phosphorothioate Oligodeoxynucleotides--Anti-Sense Inhibitors of Gene Expression? Pharmacol Ther. 52: 365-384.
[39] Xodo L, Alunni-Fabbroni M, Manzini G,Quadrifoglio F (1994) Pyrimidine Phosphorothioate Oligonucleotides Form Triple-Stranded Helices and Promote Transcription Inhibition. Nucleic Acids Res. 22: 3322-3330.
[40] Torigoe H, Shimizume R, Sarai A,Shindo H (1999) Triplex Formation of Chemically Modified Homopyrimidine Oligonucleotides: Thermodynamic and Kinetic Studies. Biochemistry. 38: 14653-14659.

Chapter 7

The Mutation of Transient Receptor Potential Vanilloid 4 (TRPV4) Cation Channel in Human Diseases

Sang Sun Kang

Additional information is available at the end of the chapter

http://dx.doi.org/10.5772/50520

1. Introduction

The transient receptor potential vanilloid 4 (TRPV4) cation channel, a member of the TRP vanilloid subfamily, is expressed in a broad range of tissues, in which it contributes to the generation of Ca^{2+} signals and/or depolarization of membrane potential. TRPV4 is a polymodal Ca^{2+}-permeable cation channel with a length of 871 amino acids. It shows very prominent outward rectification, rarely opening upon hyperpolarization. Mutational analyses suggest that outward rectification is governed by a gating mechanism independent of the main intracellular gates [1-4]. The predicted TRPV4 structure harbors six membrane-spanning domains with a pore loop, an N-terminal domain with at least three ankyrin repeats, and a C-terminal domain residue within the cytoplasm [3-5]. These characters are common features in all six TRPVs (TRPV1–6). However, although the TRPV family shows similar characteristics (Fig. 1), each member has its own distinguishable functions from other TRPVs.

The participation of TRPV4 in osmo and mechanotransduction is relevant to several important functions, including cellular and systemic volume homeostasis, arterial dilation, nociception, bladder voiding, and the regulation of ciliary beat frequency. TRPV4 channel activity can be sensitized by coapplying a variety of stimuli and by the participation of a number of cell signaling pathways, which suggests the presence of different regulatory sites. In this regard, several proteins have been proposed to modulate TRPV4 subcellular localization and/or function: microtubule-associated protein 7, calmodulin, F-actin, and pacsin3 [5, 6]. Other studies have demonstrated a functional and physical interaction between inositol trisphosphate receptor 3 and TRPV4, which sensitizes the latter to the mechano and osmotransducing messenger 5'-6'-epoxieicosatrienoic acid. TRPV4 is also responsive to temperature, endogenous arachidonic acid (AA) metabolites, and phorbol esters, including 4-α phorbol 12, 13-didecanoate(4-αPDD), and participates in receptor-operated Ca^{2+} entry; thus, showing multiple activation modes [1-7]. However, the precise manner in which TRPV4 is regulated in the cell by these protein interactions, chemicals, and stimuli remains to be clearly established.

2. Naturally occurring TRPV4 mutants and genetic disorders

Few naturally occurring TRPV4 mutants have been identified. Interestingly, most of these mis-sense and nonsense point mutations are linked with the development of genetic disorders in humans and a detailed list of naturally occurring TRPV4 mutations and related disease has been documented (Table 1 and Fig. 1). Here, I discuss some of these mutations that have gained importance in terms of genetic diseases [4-6].

2.1. Serum sodium level quantitative trait locus (hyponatremia)

Tian et al. (2009) demonstrated that the rs3742030 single nucleotide polymorphism in the TRPV4 gene (P19S) is significantly associated with serum sodium concentration. After this discovery, hyponatremia was defined as serum sodium < 135 mEq/L in non-Hispanic Caucasian male populations. In heterologous expression studies in HEK293 cells, P19S mutant channels show a diminished response to hypotonic stress and to the osmotransducing lipid epoxyeicosatrienoic acid compared to that in wild-type channels. The P19S polymorphism affects TRPV4 function *in vivo* and likely influences systemic water balance on a population wide basis [8].

2.2. Chronic obstructive pulmonary disease (COPD)

COPD is characterized by airway epithelial damage, bronchoconstriction, parenchymal destruction, and mucus hypersecretion. Upon activation by a broad range of stimuli, TRPV4 functions to control airway epithelial cell volume and epithelial and endothelial permeability; it also triggers bronchial smooth muscle contraction and participates in autoregulation of mucociliary transport [9, 10]. These TRPV4 functions may be important for regulating COPD pathogenesis; thus, TRPV4 is a candidate COPD gene. The TRPV4 P19S mutant, which is also characterized as the cause of hyponatremia, is observed in patients with COPD.

2.3. Brachyolmia type 3 (BRAC3) [MIM:113500]

BRAC3 has been characterized using linkage analysis and candidate gene sequencing. Rock et al. found that some patients affected with brachylomia have a TRPV4 missense mutation, specifically at positions R616Q or V620I [11]. These mutations are located in the fifth transmembrane region, which is part of the functional pore. Each of these two mutations increases basal level activity when compared to the wild-type TRPV4. Additionally, the response to 4-αPDD (a TRPV4 specific agonist) is greater in mutants when compared with that in the wild-type [11]. This result also indicates that these two mutations preferably stabilize TRPV4 in its "open stage", resulting in constitutive channel activity. BRAC3 constitutes a clinically and genetically heterogeneous group of skeletal dysplasias characterized by a short trunk, scoliosis, and mild short stature. BRAC3 is an autosomal dominant form in which patients have severe kyphoscoliosis and flattened, irregular cervical vertebrae[11].

BRAC3, causing a R616Q gain-of-function channel, was examined and found to increase whole-cell current densities compared with that in wild-type channels. A single-channel analysis revealed that R616Q channels maintain mechanosensitivity but have greater constitutive activity and no change in unitary conductance or rectification [12]. BRAC3 ranges from mild autosomal-dominant BO, diagnosed by a shortened spine with characteristic vertebral defects and minor defects in the long bones to metatropic dysplasia characterized by more prominent spine defects as well as pronounced abnormalities in the articular skeleton resulting in short dumbbell-shaped long bones, which leads to prenatal lethality in its severest form [13].

2.4. Metatropic dysplasia (MTD) [MIM:156530]

MTD is a clinical heterogeneous skeletal dysplasia characterized by short extremities, a short trunk with progressive kyphoscoliosis, and craniofacial abnormalities that include a prominent forehead, midface hypoplasia, and a squared-off jaw [14]. Dominant mutations in the gene encoding TRPV4, a calcium permeable ion channel, have been identified in all 10 of a series MTD cases, ranging in severity from mild to perinatal lethal [14]. MTD is also called metatropic dwarfism. Metatropic dysplasia is a severe spondyloepimetaphyseal dysplasia characterized by short limbs, enlarged joints, and usually severe kyphoscoliosis [15]. Radiological features include severe platyspondyly, severe metaphyseal enlargement, and shortening of long bones. TRPV4 I331F and P799L mutants induce MTD [16, 17]. As all the above mentioned mutants are naturally occurring, these mutants are not embryonically lethal (as most lethal mutants are naturally excluded from the population). It is also important to note that none of these mutants show complete loss of their prime function, i.e., ion conductivity [12].

Several experimental results suggest that some of these mutants even have enhanced channel opening. These results demonstrate that the lethal form of the disorder is dominantly inherited and suggest locus homogeneity in the disease. Furthermore, electrophysiological studies have shown that the mutations activate the TRPV4 channel, indicating that the mechanism of the disease may result from increased calcium in chondrocytes [12, 16, 17].

Histological studies in two cases of lethal MTD revealed markedly disrupted endochondral ossification, with reduced numbers of hypertrophic chondrocytes and the presence of islands of cartilage within the primary mineralization zone [16]. These data suggest that altered chondrocyte differentiation in the growth plate leads to the clinical findings of MTD [18].

2.5. Distal spinal muscular atrophy congenital non-progressive (DSMAC) [MIM:600175]

DSMAC (also called hereditary motor and sensory neuropathy, Type IIC; HMSN2C) is a clinically variable, neuromuscular disorder characterized by a congenital lower motor neuron disorder restricted to the lower part of the body[19]. Clinical manifestations include nonprogressive muscular atrophy, thigh muscle atrophy, weak thigh adductors, weak knee and foot extensors, minimal jaw muscle and neck flexor weakness, flexion contractures of the knees and pes equinovarus. However, tendon reflexes are normal [20].

Inheritance is autosomal dominant. The R315W mutation has been identified in an unrelated family that also had HMSN2C [21]. Auer-Grumbach et al. identified two additional TRPV4 mutations (R269H and R316C) in affected members of three additional families with these three phenotypes, indicating that they are allelic disorders [22]. All three mutations occurred at the outer helices of the ANK4 and ANK5 domains, in the N-terminal cytoplasmic domain (Fig. 1). *In vitro* functional expression studies in HeLa cells show that the mutant protein forms cytoplasmic aggregates and has reduced surface expression, as well as an impaired response to stimulus-dependent channel activity. These results suggest that the mutations interfere with normal channel trafficking and function [21, 22]. Furthermore, Auer-Grumbach et al. identified a different heterozygous mutation in the TRPV4 gene (R315W; 605427.0008) in a patient with congenital distal SMA whose other family members with the same mutation had phenotypes consistent with hereditary motor and sensory neuropathy-2 or scapuloperoneal spinal muscular atrophy; thus, proving that these are allelic disorders with overlapping phenotypes[21, 22].

2.6. Spondyloepiphyseal dysplasia Maroteaux type (SEDM) [MIM:184095]

SEDM is a clinically variable spondyloepiphyseal dysplasia with manifestations limited to the musculoskeletal system [23]. Clinical features of SEDM include short stature, brachydactyly, platyspondyly, short and stubby hands and feet, epiphyseal hypoplasia of the large joints, and iliac hypoplasia; however, the patients have normal intelligence [23, 24]. Genetic mapping of patients affected with this disease show a missense mutation in TRPV4, either E183K, Y602C, or E797K [25]. Channel activity of the TRPV4 E797K mutant in HEK293 cells is constitutively active, consistent with the argument that the effects in TRPV4 are the cause SEDM [25]. SEDM is a clinically variable spondyloepiphyseal dysplasia with manifestations limited to the musculoskeletal system. Clinical features include short stature, brachydactyly, platyspondyly, short and stubby hands and feet, epiphyseal hypoplasia of the large joints, and iliac hypoplasia [26]. Both SEDM and parastremmatic dysplasia are part of the TRPV4 dysplasia family and TRPV4 mutations show considerable variability in phenotypic expression resulting in distinct clinical-radiographic phenotypes.

2.7. Parastremmatic dwarfism (PSTD) [MIM:168400]

PSTD is also characterized by defects in TRPV4 which is a bone dysplasia characterized by severe dwarfism, kyphoscoliosis, distortion, and bowing of the extremities, and contractures of the large joints [27]. The disease is radiographically characterized by a combination of decreased bone density, bowing of the long bones, platyspondyly, and striking irregularities of endochondral ossification with areas of calcific stippling and streaking in radiolucent epiphyses, metaphyses, and apophyses [27].

In a 7-year-old girl with PSTD, Nishimura et al. (2010) analyzed the TRPV4 candidate gene and identified heterozygosity for a missense mutation (R594H; 605427.0003), which had previously been found in patients with the Kozlowski type of spondylometaphyseal dysplasia **(SMDK;** 184252)[25]. However, in patients with the Kozlowski type of spondylometaphyseal dysplasia (SMDK; 184252), Krakow et al. (2009) identified a 1781G-A transition in exon 11 of the TRPV4 gene, resulting in an arg594-to-his (R594H) substitution in the cytoplasmic S4 domain [12]. Thus, both PSTD and SMDK, which are caused by a TRPV4 mutation, seem to be associated with increased basal intracellular calcium ion concentration and intracellular calcium activity [12, 16, 25]. However, the Kozlowski type of spondylometaphyseal dysplasia **(SMDK;** 184252) is different from SEDM at TRPV4 mutation sites (E183K Y602C or E797K) [23-25].

2.8. Charcot–Maries–Tooth disease type 2C (CMT2C) and scapuloperoneal spinal muscular atrophy (SPSMA) [MIM:606071]

CMT2C is an axonal form of Charcot–Marie–Tooth disease, a disorder of the peripheral nervous system, characterized by progressive weakness and atrophy, initially of the peroneal muscles and later of the distal muscles of the arms [28]. Charcot–Marie–Tooth disease is classified into two main groups based on electrophysiological properties and histopathology: primary peripheral demyelinating neuropathies (designated CMT1 when they are dominantly inherited) and primary peripheral axonal neuropathies (CMT2) [29]. CMT2 group neuropathies are characterized by signs of axonal regeneration in the absence of obvious myelin alterations, normal or slightly reduced nerve conduction velocities, and progressive distal muscle weakness and atrophy [28, 29]. Nerve conduction velocities are normal or slightly reduced. CMT2C and SPSMA are also known as hereditary motor and sensory neuropathy type 2 (HMSN2C) [30, 31]. Patients with SPSMA are characterized by weakness of the scapular muscle and bone abnormalities. CMT2C leads to weakness of distal limbs, vocal cords, and often impairs hearing and vision [32]. Genetic analyses of these patients show the presence of TRPV4 missense mutations, particularly at the R269H, R315W, and R316C positions. [31, 33]

2.9. Familial digital arthropathy-brachydactyly (FDAB)

FDAB is a dominantly inherited condition that is characterized by aggressive osteoarthropathy of the fingers and toes and consequent shortening of the middle and distal phalanges [34]. Lamandé *et al.* showed that FDAB is caused by mutations encoding p.Gly270Val, p.Arg271Pro, and p.Phe273Leu substitutions in the intracellular ankyrin-repeat domain of the TRPV4 cation channel. The TRPV4 mutant in HEK-293 cells shows that the mutant proteins have poor cell-surface localization. TRPV4 mutations that reduce channel activity cause a third phenotype, inherited osteoarthropathy, and show the importance of TRPV4 activity in articular cartilage homeostasis. Thus, the TRPV4 mutant (G270V, R271P, Y273L) also seems to be related with FDAB [34].

3. Conclusions and perspective

The TRPV4 functional Ca2+ channel consists of homo tetramer subunits [35]. TRPV4 and TRPC1 can coassemble to form heteromeric TRPV4–C1 channels [36, 37]. Because the TRPV4 ankyrin repeat is responsible for its channel selfassembly in the cell line, mutations in the TRPV4 ankyrin domain also seem to affect channel assembly in humans, as shown in the many genetic disorders (Fig. 1 and Table 1).

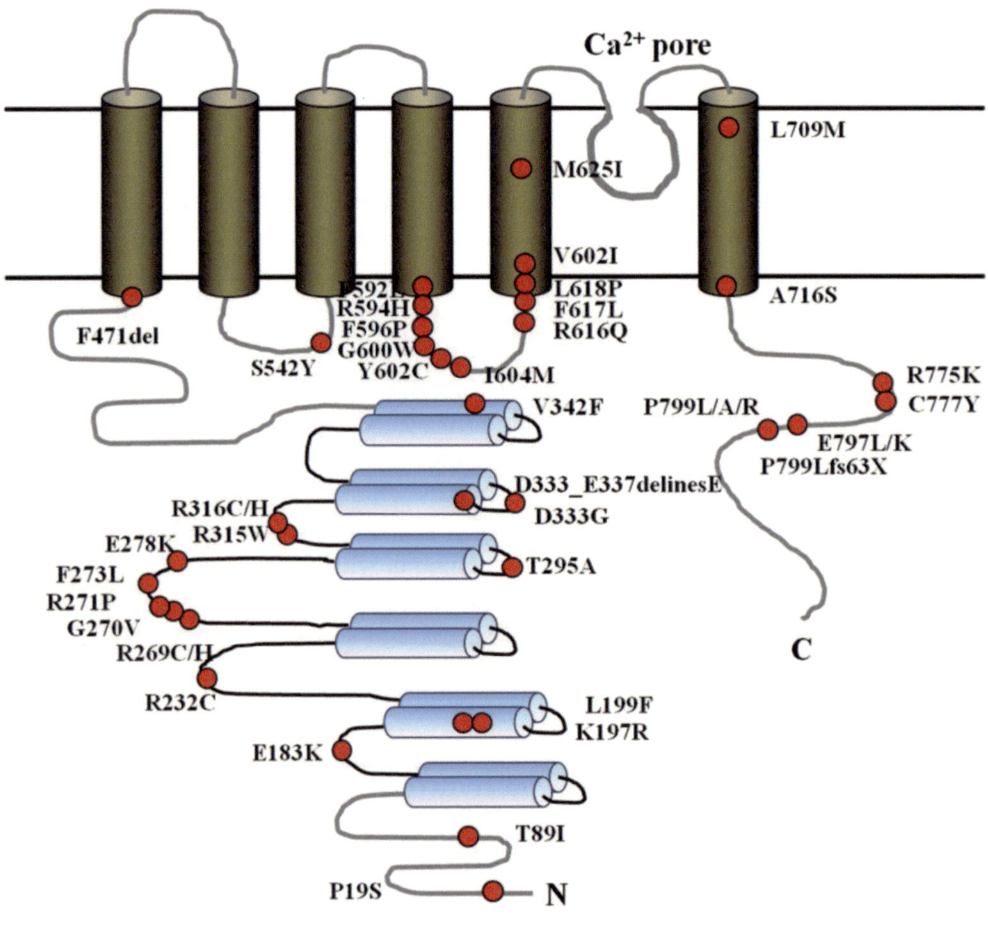

Figure 1. The naturally mutation sites on human TRPV4.

Mutation		Residue	Change in charge	Domain/motif effected	Effects on ion conductivity	Genetic disorder
1	C144T (exon 2)-	P19S	Nonpolar to polar	N-terminal	Less conductivity	Hyponatermia COPD
2	C366T (exon 2)	T89I	Polar (uncharged) to nonpolar	N-terminal	Not done	Metatropic dysplasia
3	G547A (exon 3)	E183K	Negative to plus	ARD1	Not done	SEDM-PM2
4	A590G (exon 4)	K197R	Plus to plus	ARD2	Not done	Metatropic dysplasia
5	-	L199F	Nonpolar to aromatic	ARD2	Not done	Metatropic dysplaisa
6	G806A (exon 5)	R269H	Plus to plus	ARD3	Less conductivity	SMA
7	G806A (exon 5)	R269H	Plus to plus	ARD3	More conductivity	CMT2C
8	G806A (exon 5)	R269H	Plus to plus	ARD3	More conductivity	CMT2C
9	G806A (exon 5)	R269C	Plus to polar un charged	ARD3	More conductivity	CMT2C
10		G270V	Nonpolar to polar	ARD3	Not done	FDAB
11		R271P	Plus to nonpolar	ARD3	Not done	FDAB
12		F273L	Aromatic to nonpolar	ARD3	Not done	FDAB
10	-	E278K	Negative to plus	ARD3	Not done	SMDK
11	-	T295A	Polar (uncharged) to nonpolar	ARD4	Not done	Metatropic dysplaisa
12	C943T (exon 6)	R315W	Plus to aromatic	ARD4	Less conductivity	HMSN2C
13	C946T (exon 6)	R316C	Plus to polar (uncharged)	ARD4	Less conductivity	HMSN2C
14	A1080T (exon 6)	I331F	Nonpolar to aromatic	ARD5	Not done	Metatropic dysplasia
15	-	I331T	Nonpolar to polar (uncharged)	ARD5	Not done	Metatropic dysplasia
16	A992G (exon 6)	D333G	Negative to nonploar	ARD4	More conductivity	SMDK
17	-	V342F	Nonpolar to aromatic	ARD5	Not done	Metatropic dysplasia
18	-	F592L	Aromatic to nonpolar	TM4	Not done	Metatropic dysplasia
19	G1781A (exon 11)	R594H	Plus to plus	TM4	More conductivity	SMDK
20	A1805G (exon 11)	Y602C	Aromatic to polar	TM4-TM5	Not done	SEDM-PM2
21	C1812G (exon 11)	I604M	Nonpolar to nonpolar	TM4-TM5	Not done	Metatropic dyslpasia

Mutation		Residue	Change in charge	Domain/motif effected	Effects on ion conductivity	Genetic disorder
22	G1847A (exon 12)	R616Q	Plus to polar uncharged	TM5, pore region	More conductivity	Brachylomia
23	C1851A (exon 12)	F617L	Aromatic to nonpolar	TM5, pore region	Not done	Metatropic dysplasia
24	T1853C (exon 12)	L618Q	Nonpolar to polar (uncharged)	TM5, pore region	Not done	Metatropic dysplasia
25	G858A (exon 12)	V620I	Nonpolar to nonpolar	TM5, pore region	More conductivity	Brachylomia
26	-	M625I	Nonpolar to nonpolar	TM5, pore region	Not done	SMDK
27	-	L709M	Nonpolar to nonpolar	TM5, pore region	Not done	SMDK
28	C2146T (exon 13)	A716S	Nonpolar to polar	Cytoplasmic side of TM6	Same as wild type	SMDK
29	-	R775K	Plus to plus	C-terminal region	Not done	Metatropic dysplasia
30	-	C777Y	Polar (uncharged) to aromatic	C-terminal region	Not done	SMDK
31	-	E797K	Negative to plus	C-terminal region	Not done	SEDM-PM2
32	-	P799R	Nonpolar to plus	C-terminal region	Not done	Metatropic dysplasia
33	-	P799S	Nonpolar to polar (uncharged)	C-terminal region	Not done	Metatropic dysplasia,
34	-	P799A	Nonpolar to non polar	C-terminal region	Not done	Metatropic dysplasia
35	C2396T (exon 15)	P799L	Nonpolar to nonpolar	C-terminal	Not done	SMDK

The disease abbreviation means below:

Serum Sodium Level Quantitative Trait Locus (Hyponatermia) (the # of MIM is not available)

Chronic obstructive pulmonary disease (COPD) (the # of MIM is not available)

Brachyolmia type 3 (BRAC3) [MIM:113500]

Metatropic dysplasia (MTD) [MIM:156530];

Distal spinal muscular atrophy congenital non-progressive (DSMAC) [MIM:600175]. (DSMAC is also called as Hereditary Motor and Sensory Neuropathy, Type IIC; HMSN2C)

Spondyloepiphyseal dysplasia Maroteaux type (SEDM) [MIM:184095].

Parastremmatic dwarfism (PSTD) [MIM:168400]

Charcot-Maries-Tooth disease type 2C (CMT2C) and Scapuloperoneal Spinal Muscular Atrophy (SPSMA) [MIM:606071]

Familial digital arthropathy-brachydactyly (FDAB): (the # of MIM is not available)

Kozlowski type of spondylometaphyseal dysplasia (SMDK): [mim:184252]

*MIM : Mendelian Inheritance in Man

Table 1. The Summary of the naturally occurring TRPV4 mutations and human diseases.

Our recent observations indicate that TRPV4 is modulated by phosphorylation of the Ser824 residue as a positive regulation loop [7, 38]. However, the TRPV4 C-terminal domain near serine residue 824 seems to regulate its function by an unknown controlling mechanism beyond a phosphorylation modification, such as a protein-protein interaction with CaM. TRPV4 C-terminal domain mutations also seem to affect protein-protein interactions, resulting in the genetic disorders listed in Fig. 1 and Table 1. In the future, TRPV4 mutant knockdown in an animal model will be helpful to elucidate how the TRPV4 mutations cause the genetic disorders.

TRPV4 was originally shown to be activated by hypotonicity, but later studies have demonstrated that activation can also be achieved by phorbol esters, AA, and moderate heat. TRPV4 appears to be an important player in pathological sensory perception and bone growth [1-6]. The potential effect of mutations on TRPV4 function, which are related to human diseases through its altered function, remains to be elucidated. Furthermore, the role of TRPV4 in the pathogenesis of several diseases should be characterized and how the channel protein contributes to the specific disease must be understood. This information may be useful to cure or alleviate the human diseases caused by TRPV4 mutations.Transmembrane topology of the human TRPV4 (871aa length). Indicated are the three ankyrin-binding repeats (ANK; blue bar), the six trans-membrane regions (TM1–TM6), the Ca^2 pore and the mutation site (WT; Gene Bank #. BC127052). The putative cytoplasmic region of N-terminal (1-471 aa) and C-terminal (718-871aa) of TRPV4 are indicated with N and C. Two "hot spots" in TRPV4 sequences are prominent, one at the pore region and the other one in the ARDs. (del: deletion, delines: deletion or insertion extra sequence, fs: fame shift)

Author details

Sang Sun Kang

Department of Biology Education, Chungbuk National University, Seongbong, Road, Heungdok-gu, Cheongju, Chungbuk, Republic of Korea

Acknowledgement

This work was supported by National Research Foundation of Korea (NRF) grants (2009-0076024 and 2009-0069007) funded by the Korea government (MEST) to S S Kang. I also appreciated The Core Facility of Chungbuk National University.

4. References

[1] R. Strotmann, C. Harteneck, K. Nunnenmacher, G. Schultz, T.D. Plant, OTRPC4, a nonselective cation channel that confers sensitivity to extracellular osmolarity, Nat Cell Biol, 2 (2000) 695-702.

[2] R. Strotmann, G. Schultz, T.D. Plant, Ca2+-dependent potentiation of the nonselective cation channel TRPV4 is mediated by a C-terminal calmodulin binding site, J Biol Chem, 278 (2003) 26541-26549.

[3] Y. Itoh, N. Hatano, H. Hayashi, K. Onozaki, K. Miyazawa, K. Muraki, An environmental sensor, TRPV4 is a novel regulator of intracellular Ca2+ in human synoviocytes, Am J Physiol Cell Physiol, 297 (2009) C1082-1090.

[4] P. Verma, A. Kumar, C. Goswami, TRPV4-mediated channelopathies, Channels (Austin), 4 (2010) 319-328.

[5] W. Everaerts, B. Nilius, G. Owsianik, The vanilloid transient receptor potential channel TRPV4: from structure to disease, Prog Biophys Mol Biol, 103 (2010) 2-17.

[6] C.D. Wee, L. Kong, C.J. Sumner, The genetics of spinal muscular atrophies, Curr Opin Neurol, 23 (2010) 450-458.

[7] S.H. Shin, E.J. Lee, S. Hyun, J. Chun, Y. Kim, S.S. Kang, Phosphorylation on the Ser 824 residue of TRPV4 prefers to bind with F-actin than with microtubules to expand the cell surface area, Cell Signal, 24 (2012) 641-651.

[8] W. Tian, Y. Fu, A. Garcia-Elias, J.M. Fernandez-Fernandez, R. Vicente, P.L. Kramer, R.F. Klein, R. Hitzemann, E.S. Orwoll, B. Wilmot, S. McWeeney, M.A. Valverde, D.M. Cohen, A loss-of-function nonsynonymous polymorphism in the osmoregulatory TRPV4 gene is associated with human hyponatremia, Proc Natl Acad Sci U S A, 106 (2009) 14034-14039.

[9] G. Zhu, A. Gulsvik, P. Bakke, S. Ghatta, W. Anderson, D.A. Lomas, E.K. Silverman, S.G. Pillai, Association of TRPV4 gene polymorphisms with chronic obstructive pulmonary disease, Hum Mol Genet, 18 (2009) 2053-2062.

[10] J. Dai, T.J. Cho, S. Unger, E. Lausch, G. Nishimura, O.H. Kim, A. Superti-Furga, S. Ikegawa, TRPV4-pathy, a novel channelopathy affecting diverse systems, J Hum Genet, 55 (2010) 400-402.

[11] M.J. Rock, J. Prenen, V.A. Funari, T.L. Funari, B. Merriman, S.F. Nelson, R.S. Lachman, W.R. Wilcox, S. Reyno, R. Quadrelli, A. Vaglio, G. Owsianik, A. Janssens, T. Voets, S. Ikegawa, T. Nagai, D.L. Rimoin, B. Nilius, D.H. Cohn, Gain-of-function mutations in TRPV4 cause autosomal dominant brachyolmia, Nat Genet, 40 (2008) 999-1003.

[12] D. Krakow, J. Vriens, N. Camacho, P. Luong, H. Deixler, T.L. Funari, C.A. Bacino, M.B. Irons, I.A. Holm, L. Sadler, E.B. Okenfuss, A. Janssens, T. Voets, D.L. Rimoin, R.S. Lachman, B. Nilius, D.H. Cohn, Mutations in the gene encoding the calcium-permeable ion channel TRPV4 produce spondylometaphyseal dysplasia, Kozlowski type and metatropic dysplasia, Am J Hum Genet, 84 (2009) 307-315.

[13] S. Unger, E. Lausch, F. Stanzial, G. Gillessen-Kaesbach, I. Stefanova, C.M. Di Stefano, E. Bertini, C. Dionisi-Vici, B. Nilius, B. Zabel, A. Superti-Furga, Fetal akinesia in metatropic dysplasia: The combined phenotype of chondrodysplasia and neuropathy?, Am J Med Genet A, 155A (2011) 2860-2864.

[14] M. Beck, M. Roubicek, J.G. Rogers, P. Naumoff, J. Spranger, Heterogeneity of metatropic dysplasia, Eur J Pediatr, 140 (1983) 231-237.

[15] S.D. Boden, F.S. Kaplan, M.D. Fallon, R. Ruddy, J. Belik, E. Anday, E. Zackai, J. Ellis, Metatropic dwarfism. Uncoupling of endochondral and perichondral growth, J Bone Joint Surg Am, 69 (1987) 174-184.

[16] N. Camacho, D. Krakow, S. Johnykutty, P.J. Katzman, S. Pepkowitz, J. Vriens, B. Nilius, B.F. Boyce, D.H. Cohn, Dominant TRPV4 mutations in nonlethal and lethal metatropic dysplasia, Am J Med Genet A, 152A (2010) 1169-1177.
[17] J. Dai, O.H. Kim, T.J. Cho, M. Schmidt-Rimpler, H. Tonoki, K. Takikawa, N. Haga, K. Miyoshi, H. Kitoh, W.J. Yoo, I.H. Choi, H.R. Song, D.K. Jin, H.T. Kim, H. Kamasaki, P. Bianchi, G. Grigelioniene, S. Nampoothiri, M. Minagawa, S.I. Miyagawa, T. Fukao, C. Marcelis, M.C. Jansweijer, R.C. Hennekam, F. Bedeschi, A. Mustonen, Q. Jiang, H. Ohashi, T. Furuichi, S. Unger, B. Zabel, E. Lausch, A. Superti-Furga, G. Nishimura, S. Ikegawa, Novel and recurrent TRPV4 mutations and their association with distinct phenotypes within the TRPV4 dysplasia family, J Med Genet, 47 (2010) 704-709.
[18] R. Lewis, C.H. Feetham, R. Barrett-Jolley, Cell volume regulation in chondrocytes, Cell Physiol Biochem, 28 (2011) 1111-1122.
[19] P. Fleury, G. Hageman, A dominantly inherited lower motor neuron disorder presenting at birth with associated arthrogryposis, J Neurol Neurosurg Psychiatry, 48 (1985) 1037-1048.
[20] C.J. Frijns, J. Van Deutekom, R.R. Frants, F.G. Jennekens, Dominant congenital benign spinal muscular atrophy, Muscle Nerve, 17 (1994) 192-197.
[21] M.E. McEntagart, S.L. Reid, A. Irrthum, J.B. Douglas, K.E. Eyre, M.J. Donaghy, N.E. Anderson, N. Rahman, Confirmation of a hereditary motor and sensory neuropathy IIC locus at chromosome 12q23-q24, Ann Neurol, 57 (2005) 293-297.
[22] M. Auer-Grumbach, A. Olschewski, L. Papic, H. Kremer, M.E. McEntagart, S. Uhrig, C. Fischer, E. Frohlich, Z. Balint, B. Tang, H. Strohmaier, H. Lochmuller, B. Schlotter-Weigel, J. Senderek, A. Krebs, K.J. Dick, R. Petty, C. Longman, N.E. Anderson, G.W. Padberg, H.J. Schelhaas, C.M. van Ravenswaaij-Arts, T.R. Pieber, A.H. Crosby, C. Guelly, Alterations in the ankyrin domain of TRPV4 cause congenital distal SMA, scapuloperoneal SMA and HMSN2C, Nat Genet, 42 (2010) 160-164.
[23] A.N. Doman, P. Maroteaux, E.D. Lyne, Spondyloepiphyseal dysplasia of Maroteaux, J Bone Joint Surg Am, 72 (1990) 1364-1369.
[24] A. Megarbane, P. Maroteaux, C. Caillaud, M. Le Merrer, Spondyloepimetaphyseal dysplasia of Maroteaux (pseudo-Morquio type II syndrome): report of a new patient and review of the literature, Am J Med Genet A, 125A (2004) 61-66.
[25] G. Nishimura, J. Dai, E. Lausch, S. Unger, A. Megarbane, H. Kitoh, O.H. Kim, T.J. Cho, F. Bedeschi, F. Benedicenti, R. Mendoza-Londono, M. Silengo, M. Schmidt-Rimpler, J. Spranger, B. Zabel, S. Ikegawa, A. Superti-Furga, Spondylo-epiphyseal dysplasia, Maroteaux type (pseudo-Morquio syndrome type 2), and parastremmatic dysplasia are caused by TRPV4 mutations, Am J Med Genet A, 152A (2010) 1443-1449.
[26] G. Nishimura, R. Kizu, Y. Kijima, K. Sakai, Y. Kawaguchi, T. Kimura, I. Matsushita, S. Shirahama, T. Ikeda, S. Ikegawa, T. Hasegawa, Spondyloepiphyseal dysplasia Maroteaux type: report of three patients from two families and exclusion of type II collagen defects, Am J Med Genet A, 120A (2003) 498-502.
[27] L.O. Langer, D. Petersen, J. Spranger, An unusual bone dysplasia: parastremmatic dwarfism, Am J Roentgenol Radium Ther Nucl Med, 110 (1970) 550-560.

[28] P.J. Dyck, W.J. Litchy, S. Minnerath, T.D. Bird, P.F. Chance, D.J. Schaid, A.E. Aronson, Hereditary motor and sensory neuropathy with diaphragm and vocal cord paresis, Ann Neurol, 35 (1994) 608-615.

[29] M. Donaghy, R. Kennett, Varying occurrence of vocal cord paralysis in a family with autosomal dominant hereditary motor and sensory neuropathy, J Neurol, 246 (1999) 552-555.

[30] C.J. Klein, Y. Shi, F. Fecto, M. Donaghy, G. Nicholson, M.E. McEntagart, A.H. Crosby, Y. Wu, H. Lou, K.M. McEvoy, T. Siddique, H.X. Deng, P.J. Dyck, TRPV4 mutations and cytotoxic hypercalcemia in axonal Charcot-Marie-Tooth neuropathies, Neurology, 76 (2011) 887-894.

[31] D.H. Chen, Y. Sul, M. Weiss, A. Hillel, H. Lipe, J. Wolff, M. Matsushita, W. Raskind, T. Bird, CMT2C with vocal cord paresis associated with short stature and mutations in the TRPV4 gene, Neurology, 75 (2010) 1968-1975.

[32] G. Landoure, A.A. Zdebik, T.L. Martinez, B.G. Burnett, H.C. Stanescu, H. Inada, Y. Shi, A.A. Taye, L. Kong, C.H. Munns, S.S. Choo, C.B. Phelps, R. Paudel, H. Houlden, C.L. Ludlow, M.J. Caterina, R. Gaudet, R. Kleta, K.H. Fischbeck, C.J. Sumner, Mutations in TRPV4 cause Charcot-Marie-Tooth disease type 2C, Nat Genet, 42 (2010) 170-174.

[33] H.X. Deng, C.J. Klein, J. Yan, Y. Shi, Y. Wu, F. Fecto, H.J. Yau, Y. Yang, H. Zhai, N. Siddique, E.T. Hedley-Whyte, R. Delong, M. Martina, P.J. Dyck, T. Siddique, Scapuloperoneal spinal muscular atrophy and CMT2C are allelic disorders caused by alterations in TRPV4, Nat Genet, 42 (2010) 165-169.

[34] S.R. Lamande, Y. Yuan, I.L. Gresshoff, L. Rowley, D. Belluoccio, K. Kaluarachchi, C.B. Little, E. Botzenhart, K. Zerres, D.J. Amor, W.G. Cole, R. Savarirayan, P. McIntyre, J.F. Bateman, Mutations in TRPV4 cause an inherited arthropathy of hands and feet, Nat Genet, 43 (2011) 1142-1146.

[35] M. Arniges, J.M. Fernandez-Fernandez, N. Albrecht, M. Schaefer, M.A. Valverde, Human TRPV4 channel splice variants revealed a key role of ankyrin domains in multimerization and trafficking, J Biol Chem, 281 (2006) 1580-1586.

[36] T. Kobori, G.D. Smith, R. Sandford, J.M. Edwardson, The transient receptor potential channels TRPP2 and TRPC1 form a heterotetramer with a 2:2 stoichiometry and an alternating subunit arrangement, J Biol Chem, 284 (2009) 35507-35513.

[37] X. Ma, B. Nilius, J.W. Wong, Y. Huang, X. Yao, Electrophysiological properties of heteromeric TRPV4-C1 channels, Biochim Biophys Acta, 1808 (2011) 2789-2797.

[38] E.J. Lee, S.H. Shin, J. Chun, S. Hyun, Y. Kim, S.S. Kang, The modulation of TRPV4 channel activity through its Ser 824 residue phosphorylation by SGK1, Animal Cells and Systems, 14 (2010) 99-114.

Chapter 8

Bacterial Systems for Testing Spontaneous and Induced Mutations

Anna Sikora, Celina Janion and Elżbieta Grzesiuk

Additional information is available at the end of the chapter

http://dx.doi.org/10.5772/50471

1. Introduction

Changes in genetic material result from introduction of mutations into DNA. Spontaneous mutations can occur because of replication errors or as a consequence of lesions introduced into DNA during normal cell growth. Induced mutations arise after treatment of the organism with an exogenous mutagen being physical or chemical agent increasing the frequency of mutations.

Bacteria are simple and widely used models for examination of mutagenesis and DNA repair processes. The advantages of bacterial systems are their availability, easy cultivation, short time of cell division, and haploidity. Many DNA damaging agents and/or mutator genes cause mutations that are readily and clearly observed in changes of phenotype. Additional observations like (i) analysis of bacterial survival after treatment with mutagenic agents; (ii) microscopic examination of bacterial cells; (iii) examination of plasmid DNAs isolated from mutagen-treated cells for their sensitivity to the specific enzymes that recognize DNA lesions; (iv) induction of the SOS system measured by induction of β-galactosidase in *Escherichia coli*; (v) sporulation of bacteria in *Bacillus subtilis*, all provide a simple and rapid yet highly informative characterization of the examined process. Results of these studies usually constitute a first step of deeper examination of the multiple processes on molecular level.

Two representatives of *Enterobacteriaceae* (Gammaproteobacteria), *E. coli* and *Salmonella typhimurium*, are commonly used in the studies on spontaneous and induced mutagenesis. Other bacterial models like representatives of *Pseudomonadaceae* (Gammaproteobacteria), *Pseudomonas putida* and *Bacillaceae* (Firmicutes), *B. subtilis* have been also used in mutagenesis studies.

2. Mutation detection systems in *Escherichia coli*

2.1. The *argE3* → Arg$^+$ reversion system in *Escherichia coli* K12

Escherichia coli K12 was isolated from the stool of a convalescent diphtheria patient in USA (Palo Alto, California) in 1922 and deposited in the strain collection of the Department of Bacteriology of Stanford University. Serological studies revealed that after many years of cultivation under laboratory conditions the strain lost the K and O antigens and became incapable of human gut colonization. A lot of mutant derivatives of strain K12 have been obtained in many laboratories around the world. One of them is AB1157 strain with relevant genotype: *thr-1 ara-14 leuB6 Δ(gpt-proA)62 lacY1 tsx-33 supE44*$_{amber}$ *galK2 hisG4 rfbD1 mgl-51 rpsL31 kdgK51 xyl-5 mtl-1 argE3 thi-1* and its derivatives (Bachman, 1987). The *argE3*(ochre), *hisG4*(ochre) and *thr-1*(amber) are nonsense point mutations in genes encoding enzymes involved in arginine, histidine and threonine biosynthesis pathways, respectively. The *supE44* encodes *supE* amber suppressor reading UAG. However, it can only weakly suppress the *thr-1* mutation.

A suppressor mutation is a mutation that counteracts the effects of another mutation. One type of suppressor mutations are mutations that appear in the tRNA encoding genes at the anticodon site. The changed tRNAs are able to recognize a nonsense codon that occur elsewhere in protein-coding genes and incorporate the amino acids specific for them into the polypeptide chain during protein synthesis.

The bacterial test system of mutation detection described here is based on reversion of the auxotrophic *argE3* mutation to prototrophy and subsequent determination of specificity of mutation with the use of a set of bacteriophage T4 amber and ochre mutants. The marker is situated in the chromosome. The *argE* gene encodes acetylornithine deacetylase, one of the enzymes of arginine biosynthesis pathway. The Arg$^+$ phenotype can be restored by (i) any point mutation at *argE3* that changes nonsense UAA codon to any sense nucleotide triplet coding for any amino acid; (ii) an AT→GC transition at *argE3* that changes the UAA nonsense codon to the UAG nonsense codon recognized by *supE44* amber suppressor; and (iii) suppressor mutations enabling reading UAA nonsense codon. The suppressors can be created *de novo* or as the result of a GC→AT transition at *supE44* (formation of *supE* ochre suppressor) (Sargentini & Smith, 1989; Śledziewska-Gójska et al., 1992).

Considering all the theoretical possibilities of the *ochre* suppressor formation in *E. coli* resulting from a single base substitution in tRNA genes it can be seen that such suppressors may arise from tRNA for tyrosine, lysine, glutamine, glutamate, leucine and serine. The following tRNA species that may produce *de novo* an ochre suppressor by a single base substitution in the anticodon site are $tRNA^{Gln}_{UUG}$, $tRNA^{Lys}_{UUU}$, and $tRNA^{Tyr}_{GUA}$. The formed suppressors are, respectively, *supB*, *supL* (*supG*, *supN*) and *supC* (*supO*, *supM*), created as a result of GC→AT (*supB*), AT→TA (*supL*, *supG*, *supN*) or GC→TA (*supC*, *supO*, *supM*) base substitutions in *gln*-tRNA, *lys*-tRNA and *tyr*-tRNA genes, respectively (Table 1). The *supX* suppressor is also found in the Arg$^+$ revertants, but it has not yet been identified. This suppressor can be formed as a result of either GC→TA or AT→TA transversions (Sargentini

& Smith, 1989; Śledziewska-Gójska et al., 1992 and cited therein). Raftery and Yarus (1987) constructed the *gltT*(SuUUA/G) gene encoding $tRNA^{Glu}_{UUA}$ as a result of GC→TA transversion in the *gltT* gene encoding $tRNA^{Glu}_{UUC}$. This construct was expected to explain the mystery of the *supX* suppressor. However, it failed to suppress the *argE3*oc mutation in *E. coli* AB1157 strain (Płachta & Janion, 1992). Moreover, Prival (1996) identified three $tRNA^{Glu}_{UUA}$ suppressors: *supY*, *supW* and *supZ* that arose from the *gltW*, *gltU* and *gltT* genes, respectively. These suppressors were found in late-arising spontaneous Arg^+ revertants.

There are also theoretical possibilities of creating ochre suppressors from $tRNA^{Tyr}_{AUA}$, $tRNA^{Ser}_{UGA}$ and $tRNA^{Leu}_{UAA}$, but these suppressors have not been identified yet (Śledziewska-Gójska et al., 1992). Figure 1 shows two schematic pictures of tRNA suppressors.

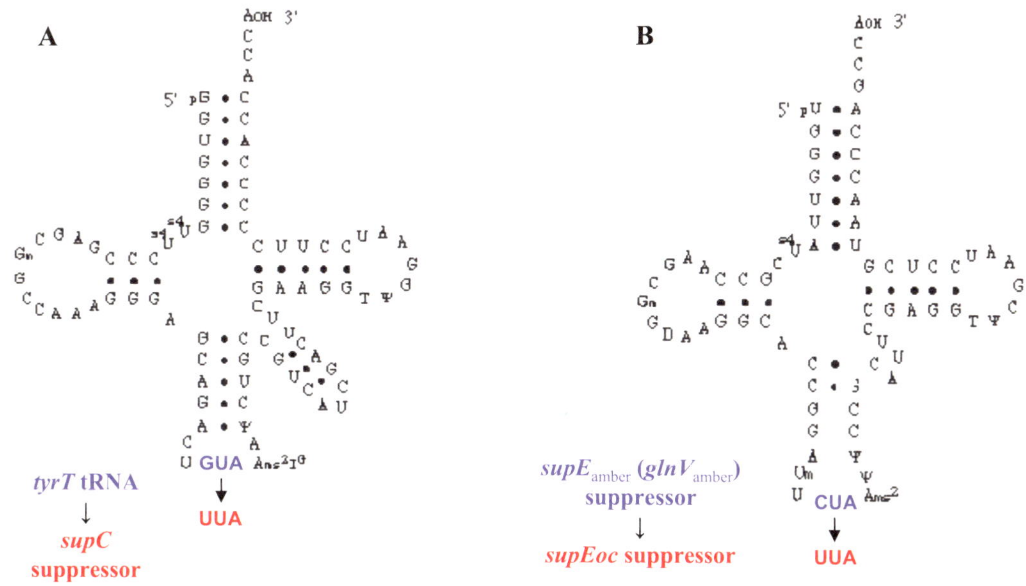

Figure 1. Two examples of tRNA particles. **A.** Tyrosine inserting tRNA into the polypeptide chain (recognizing 5′ UAC 3′ codon in the mRNA) that changes into *supC* suppressor (reading 5′ UAA 3′) in the result of GC→TA transversion in the anticodon part. **B.** Amber suppressor tRNA arisen from *glnV* gene (encoding tRNA for glutamine) inserting glutamine into the polypeptide chain (recognizing 5′ UAG 3′ codon in the mRNA) that changes into *supE*oc suppressor (reading 5′ UAA 3′) in the result of GC→AT transition in the anticodon part (from *Acta Biochimica Polonica* with permission).

Arg^+ revertants can arise spontaneously or as a result of induced mutagenesis. The first step in the analysis of the Arg^+ revertants is the examination of their requirement for histidine and threonine for growth. Arg^+ revertants have been divided into four phenotypic classes: class I: Arg^+ His^- Thr^-, class II: Arg^+ His^+ Thr^-, class III: Arg^+ His^- Thr^- and class IV: Arg^+ His^+ Thr^+. Because, as mentioned above, *supE44* suppressor to some extent suppresses the *thr-1* mutation, Thr^+ phenotype may be wrongly read, thus the revertants of class I and class II may be incorrectly classified as class III and class IV, respectively. For this reason in practice only two groups of the Arg^+ revertants have been usually considered: a sum of classes I and III, and a sum of classes II and IV (Todd et al., 1979; Śledziewska-Gójska et al., 1992).

The sensitivity of Arg^+ revertants to tester T4 phages is the second step in mutational analysis. A set of five T4 phages carrying a defined nonsense mutation includes the following phage mutants: amber B17 and NG19, ochre oc427, ps292 and ps205. Phage multiplication observed as plaque formation on a lawn of tested bacteria indicates that the host bacterium bears a specific suppressor mutation (Kato et al., 1980; Shinoura et al., 1983; Sargentini and Smith, 1989; Śledziewska-Gójska et al., 1992). A schematic procedure for determination of MMS-induced mutagenesis using *argE3* → Arg^+ reversion system is shown in Figure 2.

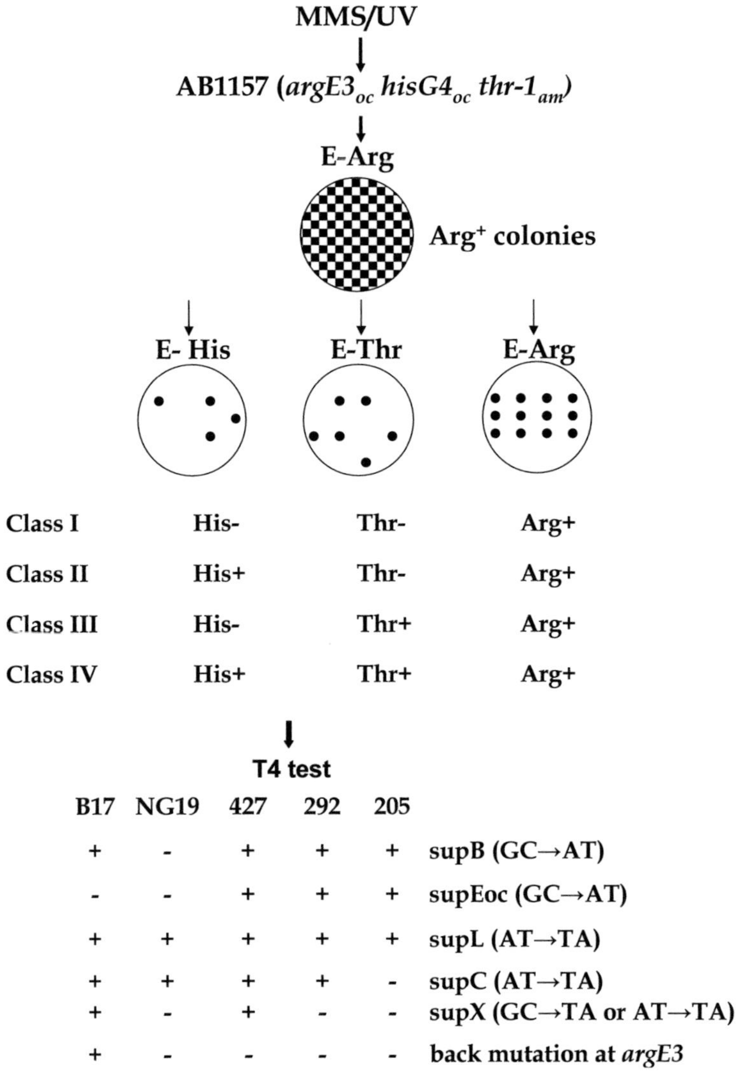

Figure 2. Schematic presentation of MMS-induced mutagenesis assay in *E.coli* AB1157 strain with the use of the *argE3* → Arg^+ reversion system (detailed description in the text).

Arg^+ revertants of class I are the result of back mutations at the *argE3* site, or *supB* or *supE*oc suppressor formation. Arg^+ revertants of class II, III and IV occur as a result of *supL*, *supX* and *supC* suppressor formation, respectively. The details of the above analysis are presented in Table 1 containing species of tRNA producing the indicated suppressor by a single base substitution in the anticodon sites.

Suppressors	**T4 phages**					**Arg^+ revertants**			**Reco-gnized codon**	**tRNA charged amino acid with**	**Gene mutation leading to recognition of UAA nonsense codon**
	amber		***ochre***			***ochre***		***amber***			
	B17	**NG19**	**oc427**	**ps 292**	**ps205**	***argE3***	***hisG4***	***thr-1***			
supB	+	-	+	+	+	+	-	-	CAA	Gln	GC → AT
***supC* (*supO*, *supM*)**	+	+	+	+	-	+	+ t^S	+ t^S	UAC	tyr	GC → TA
***supL* (*supG*, *supN*)**	+	-	+	+	-	+	+	-	AAA	Lys	AT → TA
supX	+	-	+	-	-	+	-	+	?	?	GC → TA or AT → TA
***supE*oc**	-	-	+	+	+	+	-	-	UAG	*supE*am (gln)	GC → AT
AB1157 Arg^- *supE*amber	+	-	-	-	-	-	-	-	----------	----------	----------

Table 1. tRNA suppressors counteracting effects of the *ochre* and *amber* nonsense mutations in T4 phages and Arg^+ revertants of *E. coli* AB1157. + t^S means that suppression works better at 30° C than at 37° C. (from Acta Biochimica Polonica with permission).

Sargentini and Smith (1989) constructed a set of AB1157 derivatives bearing all the mentioned suppressors: SR2151, SR2155, SR2162, SR2161, SR2154, SR2153 carrying, respectively, *supB*, *supL*, *supN*, *supM*, *supC*, *supE*oc suppressors (Sargentini & Smith, 1989; Śledziewska-Gójska et al., 1992). These strains serve to control respective phage T4 mutations.

Identification of created suppressors allows deducing the specificity of mutation without DNA sequencing. However, such analysis does not indicate the type of mutations in the *argE* gene creating a sense codon from the UAA stop codon. In this case DNA sequencing is required. The proportion of suppressor and back mutations in the *argE* gene depends on the type of mutagenic factor and bacterial background.

There is also a possibility to study the level of *hisG4*→His^+ revertants, however, only some of the suppressors may counteract the effect of *hisG4* mutation. The His^+ phenotype can be restored by (i) any point mutation at *hisG4* that changes nonsense UAA codon to any sense nucleotide triplet coding for any amino acid, or (ii) only two suppressor mutations enabling reading UAA nonsense codon, namely *supC* (*supO*, *supM*) and *supL* (*supG*, *supN*) suppressors. In this way many of the arising mutations are lost (Sargentini & Smith, 1989; Śledziewska-Gójska et al., 1992).

2.2. Studies with the use of the *argE3* → Arg^+ reversion based system

In the era of intensive development of techniques of molecular biology and genetics studies, information on reversion to prototrophy of the *argE3* mutation still provide new, interesting and valuable information on the mutagenic specificity of different mutagens and mutator genes as well as on the mechanisms of mutagenesis and DNA repair. The applications of the described genetic system are presented below. This system is particularly useful for detection of GC→TA, GC→AT and AT→TA base substitutions and examination of transcription-coupled DNA repair.

2.2.1. Specificity of mutator genes

The system confirms the mutagenic effects of mutator genes such as *mutT*, *mutY* and *fpg* (Wójcik et al., 1996; Wójcik & Janion, unpublished data; Nowosielska & Grzesiuk, 2000) or *dnaQ* (Nowosielska *et al.* 2004a; 2004b).

MutT, MutY and Fpg (MutM), proteins belonging to the GO system, defend bacteria against the mutagenic action of 8-oxoG in DNA. MutT is a pyrophosphatase that hydrolyses 8-oxo-dGTP and prevents its incorporation into DNA. MutY is a DNA glycosylase excising from DNA adenine mispaired with A, 8-oxoG or G. Among others, Fpg excises from DNA 8-oxoG when it pairs with C (or T). The level of spontaneous transversions: AT→CG in *mutT*$^-$ and GC→TA in *mutY*$^-$ and *fpg*$^-$ mutants is, respectively, about 1000 to 10000 and 10 to 100-fold higher than in the wild type strain (Michaels & Miller, 1992). We have analyzed Arg^+ revertants arising spontaneously in *mutT*$^-$, *mutY*$^-$ and *fpg*$^-$ derivatives of *E. coli* AB1157 strain. In AB1157 *mutT*$^-$ strain a 1000-fold increase in the *argE3*→Arg^+ reversions was observed. All those reversions arose due to back mutations at the *argE3* site (probably as a result of AT→CG transversions). In *mutY*$^-$ and *fpg*$^-$ mutants all of the spontaneous *argE3*→Arg^+ reversions were due to GC→TA transversions by *supC* suppressor formation (Wójcik et al., 1996; Wójcik & Janion, unpublished data; Nowosielska & Grzesiuk, 2000).

DNA polymerase III, the main replicative polymerase in *E. coli*, comprises a *dnaQ*-encoded epsilon subunit responsible for proofreading activity. Mutants defective in this subunit chronically express the SOS response and exhibit a mutator phenotype (Echols *et al.*, 1983). Using the *argE3*→Arg^+ reversion, the effects of deletions in genes *polB* and *umuDC*, encoding, respectively, the SOS-induced DNA polymerases Pol II and PolV, on the frequency and specificity of spontaneous mutations in the *dnaQ* background were studied. It was clearly shown that deletion of *umuDC* genes significantly decreased the level of spontaneous mutations in *dnaQ* strains (Nowosielska *et al.*, 2004a). The Arg^+ revertants in *mutD5* (allele of *dnaQ*) mutant occurred only as a result of tRNA suppressor formation, whereas those in *mutD5 polB* (Pol II deficient) strains arose at 81% by back mutation at the *argE3* ochre site (Nowosielska et al., 2004b).

The SOS response is a bacterial defence system enabling the survival of cells whose DNA has been damaged and replication arrested. The SOS system increases expression of over 40 genes involved in DNA repair, replication, and mutagenesis. The expression of genes of the SOS regulon is tightly regulated. The *umuD* and *umuC* genes encoding the Y-family DNA polymerase V (PolV) are expressed among the last ones. In the process of translesion synthesis (TLS), this low fidelity polymerase, composed of UmuC and two particles of shortened UmuD form, UmuD' (UmuD'$_2$C) bypasses lesions inserting a patch of several nucleotides and allowing resumption of DNA replication by PolIII, the main replicative polymerase in *E.coli* (Janion, 2008). One of the symptoms of the SOS-induction is filamentous growth of bacteria due to expression of the *sulA* gene. The *sulA* gene codes for a protein that blocks cell division by inhibiting assembling of the FtsZ protein into a ring structure leading to filament formation (Bi & Lutkenhaus, 1991). Inhibition of cell division allowed DNA repair processes to be finished before next round of division (Janion et al., 2002; Janion 2008).

It has been shown that BW535 (*nth-1, Δxth, nfo-1*), a derivative of AB1157 deficient in base excision repair (BER), chronically induces the SOS system. The *xth, nfo* and *nth* genes encode, respectively, exonuclease III (exo III), endonuclease IV (endo IV) and endonuclease III (endo) III. Exo III and endo IV account for 85% and 5% of the cell's endonuclease activity, respectively (Kow & Wallace, 1985; Cunnigham et al., 1986). Endo III is a DNA glycosylase with a broad substrate specificity which mainly excises oxidized pyrimidines, and also possesses an AP-lyase activity cleaving the sugar-phosphate backbone and generating single- and double-strand breaks in DNA (Dizdaroglu et al., 2000). The triple *nth xth nfo* mutant can not repair AP sites. The chronic induction of the SOS system is due to accumulation of AP sites that left unrepaired in DNA. A mutator phenotype measured by an increased level of spontaneous *umuDC*-dependent *argE3*→Arg$^+$ reversions was one of the symptoms of the chronic induction of the SOS system in the *nth xth nfo* mutant (Janion et al., 2003).

2.2.2. *Specificity of mutagens*

The mutagenic specificity of N^4-hydroxycytidine (oh^{4}Cyd), hydroxylamine (HA), *N*-methyl-*N*'-nitro-*N*-nitrosoguanidine (MNNG) (Śledziewska-Gójska *et al.*, 1992), ethylmethane sulfonate (EMS) (Grzesiuk & Janion, 1993), methylmethane sulfonate (MMS) (Śledziewska-Gójska & Janion, 1989; Grzesiuk & Janion, 1994) and UV light (Wójcik & Janion, 1997; 1999; Fabisiewicz & Janion, 1998) has also been confirmed with the help of the *argE3*→Arg$^+$ reversion.

Using this system it has been established that HA, a cytosine modifying agen, apart from well known GC→AT transitions may also cause a significant number of GC (or AT)→TA transversions. As much as 30% of the HA-induced Arg$^+$ revertants were formed by GC (or AT)→TA transversions (Śledziewska-Gójska et al., 1992).

Studies on *E. coli* AB1157 strain and its derivatives revealed that biological effects (survival, mutation induction and mutation specificity) of halogen light irradiation were very similar to those observed after UVC irradiation. The halogen light-induced mutations were

GC→AT transitions (*supB* or *supE* ochre suppressor formation) and back mutations at *argE3* sites resulting from T-C 6-4 photoproducts or T<>T thymine dimers, respectively. The latter damage was observed only in *uvrA* mutants defective in nucleotide excision repair (NER), constituting less than 5% of the total number of Arg$^+$ revertants (Wójcik & Janion, 1997). These results confirmed previous data showing harmful effects caused by halogen light, such as DNA damage, mutations, genotoxicity and skin cancers in mice due to emission of a broad spectrum of UV light, particularly UVC (De Flora et al., 1990; D'Agostini et al., 1993; D'Agostini & De Flora, 1994).

Analysis of Arg$^+$ revertants supplied new data on the mechanisms of mutagenesis and processes of DNA repair. The mutagenic properties of DNA damaging agents and the spectra of the induced mutations depend on the bacterial background, *i.e.*, the presence of mutations in genes encoding proteins involved in DNA repair systems.

It is known that EMS, a SN2-type alkylating agent, is an *umuDC*-independent mutagen and induces GC→AT transitions due to formation of *O*^{6}-ethylguanine in DNA. It has been shown that in the AB1157 strain, EMS-induced Arg$^+$ revertants arise by *supB* and *supE* ochre suppressor formation. However, in *mutS*$^-$, a mismatch repair-deficient strain, the specificity of the EMS-induced *argE3*→Arg$^+$ reversions was changed and formation of *supL* suppressor by AT→TA transversions was mainly observed. Moreover, these mutations were *umuDC*-dependent. It was suggested that the change in mutation specificity was due to 3meA lesions or creation of apurinic sites. These results also point to different processes of DNA repair in *mutS*$^+$ and *mutS*$^-$ strains (Grzesiuk & Janion, 1993).

MMS, another SN2-type alkylating agent, predominantly methylates nitrogen atoms in purines. This methylating agent creates the following adducts in double stranded DNA: 7-methylguanine (7meG), 3-methyladenine (3meA), 1-methyladenine (1meA), 7-methyladenine (7meA), 3-methylguanine (3meG), O^6-methylguanine (O^{6}meG), 3-methylcytosine (3meC), and methylphosphotriesters. In ssDNA, MMS induces the same lesions but in different proportions. In ssDNA, the participation of 1meA and 3meC increases significantly since the ring nitrogens at these positions are not protected by the complementary DNA strand (Wyatt & Pittman, 2006; Sedgwick *et al.*, 2007). Analysis of Arg$^+$ revertants in *E. coli* AB1157 strain without any additional mutations revealed that 70-80% of those revertants arose by AT→TA transversions in a *umuDC*-dependent process, whereas the rest occurred in a *umuDC*-independent manner either by GC→AT transitions (formation of *supB* or *supE* ochre suppressors) or by back mutations at *argE3* site. The latter ones were detected in less than 5% of the Arg$^+$ revertants. AT→TA transversions are thought to be the result of 3meA, abasic sites and 1meA, whereas GC→AT transitions come from *O*^{6}-meG and 3meC residues in DNA and from depurination of 7meG (Grzesiuk & Janion, 1994; Nieminuszczy *et al.*, 2006a; 2009; Wrzesinski et al., 2010).

The spectrum of the MMS-induced *argE3*→Arg$^+$ reversions changes in various strains deficient in DNA repair systems. In the *mutS*$^-$ mutant Arg$^+$ revertants arose mainly by GC→AT transitions (*supB* and *supE* ochre suppressor formation) or back mutations at *argE3* site. The latter group constituted a few percent of the total number of the Arg$^+$

revertants (Grzesiuk & Janion, 1998). In the *dnaQ49* derivative of the AB1157 strain about half of the MMS-induced Arg$^+$ revertants occurred by AT→TA transversions (*supL* suppressor formation). In a double *dnaQ*$^-$ *umuDC*$^-$ mutant about 90% of the revertants possessed *supB* or *supE* ochre suppressors due to GC→AT transitions (Grzesiuk and Janion, 1996).

2.2.3. Detection of mutations resulting from lesions in ssDNA

Examination of MMS-induced mutagenesis in AB1157*alkB*$^-$ derivatives indicates that the *argE3*→Arg$^+$ reversion system also enables detection of mutations arising from lesions in ssDNA (Nieminuszczy et al., 2006a; 2009; Sikora et al., 2010; Wrzesiński et al., 2010). AlkB is an α-ketoglutarate-, O_2- and Fe(II)-dependent dioxygenase that oxidatively demethylates 1meA and 3meC in ds- and ssDNA and in RNA. However, ssDNA is repaired much more effectively than dsDNA (Trewick et al., 2002; Falnes et al., 2002). It has been shown that in *alkB*$^-$ mutants the level of MMS-induced mutagenesis depends on the test system used, and is several orders of magnitude higher when measured in the *argE3*→Arg$^+$ reversion test system in *E. coli* AB1157 in comparison to *lacZ*→Lac$^+$ reversion studied in CC101-CC106 strains (Nieminuszczy et al., 2006a; 2006b; 2009; Kataoka et al., 1983; Dinglay et al., 2000). The CC101-CC106 tester strains are described in more detail in Chapter 2.4. Briefly, the *lacZ*→Lac$^+$ reversion occurs only by a back mutation at one point in the structural gene encoding the β-galactosidase that if not expressed is primarily in dsDNA form.

The *argE3*→Arg$^+$ reversion-based system has showed that in AB1157 *alkB*$^-$ strain 95-98% of the induced mutations are *umuDC* (Pol V)-dependent AT→TA transversions (*supL* suppressor formation) and GC→AT transitions (*supB* or *supE* ochre suppressor formation). Back mutations in the *argE3* site constitute only about 2-5% of all types of Arg$^+$ revertants (Nieminuszczy et al., 2006a). Genes encoding tRNA are heavily transcribed and exist mostly as ssDNA in cells. It facilitates methylation of A/C to 1meA/3meC. That is why we assume that in AB1157 *alkB*$^-$ strain the targets undergoing mutations leading do Arg$^+$ revertants are predominantly located in ssDNA. Reversion to Arg$^+$ occurs mostly by formation of a variety of *sup*tRNA ochre suppressors. The number of targets undergoing mutations and differences in the reactivity of MMS to form 1meA/3meC lesions in ss- *vs.* dsDNA are the main reasons of the great discrepancy in the frequencies of MMS-induced *argE3*→Arg$^+$ and *lacZ*→Lac$^+$ revertants observed (Nieminuszczy et al., 2009).

An extremely high level of the MMS-induced *argE3*→Arg$^+$ reversions has been observed in *E. coli* AB1157 *nfo xth alkB* strain defected in the repair of AP sites caused by invalid base excision repair system (BER) and deficiency in AlkB dioxygenase. This phenomenon can be explained by local relaxation of dsDNA structure due to the presence of AP sites in AB1157 *nfo*$^-$ *xth*$^-$ strain. We assume that under these conditions more single stranded DNA appears in the bacterial chromosome that facilitating methylation by MMS and resulting in "error catastrophy" in the triple AB1157 *nfo xth alkB* mutant (Sikora et al., 2010). Analysis of MMS-induced Arg$^+$ revertants in *alkB*$^-$ and *alkB*$^-$BER$^-$ strains clearly points to a mutagenic activity of 1meA and 3meC (Nieminuszczy et al., 2006a; Sikora et al., 2010).

2.2.4. Determination of transcription-coupled DNA repair

The *argE3*→Arg$^+$ reversion system in *E. coli* AB1157 also enables studies on preferential removal of lesions from the transcribed DNA strand. This type of DNA repair, called transcription-coupled repair (TCR), requires Mfd protein that removes transcription elongation complexes stalled at non-coding lesions in DNA and recruits to these sites proteins involved in nucleotide excision repair (NER). TCR occurs under conditions of temporary inhibition of protein synthesis and results in a decrease in the frequency of induced mutations (Selby & Sancar, 1993; Savery, 2007). This phenomenon is called mutation frequency decline (MFD) and was discovered for UV-irradiated bacteria by Evelin Witkin (for review see Witkin, 1994). The MFD phenomenon has been studied by the Janion and Grzesiuk's group on UV (or halogen light)- and MMS-induced Arg$^+$ revertants in the AB1157 strain transiently incubated under non-growth conditions (amino acid starvation) after treatment with a mutagen (Grzesiuk & Janion, 1994; 1996; 1998; Wójcik & Janion, 1997; Fabisiewicz & Janion, 1998; Wrzesiński et al., 2010).

Table 2 shows all the mutagenic targets for UV- and MMS-induced DNA damage. Potential targets for UV-modifications (T-C and T-T sequences for creation of 6-4 photoproducts and pyrimidine dimers, respectively) are underlined. Potential targets (single bases) for MMS-induced modifications are shadowed. UV- or halogen light-induced Arg$^+$ revertants occur mainly as a result of a GC→AT transition forming the *supB* and *supE* ochre suppressors, respectively, at the transcribed DNA strand of the *glnU* and the coding DNA strand of the *glnV* amber (*supE44* amber) gene. In *mfd*$^-$ strains the formation of *supB* predominated over *supE* ochre suppressors and their number, in contrast to the *mfd*$^+$ strain, did not decrease during amino acid starvation. The MFD effect observed in *mfd*$^+$ strains is a reflection of repair of premutagenic lesions in the transcribed strand of the *glnU* gene leading to *supB* suppressor formation (Wójcik & Janion, 1997; Fabisiewicz & Janion, 1998).

Studies on TCR involvement in the repair of MMS-induced lesions have included (i) an analysis of Arg$^+$ revertants, and (ii) examination of plasmid DNA isolated from MMS-treated and transiently starved bacteria for their sensitivity to the Fpg and Nth endonucleases. The decrease in the level of MMS-induced mutations during transient starvation was accompanied by repair of abasic sites in plasmid DNA. As it is shown in Table 2, potential targets for MMS damage are located on both the transcribed and coding DNA strands of *glnU*, *glnV* amber and *argE* genes and only on the transcribed strand of *lys*-tRNA genes. Lesions resulting from methylation of the transcribed DNA strand are subject to MFD repair. Previous studies on the MFD phenomenon after MMS treatment of the AB1157 strain and its derivatives focused on the preferential repair of transcribed-strand lesions of genes coding for *lys*-tRNA; this repair was manifested by a decrease in the number of *supL* suppressors (Grzesiuk & Janion, 1994; 1998). Recent studies revealed a significantly slower and completely absent MFD effect in, respectively, AB1157*mfd* and double *alkB mfd* mutants. It has been assumed that the former effect is the result of action of other DNA repair systems and the latter is a reflection of an accumulation of damage to DNA and induction of SOS response. These results again have confirmed the strong mutagenic effects of 1meA/3meC lesions (Wrzesiński *et al.*, 2010).

Interestingly, in a *dnaQ* mutant no TCR was observed indicating that in this mutant the processes of DNA repair are different, probably due to chronic induction of SOS response and the presence of Pol V and Pol IV DNA repair polymerases induced within SOS regulon (Grzesiuk & Janion, 1996).

DNA		→		tRNA
glnU gene ↓ *supB* suppressor	5′----TT ***TTG***AT----3′ 3′--- AAA***AC***TA----5′ ↓ 5′----TT ***TTA***AT----3′ 3′----AAA***AT***TA----5′	→	5'--UUG--3' ↓ 5'--UUA--3'	gln-tRNA$_{CAA}$ - tRNA anticodon for glutamine reading 5'CAA3' codon in mRNA ↓ gln-tRNA$_{UAA}$ - tRNA anticodon reading nonsense *ochre* triplet 5'UAA3' in mRNA
glnV$_{am}$ (*supE44*$_{am}$) suppressor ↓ *supE*$_{OC}$ suppressor	5′----T***CTA***----3′ 3′----A***GAT***----5′ ↓ 5′----T***TTA***----3′ 3′----A***AAT***----5′	→	5'--CUA--3' ↓ 5'--UAA--3'	gln-tRNA$_{UAG}$ - tRNA anticodon reading nonsense *amber* triplet 5'UAG3' in mRNA ↓ gln-tRNA$_{UAA}$ - tRNA anticodon reading nonsense *ochre* triplet 5'UAA3' in mRNA
lys-tRNA genes ↓ *supL* suppressor	5′---***T TT***-----3′ 3′--***AAA***----5′ ↓ 5′---***TTA***----3′ 3′--- ***AAT***----5′	→	5'--UUU--3' ↓ 5'--UAA--3'	lys-tRNA$_{AAA}$ - tRNA anticodon for lysine reading 5'AAA3' codon in mRNA ↓ lys-tRNA$_{UAA}$ - tRNA anticodon reading nonsense *ochre* triplet 5'UAA3' in mRNA
argE3 mutation in *argE* gene ↓ mutations leading to any sense nucleotide triplet or UAG nonsense codon recognized by *supE44* amber suppressor	5′----TTTAAAT----3′ 3′---AAATTTA---5′	→		No changes in tRNA encoding genes

Table 2. Potential mutagenic targets for UV and MMS modification and mechanisms of mutation creation in *glnU*, *glnV* amber, *lys*-tRNA and *argE* genes, leading to Arg$^+$ phenotype in *E. coli* K-12 AB1157. Nucleotide triplets corresponding to tRNA anticodon in *glnU*, *glnV* amber and *lys*-tRNA genes are in italics. Underlined sequences and shadowed bases show potential sites of photoproducts (6-4 photoproducts and thymine dimers) formation and targets for methylation, respectively.

2.3. The *trpE65* → Trp$^+$ and *tyrA14* → Tyr$^+$ reversion systems in *E. coli* B/r derivatives

It is thought that *E. coli* B is the clonal descendant of a *Bacillus coli* strain used by Felix d'Herelle from the Pasteur Institute in Paris, in his studies performed on bacteriophages almost 100 years ago. *B. coli* was isolated from human feces as a normal commensal of the human gut. *B. coli* was renamed to *E.coli* strain B and published by Delbrück and Luria in 1942 (Delbrück & Luria in 1942). The history of *E.coli* B was excellently presented by Daegelen and co-workers (Daegelen et al., 2009). *E. coli* B/r (B resistant to radiation) is one of the mutants obtained from *E. coli* B after irradiation with UV light by Evelyn Witkin in 1942 (Witkin, 1946). *E. coli* B was found to be very sensitive to UV irrradiation due to La protease (Lon protein, product of the *lon* gene) deficiency. *E. coli* B/r strain owns its UV-resistance to *sulA* mutation (Studier et al., 2009). SulA protein is synthesized in bacterial cell during the SOS response induction and is a substrate for the Lon protease (Goldberg et al., 1994).

Reversions of *trpE65* to Trp^+ phenotype and *tyrA14* to Tyr^+ in *E. coli* B/r WP2 (Ohta *et al.*, 2002) and WU3610 derivatives (Bockrath *et al.*, 1987), respectively, are analogous to *E.coli* K12 AB1157 mutation detection systems. Both *trpE65* and *tyrA14* are ochre mutations in genes coding for enzymes involved in tryptophane and tyrosine biosynthesis, respectively. The Trp^+ or Tyr^+ phenotype may be recovered by (i) any point mutation at *trpE65* or *tyrA14* leading to the formation of any sense nucleotide triplet, and (ii) ochre suppressor mutations. In the WP2 (*trpE65*) system the examined suppressors are *supB, supC, supG* and *supM*, formed in the genes coding for tRNA: *glnU, tyrT, lysT* and *tyrU*, respectively (Ohta *et al.*, 2002). In the WU3610 (*tyrA14*) strain *de novo* ochre suppressor mutations in glutamine tRNA are studied. The WU3610-11 derivative bears an amber suppressor created from another glutamine tRNA gene that can be converted to an ochre suppressor (Bockrath and Palmer, 1977; Bockrath *et al.*, 1987). Both systems have been used in MFD studies (Bridges *et al.*, 1967; George & Witkin, 1974; Bockrath and Palmer, 1977; Bockrath *et al.*, 1987).

Besides *E. coli* WP2 strain, its derivatives are widely used: WP2 carrying pKM101 plasmid from *S.typhimurium*, WP2 *uvrA* mutant and WP2 *uvrA* bearing pKM101. *E. coli* WP2 and its derivatives are recommended to be used in conjunction with Ames *S.typhimurium* tester strains to screen various compounds for mutagenic activity (Mortelmans & Riccio, 2000), (see chapter 2.8).

2.4. Lac^+ reversion system for determination of base substitutions and frameshift mutations

A commonly used and convenient *E. coli* K-12 *lacZ*→Lac^+ reversion system allows rapid detection of specificity of mutation. The β-galactosidase encoding *lacZ* gene is a part of the lactose operon. Mutants in *lacZ* gene are unable to grow on a medium containing lactose as the sole carbon source. A set of 11 mutants (*E.coli* K12 CC101-111 strains) with a *lacZ* deletion in the chromosome and F′ episome with cloned *lacZ* gene bearing defined mutations (six base substitutions in CC101-106 strains, and five frame shift mutations in CC107-111 strains) have been constructed (Coupples and Miller, 1989; Coupples et al., 1990)

Glutamine at 461 position is essential for β-galactosidase activity. In CC101-CC106 strains coding position 461 was changed. Reversion to the Lac^+ phenotype is due to a specific base substitution at 461 position restoring the glutamic acid codon. CC107-CC111 carry mutations in the *lacZ* gene that revert to Lac^+ *via* specific frameshifts. The altered sequences contain monotonous runs of each of the four bases or a run of –G-C- sequences on one strand. Addition or loss of a single base pair or loss of –G-C- sequence lead to reversion of the *lacZ* mutation. In the case of CC101-CC111 strains the marker is episomal, in contrast to e.g. the AB1157 strain where the marker is situated on the chromosome. Figure 3 presents the idea of the CC101-CC106 and CC107-CC111 strains construction. The *lacZ*→Lac^+ reversion system in *E.coli* K12 CC101-111 strains is very handy and used all over the world for studying specificity of mutations in genes under investigation.

A.

Glu-461

TGG TCG CTG GGG AAT GAA TCA GGC CAC GCC GCT AAT CAC GAC GCG CTG TAT CGC TGG ATC AAA TCT CTC GAT CCT TCC CGC CCG GTG CAG TAT GAA GGC GGC GGA GCC GAC ACC ACG GCC ACC GAT ATT ATT TGC CCG ATG TAC GCG CGC GTG GAT GAA GAC CAG CCC TTC CCG GCT GTG CCG AAA TGG TCC ATC AAA AAA TGG CTT

B.

	Codon at position 461 in *lacZ* gene	Mutation leading to Lac+ phenotype
wild type Lac+	-GAG-	-
CC101	-TAG-	AT → CG
CC102	-GGG-	GC → AT
CC103	-CAG-	GC → CG
CC104	-GCG-	GC → TA
CC105	-GTG-	AT → TA
CC106	-AAG-	AT → GC

C.

	The wild type and altered sequence	Mutation leading to Lac+ phenotype
wild type Lac+	GGC GGC GGA GCC	
CC107	GGC GGG GGG CC	+ 1 G
wild type Lac+	GGC GGC GGA GCC	
CC108	GGG GGG CGG AGC C	- 1 G
wild type Lac+	TAC GCG CGC GTG	
CC109	TAC GCG CGC GCG TG	- 2 CG
wild type Lac+	ATC AAA AAA TGG	
CC110	ATA AAA AAT GG	+ 1 A
wild type Lac+	ATC AAA AAA TGG	
CC111	ATC AAA AAA ATG G	- 1 A

Figure 3. The idea of the *E. coli* CC101-CC106 and CC107-CC111 strains construction. **A.** A fragment of the *lacZ* sequence - the underlined sequences have been altered to create the tester strains: Glu-461 to yield strains CC101-CC106, the remaining sequences rich in –G-, -C-G- and A, respectively, to yield strains CC107-CC111. **B.** Altered codon at position 461 in *lacZ* gene in six different strains (CC101-CC106) and base substitutions recovering the Lac+ phenotype. **C.** Altered sequences in *lacZ* gene in five strains (CC107-CC111) and frameshifts recovering the Lac+ phenotype.

Fijałkowska and Shaaper with colleagues constructed a series of *lacZ* strains allowing studies on replication fidelity based on analysis of frequency of Lac+ revertans. The entire *lacIZYA* operon from F'*pro lac* plasmid of Coupples and Miller's strains containing specific *lacZ* mutation has been inserted into the chromosome of the *E. coli* MC4100 (*lac*−) in two possible orientations with regard to the chromosomal replication origin *oriC*. This system enables investigation of frequencies of base pair substitutions and frame-shift mutations. It has been used to show that during chromosomal DNA replication in *E. coli* two DNA strands, the leading and the lagging, are replicated with different accuracy in *w.t.* as well as in various mutants in genes involved in replication or DNA repair (Fijałkowska et al., 1998; Maliszewska-Tkaczyk et al., 2000; Gaweł et al., 2002).

2.5. Forward mutation system with the use of *E. coli lacI* strain

A forward mutational system, in contrary to reversion systems, monitors the mutation of a wild type gene. The *E.coli lacI* nonsense system described by Miller and Coulondre (Coulondre & Miller, 1977; Miller, 1983) is a forward system playing an important role in the examination of specificity of numerous mutagens. The *lacI* gene encodes the repressor of *lac* operon required to metabolize lactose. The base of *lacI* system is the analysis of nonsense mutations in the *lacI* gene (present on an F′ episome). The system also involves several techniques to identify each nonsense mutation. There are over 80 sites within *lacI* gene where a nonsense mutation can arise by a single base change. Nonsense mutations constitute 20-30% of all mutations induced by many mutagens. Since the *lacI* gene encodes the repressor of the *lac* operon, *E.coli lacI*$^-$ cells express the operon constitutively. In this way *lacI* mutants can be selected on the plates containing phenyl-β-galactosidase (a lactose analog) as the only source of carbon. These mutants can metabolize the analog but cannot induce the operon. Further mutant analysis involves ability to be suppressed by various tRNA suppressors and subsequent genetic analyses.

The distribution of *lacI* mutations can be arranged according to base substitution generated by each nonsense mutation, creating a map of mutational hot and cold spots, for places where the number of mutations exceeds or is smaller, respectively, in comparison to other sites. Created map of mutational spectra is characteristic for different mutagens. The *lacI* nonsense system shows limitations of detecting only base substitution mutations and not detecting AT→GC transitions. Nevertheless, it became possible to determine the nature of the *lacI* mutations directly by DNA sequencing.

2.6. The *trpA* → Trp$^+$ reversion system in *E.coli* K12

An important approach to determine mutagen specificity based on reversion of an auxotrophic *trpA* mutation to the Trp$^+$ phenotype in *E. coli* K12 was developed by Yanofsky and co-workers (Berger et al., 1968). The *trpA* gene is a part of the *trp* operon and codes for the tryptophan synthetase α chain in *E.coli*. There is a set of the *trpA* alleles that enable to monitor all possible base substitutions (*trp88, trp46, trp23, trp3, trp223, trp58, trp78, trp11, trp446*) and frame shifts (*trpE9777, trpA21, trpA540, trpA9813* alleles) and allow studying mutagen specificity. The Trp$^+$ revertants are divided into classes based upon colony size and two physiological tests: 5-methyl tryptophan (5-MT) sensitivity and indole glycerol phosphate (IGP) accumulation. Moreover, full Trp$^+$ revertants (FR) and partial Trp$^+$ revertants (PR) are distinguished. The PR group is divided into 3 more classes: PRI, PRII, PRIII. To enhance the frequiences of spontaneous and induced Trp$^+$ revertants pKM101 plasmid from *S. typhimurium* (described in chapter 2.8) was introduced to *E. coli trp*⁻ strains by conjugation (Fowler et al., 1979; McGinty & Fowler, 1982; Persing et al., 1981; Fowler & McGinty, 1981). Table 3 shows characterization of UV-induced Trp^+ revertants.

***trpA* allele**	**Revertant class**	**5-MT sensitivity**	**IGP accumulation**	**Inferred base-pair substitutions**
3	I FR PR	R S	- +	AT → TA -
11	I FR PR	R S	- +	GC → CG - AT → TA AT→ GC
23	I FR II PR III PR IVPR	R S S S	- + + +	AT → CG GC → CG CG → AT -
46	I FR II PR III PR	R S S	- + +	AT → GC AT → CG AT → TA -
58	I FR II PR III PR	R S S	- + +	AT → GC AT → CG GC → AT
78	I FR PR	R S	- +	AT → CG -
88	I FR PR	R S	- +	AT → CG -
223	I FR PR	R S	- +	AT → GC AT → CG -

Table 3. The characterization of UV-induced *trpA* base pair substitutions based on Fowler et al., 1981

2.7. Adaptive mutations

Adaptive mutations (also called "directed", "stationary phase" or "starvation associated") are a special kind of spontaneous mutations that occur in non-dividing or slowly-growing stationary-phase cells. Mutations of this type are detectable after exposure to a non-lethal selection and allow growth under these conditions.

As a tool for studying stationary phase mutations, *lacI33* →Lac$^+$ reversion has been used. The *lacI33* marker (+1 G frame shift) is carried on the F'sex plasmid in the FC40 *E.coli* K12 strain and its derivatives. The reversion to Lac$^+$ phenotype is due to a -1G frameshift mutation (Foster and Trimarchi, 1995; Rosenberg et al., 1994). The *lacI33* →Lac$^+$ reversion, if plasmid born, depends on *recABC* encoded proteins (Foster and Trimarchi, 1994). When *lacI33* mutation is localized on the chromosome it reverts adaptively at a much lower rate and the event is *recA* independent.

The systems searching for stationary phase mutations operating on bacterial chromosomal loci use reversion to prototrophy of auxotrophic *E.coli* strains. Prototrophic revertants are able to grow on minimal plates lacking required compounds. These systems often use reversions in genes such as *trpE* (Bridges, 1993) or *tyrA* (Bridges, 1996) in *E.coli* B/r (see chapter 2.3) and *trp* operon in *E.coli* K12 (Hall, 1990).

The *argE3*→Arg$^+$ reversion system in *E. coli* K12 AB1157 strain can be also used for selection bacteria mutated adaptively on minimal plates lacking arginine. Colonies of stationary phase mutations appear on these plates after four and more days of incubation at 37°C. Further phenotypic analysis and susceptibility to a set of amber and ochre T4 phages allowed the identification of stationary phase mutations in AB1157 *mutY*$^-$ strain defective in the ability to remove adenine from A-8-oxo-G and A-G mispairing (Nowosielska and Grzesiuk, 2000), and in AB1157*dnaQ*$^-$ mutated in proofreading subunit of *E.coli*, main replicative DNA polymerase PolIII (Nowosielska et al., 2004a; 2004b). It has been shown that in the *dnaQ*$^-$ strain two repair polymerases, PolIV and PolV, influence the frequency and specificity of starvation-associated mutations.

In bacteria there is no single mechanism for the generation of stationary-phase mutations. Under starvation conditions mutations can arise as a result of oxidative and other DNA damage, errors occurring during DNA replication, defects or inefficiency of DNA repair systems but also DNA repair synthesis by itself may be a source of mutagenesis under conditions restricted for growth. *Pseudomonas* possesses a different set of specialized DNA polymerases compared with enetrobacteria, also its DNA repair systems involved in stationary-phase mutagenesis differ. To study stationary-phase mutagenesis in *Pseudomonas* the *pheA*→Phe$^+$ reversion system described in chapter 2.9. is used the most.

2.8. The Ames test with the use of *Salmonella* strains

The *Salmonella* mutagenicity assay was introduced by Bruce Ames and co-workers in the early 1970s, later modified (Ames *et al.*, 1975) and constantly improved. The test involves reversion of histidine auxotrophs of *Salmonella enterica* serovar Typhimurium mutation to prototrophy by base substitutions in the *hisG46* allele or by frame-shift in the *hisD3052* allele. In addition, the tester strains carry: (i) additional mutations, such as *rfa*, increasing the permeability of the bacterial cell wall and enabling better penetration of mutagenic agents to the cell, or *uvrB*, disturbing DNA repair; (ii) plasmid pKM101 – a mutagenesis-enhancing plasmid bearing *mucA* and *mucB* genes that code for proteins corresponding to *E.coli* UmuC and UmuD encoding SOS-induced, repair polymerase V (PolV), responsible for translesion synthesis (TLS) (Mortelmans, 2006).

Ames *Salmonella* test has been developed to screen chemicals for their potential mutagenicity and genotoxicity. It is used routinely as a screening assay to predict animal carcinogens. Since many compounds show genotoxicity only after enzymatic conversion to active form, a method was discovered to imitate mammalian metabolism in a bacterial system by adding an extract of rat liver. Rats are first injected with a polychlorinated biphenyl (PCB) mixture, Aroclor 1254, inducing expression of enzymes involved in activation of chemicals. The livers of these rats are homogenized, centrifuged, and supernatans (microsomal fraction), termed S9 mix, are collected. Test method involves histidine auxotroph *S. typhimurium* TA98 for testing frameshift mutations and TA100 for testing base-pair substitutions (several other strains have also been constructed). Cultures of these strains are mixed with a chemical tested, incubated in the presence (or absence) of S9 mix, and plated on solid minimal medium lacking histidine. A two-day incubation allows His^+ revertants to form colonies on minimal plates. The frequency of His^+ revertants indicates mutagenic potency of tested chemical (Figure 4).

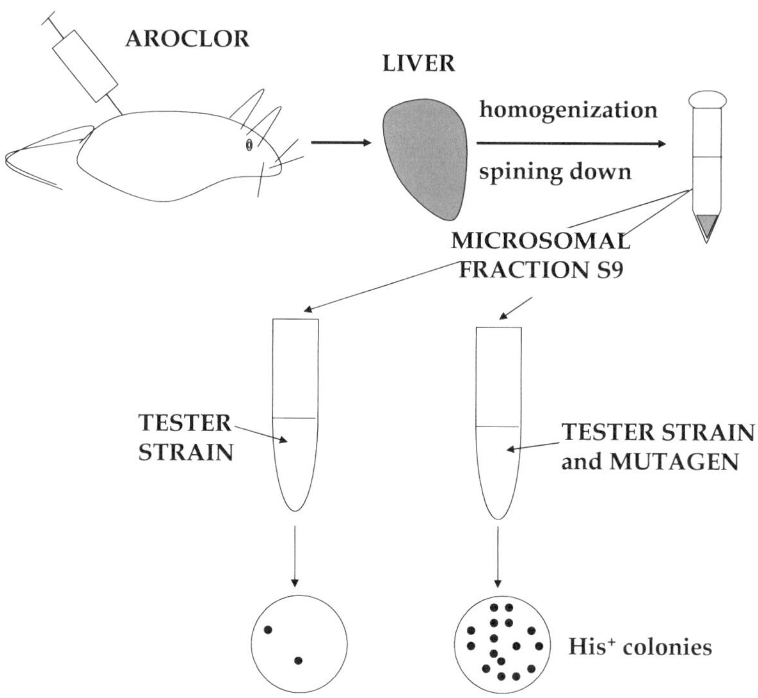

Figure 4. An idea of Ames test – description in the text.

The following *S. typhimurium* strains are recommended for general mutagenesis: TA97 (*hisD6610, rfa, ΔuvrB,* +R), TA98 (*hisD3052, rfa, ΔuvrB,* +R), TA100 (*hisG46, rfa,* del*uvrB,* +R), and TA102. The last strain, except the plasmid pKM101bearing R-factor, contains multicopy plasmid pAQ1 carrying the *hisG428* mutation and a tetracycline resistance gene. Strains containing R-factor are much better tester strains for a number of mutagenes that weakly or not at all revert strains devoid of pKM101plasmid. In *E.coli* and *S. typhimurium* the presence of pKM101 increases mutagenesis by inducing SOS response leading to error-prone DNA repair.

The *hisG* gene encodes for the first enzyme of histidine biosynthesis. Mutation in this gene, *hisG46,* present in TA1535 strain and its R-factor⁺ derivative, TA100, substitutes $\frac{-GGG-}{-CCC-}$ (proline) for $\frac{-GAG-}{-CTC-}$ (leucine). Both strains serve to detect mutagens that cause base-pair substitutions. Mutation in *hisD* gene, *hisD3052,* present in TA1538 strain and its R-factor⁺ derivative, TA98 detect various frameshift mutagens. Frameshift mutagens can stabilize the shifted pairing occurring in repetitive sequences or hot spots resulting in frameshift mutation which restores the correct for histidine synthesis reading frame. The *hisD3052* mutation has 8 repetitive –GC- residues next to -1 frameshift in the *hisD* gene. This mutation is reverted by such mutagens as 2-nitrosofluorene and daunomycin.

The Ames II assay is a liquid microtiter modification of the Ames test. It involves new set of *S. typhimurium* strains TA7001-6 (Gee *et al.*, 1994) and also a mixture of these strains called TAMix (Fluckiger-Isler *et al.*, 2004). Table 4 contains genotypes of mentioned above strains, and mutation specificity of indicated mutagens.

Strain	Genotypes	Mutagen	Mutation detected
TA98	*hisD3052 Δara9 Δchl008 (bio chl uvrb gal) rfa1004*/pKM101		frameshifts
TA7001	*hisG1775 Δara9 Δchl004 (bio chlD uvrb chlA) galE503 rfa1041*/pKM101	STN, N4AC	A:T → G:C
TA7002	*hisG9138 Δara9 Δchl004 (bio chlD uvrb chlA) galE503 rfa1041*/pKM101	STN, MMS	T:A → A:T
TA7003	*hisG9074 Δara9 Δchl004 (bio chlD uvrb chlA) galE503 rfa1041*/pKM101	STN, ANG/UVA	T:A → G:C
TA7004	*hisG9133 Δara9 Δchl004 (bio chlD uvrb chlA) galE503 rfa1041*/pKM101	NQNO, MNNG	G:C → A:T
TA7005	*hisG9130 Δara9 Δchl004 (bio chlD uvrb chlA) galE503 rfa1041*/pKM101	NQNO, MMS	C:G → A:T
TA7006	*hisG9070 Δara9 Δchl004 (bio chlD uvrb chlA) galE503 rfa1041*/pKM101	NQNO, 5azaC	C:G → G:C

Table 4. Detection of mutation specificity of selected mutagen with the use of indicated *S. typhimurium* strains. STN, streptonigrin; N4AC, N^4-aminocytidine; MMS, methyl methanesulfonate; ANG, angelicin; NQNO, 4-nitroquinoline-1-oxide; MNNG, N-methyl-N′-nitro-N-nitrosoguanidine; 5azaC, 5-azacytidine.

As mentioned in chapter 2.3, *E. coli* B/r WP2 system complements the *Salmonella* assay and both strains should be used in combination when potential mutagens are examined. It has appeared that some chemicals such as nitrofurans are nonmutagenic in the *Salmonella* strains, whereas caused mutation in *E. coli* WP2 and in human cell cultures. The procedures described for the Ames *Salmonella* assay and *E. coli* WP2 tryptophan reverse mutation test are similar (Mortelmans & Riccio, 2000).

In *S. typhimurium* an another test for mutation detection was developed. It includes examination of forward mutations leading to resistance to the purine analog 8-azaguanine. The mutants are unable to convert enzymatically 8-azaguanine to the toxic metabolite. Base-pair mutations, frameshift mutations and deletions in different genes are expected to render 8-azaguanine resistance (Skopek et al., 1978).

2.9. Test system to study mutations in *Pseudomonas putida*

The genus *Pseudomonas* constitutes a large and diverse group of mostly saprophytic bacteria occurring in soil, water, plants and animals. They are known for ability to metabolize toxins, antibiotics, organic solvents and heavy metals present in environment, thus playing an important role in the development of soil microorganism community. The metabolic versatility of these organisms has been used to degrade waste products (bioremediation) and to synthesize added-value chemicals (biocatalysis) (Pieper and Reineke, 2000; Wackett *et al.*, 2002).

There is only a limited number of test systems that allow studying mutagenic processes in *Pseudomonas*. Rifampicine (Rif) resistance is used as one out of two such systems. The antibacterial action of Rif bases on binding to β subunit of RNA polymerase (RNA pol) and blocking the RNA elongation. Rifampicine resistance (Rif ^{r}) mutants harbor substitutions in β subunit of RNA pol that either make direct contacts with Rif or are located near the binding pocket (Campbell *et al.*, 2001). Mutations decrease the binding of Rif to RNA pol, making the enzyme, to different degree, not sensitive to the antibiotic (Jin and Gross, 1988). Except for few mutations located at the 5′ end, other Rif ^{r} mutations in *E.coli* are found in three clusters in a central region of *rpoB* gene. Sequence analysis in the *rpoB* region harboring Rif ^{r} mutations indicate a high level of conservation among prokaryotes (Campbell *et al.*, 2001). For that reason *rpoB*/Rif ^{r} system can be used as a mutation testing system in distinct bacterial species including *Pseudomonas*.

Using *rpoB*/Rif ^{r} system it has been found that isolated mutants of *P. putida* and *P. aeruginosa* express different levels of resistance to Rif, depending on the localization of mutations in the *rpoB* sequence (Jatsenko *et al.*, 2009). The spectrum of mutations strongly depends on temperature of growth. Thus, the usage of the same growth temperature is very important in mutation research in *Pseudomonas* while employing the *rpoB*/Rif ^{r} system.

Except *rpoB*/Rif ^{r} system, phenol utilization as growth substrate has been used to measure different types of point mutations in *P. putida*. The presence of *pheA* gene encoding phenol monooxygenase enables bacteria to utilize phenol as growth substrate and form colonies on selective plates. The reporter gene *pheA* was modified in RSF1010-derived plasmids by +1 frameshift mutation (Saumaa *et al.*, 2007). Assay system that allows base substitution detection uses two steps PCR technique. Mutant oligonucleotides contained specific base substitutions replacing the CTG for Leu-22 in the *pheA* gene with a TGA, TAA, or TAG stop codons. In the first step, the PCR with oligonucleotides pheAup and pheA22TGA, pheA22TAA or pheA22TAG, complementary to the positions 42 and 71 relative to the coding sequence of the *pheA* gene was performed. After treatment with restriction enzyme ExoI, PCR products were purified and used in a second PCR with the oligonucleotide pheAts complementary to *pheA* nucleotides 295 to 313. The amplified DNA fragments were cloned into the EcoRV site of pBluscript KS (+) (the mutations were verified by DNA sequencing). As a result, plasmids: pKTpheA22TGA, pKTpheA22TAA, and pKTpheA22TAG were obtained (Tegowa *et al.*, 2004).

2.10. *Bacillus subtilis* as a model for mutation detection

Bacillus subtilis is a Gram-positive bacterium commonly occuring in soil. It shows the ability to form endospores allowing the organism to tolerate extreme environmental conditions. The sporulation of *B. subtilis* provides a system for the detection of forward mutations in many genes whose products are responsible for spore formation. Mutants form non-sporulating or oligosporogenous colonies lacking brown pigment presented in normally sporulating cells. Using this mutation detection system mutagenic activity of many compounds, e.g. acridine orange, acriflavin, nitrous acid, 2-nitrosofluorene, nitrogen mustard, aflatoxin B1, 4-nitroquinoline-*N*-oxide, ethidium bromide, have been shown (Macgregor & Sacks, 1976).

B. subtilis HA101 strain and its derivatives (e.g. excision-repair deficient derivative, TKJ5211 strain) are other mutagen-tester bacteria. They carry suppressible nonsense mutations in *his* and *met* genes. The presence of suppresors is detected by examination of sensitivity of His^+ and Met^+ revertants to tester SPO1 phages (sus-5 and sus-11). This system has been used to examine mutagenic properties of various compounds, e.g. nitrofurazone, 4-nitroquinoline 1-oxide (4NQO), 4-aminoquinoline 1-oxide (4AQO), α- and β- naphthylamine (Tanooka, 1977) or triethanolamine (Hoshino & Tanooka, 1978).

2.11. Resistance of bacteria to antibiotics

Antibiotic resistance to streptomycin, rifampicin or nalidixic acid is often used for determination of spontaneous and induced mutagenesis in bacteria. It is a universal, rapid and simple test used in many species. The frequency of an antibiotic resistant bacteria usually is determined on plates supplemented with respective antibiotic. Rifampicin-, streptomycin- and nalidixic acid-resistant mutants arise due to spontaneous or induced mutations in chromosomal DNA. The mechanisms leading to rifampicin resistance was described in chapter 2.9. Resistance to streptomycin involves mutational changes in the 30S subunit of the ribosome, whereas resistance to nalidixic acid results from point mutations in structural genes encoding gyrase (topoisomerase II) subunit A (*gyrA E. coli*) or topoisomerase IV (*parC E. coli*). Further analysis of antibiotic resistant mutants is also possible, e.g., by sequencing of *gyrA* gene from nalidixic acid-resistant mutants of *B. subtilis* spores (Munakata et al., 1997). Genes responsible for resistance to ampiciline, tetracycline, carbenicyline, chloramphenicol, spectinomycine are usually harbored by plasmids or transposons and used as markers of new features/mutations introduced to the bacterial strains.

Resistance to other agents such as mentioned 8-azaguanine (see chapter 2.8) or phages, e.g. T1 phages, has been also used in mutagenesis tests, particularly in the 20th century studies. Resistance to T1 is due to *tonA* or *tonB* mutation (Miller, 1972).

3. Conclusions

Living organisms are continuously exposed to damaging agents both from the environment and from endogenous metabolic processes, whose action results in modification of proteins, lipids, carbohydrates and nucleic acids. The knowledge on DNA modifications leading to mutations is critical to our understanding of how and why the genome is affected during the lifespan of the organism, and how the DNA repair systems efficiently work *via* several different pathways. Bacterial systems for testing mutations are informative, cheap, and quick methods to study these processes. Connected with modern molecular biology methods, such as sequencing, RT PCR, side directed mutagenesis etc., bacterial systems constitute extremely valuable tools for studying metabolic processes, DNA repair systems, mutagenic and anticancer properties of chemicals etc. Two of the systems described here are of special value: *lacZ*$^-$→Lac$^+$ reversion system for determination the specificity of mutations and Ames test for studying mutagenic properties of chemicals. The latter, improved constantly e.g. by construction of new indicator strains and modern techniques, is extremely important in searching for new anticancer drugs.

Author details

Anna Sikora, Celina Janion and Elżbieta Grzesiuk
Institute of Biochemistry and Biophysics, Polish Academy of Sciences, Poland

Acknowledgement

We thank to the Editorial Board of *Acta Biochimica Polonica* for permission to published parts of the article entilted Reversion of *argE3* → Arg$^+$ in *Escherichia coli* AB1157 – a simple and informative bacterial system for mutation detection by Anna Sikora and Elzbieta Grzesiuk (*Acta Biochimica Polonica,* 2010, 57: 479-485).

4. References

Ames, B.N.; McCann, J. &Yamasaki, E. Methods fordetecting carcinogens and mutagens with the *Salmonella*-microsome mutagenicity test. *Mutat. Res.* Vol.31, (1975), pp. 347-364
Bachmann, B.J. (1987). Derivation and genotype of some mutant derivatives of *Escherichia coli* K-12, In *Escherichia coli and Salmonella typhimurium: Cellular and molecular biology*, F.C.

Neidhardt; J. Ingraham; K.B. Low; B. Magasanik; M. Schaechler & H.E. Umbarger HE, (Eds) vol 2, 1190-1219. ASM Press, Washington, DC.

Berger, H.; Brammar, W.J. & Yanofsky, C. Analysis of amino acid replacement resulting from frameshift and missense mutations in the tryptophan synthetase A gene of *Escherichia coli. J. Mol. Biol.* Vol.34, (1968), pp. 219-238.

Bi, E. & Lutkenhaus, J. FtsZ ring structure associated with division in *Escherichia coli. Nature* Vol.354, (1991), pp. 161 – 164

Bockrath, R.C. & Palmer, J.E. Differential repair of premutational UV-lessions at tRNA genes in *E. coli. Molec. Gen. Genet.* Vol. 156, (1977), pp. 133-140

Bockrath, R.C.; Barlow. A. & Engstrom, J. Mutation frequency decline in *Escherichia coli* B/r after mutagenesis with ethyl ethanesulfonate. *Mutat. Res.* Vol. 183, (1987), pp. 241-247

Bridges, B.A.; Dennis, R.E. & Munsen, R.J. Mutation in *Escherichia coli* B/r WP2 *try-* by reversion or suppression of a chain-terminating codon. *Mutat. Res.* Vol. 4, (1967), pp. 502-504

Bridges, B.A. Spontaneous mutation in stationary-phase *Escherichia coli* WP2 carrying various DNA repair alleles. *Mutat. Res.* Vol.302, (1993), pp. 173-176

Bridges, B.A. Elevated mutation rate in *mutT* bacteria during starvation: evidence for DNA turnover? *J. Bacteriol.* Vol.178, (1996), pp. 2709-2711

Campbell, E.A.; Korzheva, N.; Mustaev, A.; Murakami, K.; Nair, S.; Goldfarb, A. & Darst, S.A. Structural mechanism for rifampicin inhibition of bacterial RNA polymerase. *Cell* Vol.104, (2001), pp. 901-912.

Coulondre, C. & Miller, J.H. Genetic studies of the *lacI* gene of *Escherichia coli. J. Mol. Biol.* Vol. 117, (1977), pp. 577-606

Coupples, C.G. & Miller, J.H. A set of *lacZ* mutations in *Escherichia coli* that allow rapid detection of each of the six base substitutions. *Proc. Natl. Acad. Sci. USA* Vol. 86, (1989), pp. 5345-5349

Coupples, C.G.; Cabrera, M. Cruz, C. & Miller, J.H . A set of *lacZ* mutations in *Escherichia coli* that allow rapid detection of specific frameshift mutations. *Genetics* Vol.125, (1990), pp. 275-280

Cunnigham, R.P.; Saporito, S.M.; Spitzer, S.G. & Weiss,B. Endonuclease IV (*nfo*) mutant of *Escherichia coli. J. Bacteriol.* Vol. 168, (1986), pp. 1120-1127

D'Agostini, F. & De Flora, S. Potent carcinogenicity of uncovered halogen lamps in hairless mice. *Cancer Res.* Vol. 54, (1994), pp. 5081-5085

D'Agostini, F.; Izzoti, A. &, De Flora, S. Induction of micronuclei in cultured human lymphocytes exposed to quartz halogen lamps and its prevention by glass covers. *Mutagenesis* Vol.8, (1993), pp. 87-90

Daegelen, P.; Studier, F.W.; Lenski, R.E.; Cure, S. & Kim, J.F. Tracing ancestors and relatives of *Escherichia coli* B, and the derivation of B strains REL606 and BL21(DE3). *J. Mol. Biol.* Vol.394, (2009), pp. 634-643

De Flora, S.; Camoirano, A.; Izzoti, A. & Bennicelli, C. Potent genotoxicity of halogen lamps, compared to fluorescent light and sunlight. *Carcinogenesis* Vol.11, (1990), pp. 2171-2177

Delbruck, M & Luria, S.E. Interference between bacterial viruses: I. Interference between two bacterial viruses acting upon the same host, and the mechanism of virus growth. *Arch. Biochem.*, Vol.1, (1942), pp. 111–141

Dinglay, S.; Trewick, S.C.; Lindahl, T. & Sedgwick, B. Defective processing of methylated single-stranded DNA by *E. coli* AlkB mutants. *Genes. Dev.* Vol.14, (2000), pp. 2097-2105

Dizdaroglu, M.; Bauche, C.; Rodriguez, H. & Laval, J. Novel substrates of Escherichia coli nth protein and its kinetics for excision of modified bases from DNA damaged by free radicals. *Biochemistry*. Vol.39, (2000), pp. 5586-5592.

Fabisiewicz, A. & Janion, C. DNA mutagenesis and repair in UV-irradiated *Escherichia coli* K-12 under condition of mutation frequency decline. *Mutat Res* Vol.402, (1998), pp. 59-66

Falnes, P.O.; Johansen, R.F. & Seeberg, E. AlkB-mediated oxidative demethylation reverses DNA damage in *Escherichia coli*. *Nature* Vol.419, (2002), pp. 178-182

Fijalkowska, I.J.; Jonczyk, P.; Tkaczyk, M.M.; Bialoskorska, M. & Schaaper, R.M. Unequal fidelity of leading strand and lagging strand DNA replication on the *Escherichia coli* chromosome. *Proc. Natl. Acad. Sci. USA*. Vol.95, (1998), pp. 10020-10025.

Fluckiger-Isler, S.; Baumeister, M.; Braun, K.; Gervais, V.; Hasler-Nguyen N; Reimann, R.; Van Gompel, J.; Wunderlich, H.-G. & Engelhardt, G. Assessment of the performance of the AmesII™ assay: a collaborative study with 19 coded compounds. *Mutat Res* Vol.558, (2004), pp. 181-197.

Foster, P. & Trimarchi, J.M. Adaptive reversion of an episomal frameshift mutations in *Escherichia coli* requires conjugal functions but not actual conjugation. *Proc Natl Acad Sci USA* Vol.92, (1995), pp. 5487-5490

Foster, P. & Trimarchi, J.M. Adaptive reversion of a frameshift mutations in *Escherichia coli* by simple base substitutions in homopolimeric runs. *Science* Vol.265, (1994), pp. 407-409

Fowler, R.G.; McGinty, L. & Mortelmans, K.E. Spontaneous mutational specificity of drug resistance plasmid pKM101 in *Escherichia coli*. *J. Bacteriol.* No.140, (1979), pp. 929-937

Fowler, R.G. & McGinty, L. Mutational specificity of ultraviolet light in *Escherichia coli* with and without the R plasmid pKM101. *Genetics,* Vol.99, (1981), pp. 25-40

Gawel, D.; Maliszewska-Tkaczyk, M.; Jonczyk, P.; Schaaper, R.M. & Fijalkowska, I.J. Lack of strand bias in UV-induced mutagenesis in *Escherichia coli*. *J Bacteriol.*Vol.184, (2002), pp. 4449-4454.

Gee, P.; Maron, D.M. & Ames, B.N. Detection and classification of mutagens: a set of base-specific *Salmonella* tester strains. *Proc. Natl. Acad. Sci. USA* Vol.91, (1994), pp. 11606-11610

George, D.L. & Witkin, E.M. Slow excision repair in an *mfd* mutant of *Escherichia coli* B/r. *Mol. Gen. Genet*. Vol.133, (1974), pp. 283-291

Goldberg, A.L.; Moerschell, R.P; Chung, C.H. & Maurizi, M.R. ATP-dependent protease La (Lon) from *Escherichia coli*. *Methods Enzymol.* Vol.244, (1994), pp. 350-375.

Grzesiuk, E. & Janion, C. Some aspects of EMS-induced mutagenesis in *Escherichia coli*. *Mutat. Res.* Vol. 297, (1993), pp. 313-321

Grzesiuk, E. & Janion, C. The frequency of MMS-induced, *umuDC*-dependent, mutations declines during starvation in *Escherichia coli*. *Mol. Gen. Genet.* Vol.245, (1994), pp. 486-492

Grzesiuk, E. & Janion, C. MMS-induced mutagenesis and DNA repair in *Escherichia coli dnaQ49*: contribution of UmuD' to DNA repair. *Mutat. Res.* Vol.362, (1996), pp. 147-154

Grzesiuk, E. &, Janion, C. Mutation frequency decline in MMS-treated *Escherichia coli* K-12 *mutS* strains. *Mutagenesis* Vol..13, (1998), pp. 127-132

Hall, B.G. Spontaneous point mutations that occur more often when they are advantageous than they are neutral. *Genetics* Vol.90, (1990), pp. 673-691

Hoshino, H. & Tanooka, H. Carcinogenicity of triethanolamine in mice and its mutagenicity after reaction with sodium nitrite in bacteria. *Cancer Res.* Vol.38, (1978), pp. 3918-3921

Janion, C. Some Provocative Thoughts on Damage and Repair of DNA. *J. Biomed. Biotechnol.* Vol.1, (2001), pp. 50-51.

Janion, C. Inducible SOS response system of DNA repair and mutagenesis in *Escherichia coli. Int J Biol Sci.* Vol.4, (2008), pp. 338-344.

Janion, C., Sikora, A., Nowosielska, A. & Grzesiuk E. Induction of the SOS response in starved *Escherichia coli. Environ. Mol. Mutagen.* Vol.40, (2002), pp. 129-133.

Janion, C.; Sikora, A.; Nowosielska, A. & Grzesiuk, E. *E. coli* BW535, a triple mutant for the DNA repair genes *xth, nth,* and *nfo,* chronically induces the SOS response. *Environ. Mol. Mutagen.* Vol.41, (2003), pp. 237-242

Jatsenko, T.; Tover, A.; Tegova, R. & Kivisaar, M. Molecular characterization of Rif ^{r} mutations in *Pseudomonas aeruginosa* and *Pseudomonas putida. Mutat Res.* Vol.683, (2010), 106-114

Jin, D.J. & Gross, C.A. Mapping and sequencing of mutations in the *Escherichia coli rpoB* gene that lead to rifampicin resistance. *J. Mol. Biol.* Vol. 202, (1988), pp. 45-58

Kataoka, H.; Yamamoto, Y. & Sekiguchi, M. A new gene (*alkB*) of *Escherichia coli* that controls sensitivity to methyl methane sulfonate. *J. Bacteriol.* Vol.153, (1983), pp. 1301-1307

Kato, T.; Shinoura, Y.; Templin, A. & Clark, A.J. Analysis of ultraviolet light-induced suppressor mutations in the strain of *Escherichia coli* K-12 AB1157. An implication for molecular mechanisms of UV mutagenesis. *Mol. Gen. Genetic.* Vol.180, (1980), pp. 283-291

Kow, Y.W. & Wallace, S.S. Exonuclease III recognizes urea residues in oxidized DNA. *Proc. Natl. Acad. Sci. USA* Vol.82, (1985), pp. 8354-8358.

Macgregor, J.T. & Sacks, L.E. The sporulation system of *Bacillus subtilis* as the basis of a multi-gene mutagen screening test. *Mutat. Res.* Vol.38, (1976), pp. 271-286

Maliszewska-Tkaczyk, M.; Jonczyk, P.; Bialoskorska, M.; Schaaper, R.M. & Fijalkowska, I.J. SOS mutator activity: unequal mutagenesis on leading and lagging strands. *Proc. Natl. Acad. Sci. USA.* Vol.97, (2000), pp. 12678-12683.

McGinty, L.D. & Fowler, R.G. Visible light mutagenesis in *Escherichia coli. Mutat. Res.* Vol.95, (1982), pp. 171-181

Michaels, M.L. & Miller, J.H. The GO system protects organisms from the mutagenic effect of the spontaneous lesion 8-hydroxyguanine (7,8-dihydro-8-oxoguanine). *J. Bacteriol.* Vol.174, (1992), pp. 6321-6325

Miller, J.H. (1972) *Experiments in molecular genetics*. Cold Spring Harbor Laboratory, New York.

Miller, J.H. Mutational specificity in bacteria. *Annu. Rev. Genet.* Vol.17, (1983), pp. 215-238

Mortelmans, K. Isolation of plasmid pKM101 in the Stocker laboratory. *Mutat. Res.* Vol.612, (2006), pp. 151–164

Mortelmans, K. & Riccio, E.S. The bacterial tryptophan reverse mutation assay with *Escherichia coli* WP2. *Mut. Res.* Vol.455, (2000), pp.61-69

Munakata, N.; Saitou, M.; Takahashi, N.; Hieda, K. & Morohoshi, F. Induction of unique tandem-base change mutations in bacterial spores exposed to extreme dryness. *Mut. Res.* Vol.390, (1997), pp. 189-195

Nieminuszczy, J.; Sikora, A.; Wrzesinski, M.; Janion, C. & Grzesiuk, E. AlkB dioxygenase in preventing MMS-induced mutagenesis in *Escherichia coli*: effect of Pol V and AlkA proteins. *DNA Repair* Vol.5, (2006a), pp. 181-188

Nieminuszczy, J.; Janion, C.; Grzesiuk, E. Mutator specificity of *Escherichia coli alkB117* allele. *Acta Biochim. Polon.* Vol.53, (2006b), pp. 425-428

Nieminuszczy, J.; Mielecki, D.; Sikora, A.; Wrzesinski, M.; Chojnacka, A.; Krwawicz, J.; Janion, C. & Grzesiuk, E. Mutagenic potency of MMS-induced 1meA/3meC lesions in *E. coli*. *Environ. Mol. Mutagen.* Vol.50, (2009), pp. 791-799

Nowosielska, A. & Grzesiuk, E. Reversion of *argE3* ochre strain *Escherichia coli* AB1157 as a tool for studying the stationary-phase (adaptive) mutations. *Acta Biochim. Pol.* (2000), Vol.47, (2000), pp. 459-467

Nowosielska, A.; Nieminuszczy, J. & Grzesiuk, E. Spontaneous mutagenesis in exponentially growing and stationary-phase, *umuDC*-proficient and -deficient, *Escherichia coli dnaQ49*. *Acta Biochim. Pol.* Vol.51, (2004a), pp. 683-692

Nowosielska, A.; Janion, C. & Grzesiuk, E. Effect of deletion of SOS-induced polymerases, pol II, IV, and V, on spontaneous mutagenesis in *Escherichia coli mutD5*. *Environ. Mol. Mutagen.* Vol.43, (2004b), pp. 226-234

Ohta, T.; Tokishita, S.; Tsunoi, R.; Ohmae, S. & Yamagata, H. Characterization of Trp$^+$ reversions in *Escherichia coli* strain WP2*uvrA*. *Mutagenesis* Vol.17, (2002), pp. 313-316

Persing, D.H.; McGinty, L.; Adams, C.W. & Fowler, R.G. Mutational specificity of the base analogue, 2-aminopurine, in *Escherichia coli*. *Mutat. Res.* Vol.83, (1981), pp. 25-37

Pieper, D.H. & Reineke, W. Engineering bacteria for bioremediation. *Curr. Opin. Biotech.* Vol.11, (2000), pp. 262-270

Płachta, A. & Janion, C. Is the tRNA *ochre* suppressor *supX* derived from *gltT* ? *Acta Biochim. Polon.* Vol.39, (1992), pp. 265-269

Prival, M.J. Isolation of glutamate-inserting ochre suppressor mutants of *Salmonella typhimurium* and *Escherichia coli*. *J. Bacteriol.* Vol.178, (1996), pp. 2989-2990

Raftery, L.A. & Yarus, M. Systematic alterations in the anticodon arm make tRNAGlu-Suoc a more efficient suppressor. *EMBO J.* Vol.6, (1987), pp. 1499-1506

Rosenberg, S.M.; Longerich, S.; Gee, G. & Harris, R.S. Adaptive mutation by deletion in small mononucleotide repeats. *Science* Vol.265, (1994), pp. 405-407

Sargentini, N.J. & Smith, K.C. Mutational spectrum analysis of *umuC*-independent and *umuC*-dependent γ-radiation mutagenesis in *Escherichia coli*. *Mutat. Res.* Vol.211, (1989), pp. 193-203

Saumaa, S.; Tover, A.; Tark, M.; Tegova, R. & Kivisaar, M. Oxidative DNA damage defense systems in avoidance of stationary-phase mutagenesis in *Pseudomonas putida*. *J. Bacteriol.* Vol.189, (2007), pp. 5504-5514

Savery, N.J. The molecular mechanism of transcription-coupled DNA repair. *TRENDS Microbiol.* Vol.15, (2007), pp. 326-333

Sedgwick, B.; Bates, P.A.; Paik, J.; Jacobs, S.C. & Lindahl, T. Repair of alkylated DNA: Recent advances. *DNA Repair* Vol.6, (2007), pp. 429-442

Selby, C.P. & Sancar, A. Molecular mechanism of transcription-repair coupling. *Science* Vol.260, (1993), pp. 53-58

Shinoura, Y.; Kato, R. & Glickman, B.W. A rapid and simple method for the determination of base substitution and frameshift specificity of mutagens. *Mutat. Res.* Vol.111, (1983), pp. 43-49

Sikora, A.; Mielecki, D.; Chojnacka, A.; Nieminuszczy, J.; Wrzesinski, M. & Grzesiuk, E. Lethal and mutagenic properties of MMS-generated DNA lesions in *Escherichia coli* cells deficient in BER and AlkB-directed DNA repair. *Mutagenesis* Vol.25, (2010), pp. 139-147

Skopek, T.R.; Liber, H.L., Krolewski, J.J. & Thilly, W.G. Quantitaive forward mutation assay in *Salmonella typhimurium* using 8-azaguanine resistance as a genetic marker. *Proc. Natl. Acad. Sci. USA*, Vol.75, (1978), pp.410-414.

Studier, F.W.; Daegelen, P.; Lenski, R.E.; Maslov, S. & Kim, J.F. Understanding the differences between genome sequences of *Escherichia coli* B strains REL606 and BL21(DE3) and comparison of the *E. coli* B and K-12 genomes. *J. Mol. Biol.*Vol.394, (2009), pp. 653-680

Śledziewska-Gójska, E. & Janion, C. Alternative pathways of methyl methanesulfonate-induced mutagenesis in *Escherichia coli*. *Mol. Gen. Genet.* Vol.216, (1989), pp. 126-131

Śledziewska-Gójska, E.; Grzesiuk, E.; Płachta, A. & Janion, C. Mutagenesis of *Escherichia coli*:A method for determining mutagenic specifity by analysis of tRNA suppressors. *Mutagenesis* Vol.7, (1992), pp. 41-46

Tanaoka, H. Development and applications of *Bacillus subtilis* test systems for mutagens, involving DNA-repair deficiency and suppressible auxotrophic mutations. *Mut. Res.* Vol.42, (1977), pp. 19-32

Tegova, R.; Tover, A.; Tarassova, K.; Tark, M. & Kivisaar, M. Involvement of Error-Prone DNA Polymerase IV in Stationary-Phase Mutagenesis in *Pseudomonas putida*. *J. Bacteriol.* Vol.186, (2004), pp. 2735-2744

Todd, P.A.; Monti-Bragadin, C. & Glickman, B.W. MMS mutagenesis in strains of *Escherichia coli* carrying the R46 mutagenic enchancing plasmid: Phenotypic analysis of Arg^+ revertants. *Mutat. Res.* Vol.62, (1979), pp. 227-237

Trewick, S.C.; Henshaw, T.F.; Hausinger, R.P.; Lindahl, T. & Sedgwick, B. Oxidative demethylation by *Escherichia coli* AlkB directly reverts DNA base damage. *Nature* Vol.419, (2002), pp. 174-178

Wackett, L.P.; Sadowsky, M.J.; Martinez, B. & Shapir, N. Biodegradation of atrazine and related s-triazine compounds: from enzyme to field studies. *Appl. Microbiol. Biot.* Vol.58, (2002), pp. 39-45

Witkin, E.M. Inherited differences in sensitivity to radiation in *Escherichia coli*. *Proc. Natl. Acad. Sci. USA* Vol. 32, (1946), pp. 59-68

Witkin, E.M. Mutation frequency decline revisited. *BioEssays* Vol.16, (1994), pp. 437-444

Wójcik, A. & Janion, C. Mutation induction and mutation frequency decline in halogen light-irradiated *Escherichia coli* K-12 AB1157 strains. *Mutat. Res.* Vol.390, (1997), pp. 85-92

Wójcik, A.; Grzesiuk, E.; Tudek, B. & Janion, C. Conformation of plasmid DNA from *Escherichia coli* deficient in the repair systems protecting DNA from 8-oxoguanine lesions. *Biochimie* Vol.78, (1996), pp. 85-89

Wójcik, A. & Janion, C. Effect of Tn*10*/Tn*5* transposons on the survival and mutation frequency of halogen light-irradiated AB1157 *Escherichia coli* K-12. *Mutagenesis* Vol. 14, (1999), pp. 129-134.

Wrzesiński, M.; Nieminuszczy, J.; Sikora, A.; Mielecki, D.; Chojnacka, A.; Kozłowski, M.; Krwawicz, J. & Grzesiuk, E. Contribution of transcription-coupled DNA repair to MMS-induced mutagenesis in *E. coli* strains deficient in functional AlkB protein. *Mutat Res* Vol.688, (2010), pp. 19-27

Wyatt, M.D. & Pittman, D.L. Methylating agents and DNA repair responses: methylated bases and sources of strand breaks. *Chem. Res. Toxicol.* Vol.19, (2006), pp. 1580-1594

Chapter 9

ENU Mutagenesis in Mice – Genetic Insight into Impaired Immunity and Disease

Kristin Lampe, Siobhan Cashman, Halil Aksoylar and Kasper Hoebe

Additional information is available at the end of the chapter

http://dx.doi.org/10.5772/50334

1. Introduction

Over the last decade biomedical research has seen tremendous advancements in the field of genetics that enables unlimited access to >60 vertebrate genomes—including the human and mouse genomes, two of the most widely studied species in biomedical research. These advancements are largely due to rapid development of high throughput sequencing technologies such as next-generation sequencing (NGS) technologies that allow for more affordable and efficient sequencing compared to traditional Sanger technology. The availability of the entire human genome sequence has accelerated our efforts to gain insight into the genetics underlying human disease. Such efforts include Genome-Wide Association Studies (GWAS) – a widely used approach that examines the association between common genetic variants and specific human disease traits. GWAS has led to the successful identification of a large number of SNPs that are linked with chronic diseases ranging from Crohn's disease, systemic lupus erythermatosus (SLE), type I diabetes (T1D), and many other common western world diseases(reviewed by Visscher et al.[1]). On the other hand, genetic deficiencies that cause severe disease—such as primary immunodeficiency diseases associated with a poor survival— represent mostly rare mutations within the human population[2]. Such patients can be found in pediatric clinics and more often than not, the genetic deficiencies underlying disease remain elusive. The availability of NGS, however, offers exciting new opportunities in that it enables the identification of all genome-wide variants in individual patients for limited costs. Nonetheless, both approaches are faced with a significant challenge to identify the causal variants. First of all, most GWAS identify loci that contain more than one SNP but more importantly, SNP maps are incomplete and require in depth probing of the identified genetic region (reviewed by Visscher et al.[1]). Thus the approach is generally not limited to a single SNP, but rather uncovers multiple gene candidates for a single locus and researchers are often left with the critical question to identify variant causality. This is further complicated by the fact that GWAS is often used for the analysis of complex polygenic traits where gene variants need to exist in combination with one other to assert an effect. In the case of monogenic traits underlying severe disease phenotypes, linkage analysis is rarely an option, whereas whole genome sequencing likely results in the identification of numerous "unique" variants. The biological consequences of such variants would again need to be confirmed and candidate gene selections are guided by *a priori* knowledge of gene function. Thus the challenges have rather shifted from identifying genetic changes to understanding gene-function and identifying gene causality.

Providing insight into the functional genome is not just limited to understanding gene or protein function, but also includes gene regulation and complex interactions with other genes within the context of cellular or organismal function. The mammalian genome is believed to consist of ~22,000 annotated genes—most of which have been poorly described. In addition, there is almost an unlimited number of phenotypes to be probed, making this an even more daunting task. Nonetheless, experimental models, including fruit fly and mouse models, have been extremely valuable in revealing unique insight into gene function. Typically, forward and reverse genetic approaches have been applied in parallel to uncover gene function. Reverse genetics begins with the creation of a genetic change and ends with the identification of a phenotype. This approach is hypothesis-based and assumes a specific gene function up front. On the other hand, forward (or classical) genetics proceeds from phenotype to the identification of a causal genetic change (SNP or mutation). This approach has led to important discoveries in the field of immunology most notably the identification of TLR4 as the sole LPS receptor[3]— a discovery recently awarded with the Nobel Prize. Until a few years ago, identification of such genetic variants required positional cloning. This was once considered an arcane art, requiring significant effort, time and financial resources. However, the current availability of the genome sequence for most inbred mouse strains has eliminated the need for contig construction and trivialized the identification of informative markers for high-resolution mapping and/or the identification of existing variants within an associated chromosomal region. Moreover, low cost high-throughput DNA sequencing has accelerated the process of finding unique mutations either introduced spontaneously or by following treatment with mutagens. The current limitation for forward genetics is rather the restricted number of strong monogenic phenotypes, something also referred to as the "phenotype gap"[4]. To overcome this limitation, germline mutagenesis— in which random mutations are introduced in spermatogonial stem cells— has proven to be an effective approach to expand the number of phenotypes.

2. N-ethyl-N-nitrosourea mutagenesis

In mice, a widely used mutagen to create and expand the number of phenotypes is the alkylating agent N-ethyl-N-nitrosourea (ENU). ENU is a powerful mutagen that according to our latest estimates can introduce more than 3 base-pair changes per million base-pairs of genomic DNA[5]. ENU introduces point mutations in spermatogonial stem cells, predominantly affecting A/T base pairs (44% A/T→T/A transversions and 38% A/T→G/C transitions), whereas at the protein level, ENU primarily results in missense mutations (64% missense, 26% splicing errors and 10% nonsense mutations)[6]. With three bp changes per million bps and a total length of ~2,717 Mb for the mouse genome, one can calculate that each G1 male carries ~8,000 bp changes genome-wide. With the coding region being 1.3% of total genomic sequence and 76% of random bp changes creating a coding change, it follows that each G1 mouse carries about 80 coding changes genome-wide, according to our latest estimates. These exist in a heterozygous form and do not necessarily cause a phenotype. In our experience, the majority of ENU-induced mutations, behave as recessive traits or are codominant at best. The approach entails a weekly injection of ~90mg/kg ENU for 3 weeks that is followed by a brief period of sterility for up to 12 weeks. After the recovery period, each G0 male is bred to untreated, wild type C57BL/6 female mice to generate G1 offspring. These G1 animals are then either used for phenotypic screens or can be used to produce G2 mice, which in turn are backcrossed to the G1 male to generate G3 offspring. While screening the G1 population for phenotypes is limited to the identification of dominant mutations, screening of G3 mice allows for the discovery of recessive mutations. Although the total number of base-pair changes in G3 mice will be reduced—each mouse will carry ~11 coding changes in homozygous form—this has proven to be the more powerful approach to capture mice with phenotypes of interest and more importantly allows for the retrieval of lethal phenotypes.

The rate-limiting step in ENU mutagenesis has long been the identification of causative mutations. Until recently, identified mutant lines were outcrossed to genetically different inbred strains and often the analyses of hundreds if not thousands of meiosis were needed to obtain a small enough critical region that could be sequenced. However, the availability of NGS has significantly facilitated the process of variant identification. Currently, targeted exon-enrichment—i.e. targeting exonic sequence within a critical region using sequence capture probes—, whole-exome and whole-genome sequencing are all proven strategies to effectively uncover mutations. The coverage of (targeted) genomic DNA is often exceptional, particularly for the exon-enrichment approach, where generally high quality sequence (minimal depth >10) for more than >98 % of the targeted region can be obtained[5]. Nonetheless, causality of the identified mutations remains a critical aspect of this approach and low resolution mapping (generally < 30 meioses) and/or genetic confirmation are still integral parts of the ENU mutagenesis approach. In addition, the availability of NGS also provides further opportunities for the phenotypic probing of ENU germline mutants. Often, phenotypes identified in ENU germline mice are lost or significantly influenced by modifier loci located on outcross strains carrying a high degree of genetic variation. For example, identification of genes required for optimal NK cell function has been difficult because of the large variation in NK cell ligands/receptors existing on different mouse backgrounds (Hoebe, unpublished results). By being able to analyze and sequence large genomic regions, fine mapping is superfluous and the exploration of subtle phenotypes can be traced following an outcross to strains with minimal genetic variation between the outcross and parent ENU strain. Ultimately, the genetic diversity should be just enough to allow low-resolution linkage analysis—a prime example being the genetic diversity between C57BL/6J and C57BL/10J strains.

3. Unraveling lymphocyte immune function using ENU mutagenesis

As referred to above, a critical aspect of ENU mutagenesis is the (biological) field of interest to be probed. ENU mutagenesis has been used to define the genetic footprint of a wide variety of phenotypes, including visible, behavioral, developmental and immunological phenotypes[7]. Nonetheless, its success is depending on: 1) the use of reliable screening assays with limited biological variation, 2) targeting large genomic footprints, and 3) probing a biological phenotype that is poorly defined. Our laboratory has used ENU mutagenesis to identify genes with non-redundant function in lymphocyte development, priming or effector function. Among the biological screens we apply is an *in vivo* cytotoxicity assay in which we test the ability of G3 mice to induce an antigen-specific $CD8^+$ T cell response following immunization with irradiated cells containing antigen. In parallel, we test the ability of Natural Killer (NK) cells to recognize and eliminate "missing self" target cells *in vivo*—a process involving complex balancing interactions between activating and inhibitory NK cell receptors. Such screens are not just limited to identifying the presence/absence of NK cells and/or $CD8^+$ T cells *in vivo* but challenges the host response to undergo NK cell recognition/killing, antigen uptake/ processing and presentation by Dendritic Cells (DCs), ultimately causing T cell priming, expansion and T cell cytolytic effector function. The *in vivo* immune responses assess the capacity of ENU mice to induce sterile inflammatory responses mediated by self-molecules that activate either NK cells and/or Toll-like receptor-independent sensing pathways—the latter presumably activated by cell-death- or "danger-" associated molecular patterns (DAMPs). Importantly, the induction of type I or II IFNs are essential for the generation of antigen-specific T cell responses mediated via cell-death induced immune responses[8]. Whereas IFNs have been shown to promote the maturation of DCs and stimulate T cell priming, the underlying pathways inducing type I IFN following exposure of DCs to dying cells are less well defined. It is well established that host molecules such as DNA and/or RNA in apoptotic cells can cause sustained and systemic type I IFN production when they escape degradation in macrophages[9-11]. The pathways by which such nucleotide structures drive type I IFN production following administration of apoptotic cells remains still elusive to date. Thus, the *in vivo* cytotoxicity screen performed in our laboratory presents a large genetic footprint, not only comprising lymphocyte development but also targeting NK-, DC- and T cell biological function. As a result, we have identified a number of germline mutants that are either deficient in the IFN pathways, but also includes germline mutants that exhibit impaired lymphocyte survival, T cell activation and/or actin-polymerization. Here we will provide two examples how ENU germline mutants can provide new insight into gene function, immunological pathways and/or disease development.

4. Gimap5 and loss of immunological tolerance driving auto-immune diseases

Using N-ethyl-N-nitrosourea (ENU) germline mutagenesis, our laboratory previously identified Gimap5-deficient mice—designated *sphinx*—that exhibit reduced lymphocyte survival and develop severe colitis around 10-12 weeks of age[12]. Specifically, these mice lack NK or $CD8^+$ T cell populations in peripheral lymphoid organs, whereas relatively normal thymocyte development occurs, including the $CD4^+$ T cell, $CD8^+$ T cell, and $Foxp3^+$ regulatory T cell lineages. Coarse mapping and sequencing of the critical region revealed a single G→T point mutation in *Gimap5* to be the causal mutation. This mutation resulted in a G38C amino acid substitution in the predicted GTP-binding domain of Gimap5, destabilizing Gimap5 protein expression[12].

Gimap5 is part of the family of Gimap genes which are predominantly expressed in lymphocytes and regulate lymphocyte survival during development and homeostasis[13]. Gimap proteins contain a GTP-binding AIG1 homology domain, first identified in disease-resistance genes in higher plants[9, 10]. More recent crystallographic studies showed that the Gimap proteins resemble a nucleotide coordination and dimerization mode previously observed for dynamin GTPase—a component essential for the scission and fusion of cellular vesicular compartments such as endosomes at the cell surface or the Golgi apparatus in the cytosol[14]. Members of the Gimap family appear to be localized to different subcellular compartments with Gimap5 reported to localize in lysosomes based on studies in human, mouse and rat lymphocytes[15]. Overall, the function of these proteins and their role in disease development remain poorly defined.

Genetic aberrancies in Gimap5 have been strongly linked to reduced lymphocyte survival and homeostasis, but importantly have also been associated with autoimmune diseases. In humans, polyadenylation polymorphisms in GIMAP5—causing relative modest changes in GIMAP5 RNA expression—were associated with increased concentrations of IA2 auto-antibodies in type 1 diabetes (T1D) patients and an increased risk of systemic lupus erythematosus (SLE)[16, 17]. Studies using biobreeding (BB) rats— carrying a mutation (*lyp/lyp*) in Gimap5— show marked lymphopenia and predisposition to the development of T1D[18-22] and intestinal inflammation[23]. Together these observations suggest that, beyond lymphocyte survival, Gimap5 is essential for maintaining immunological tolerance.

Although in $Gimap5^{sph/sph}$ mice no auto-antibodies can be detected, males and females developed severe colitis around 8-12 weeks of age, which was dependent on the microbiome and is $CD4^+$ T cell driven[12]. Interestingly, inflammatory bowel disease (IBD) such as Crohn's disease, ulcerative colitis and indeterminate colitis[24, 25] manifest generally in adolescence or adulthood and they behave as complex, polygenic diseases often sharing common risk factors with other autoimmune diseases[26, 27]. Previous work suggests that impaired lymphocyte survival and consequent lymphopenia may be linked to the loss of immunological tolerance. Specifically, $CD4^+$ T cells in a lymphopenic environment can undergo thymic independent expansion in the periphery. This process—also referred to as lymphopenia-induced proliferation (LIP)—is accompanied by marked alterations in T cell phenotype and is linked to auto-immunity[28-30]. Most notably, $CD4^+$ T cells more readily adopt an effector phenotype, including the ability to robustly produce cytokines and can drive the development of colitis[31-33]. Importantly, the absence of Treg cells is an important determinant of immune-mediated sequelae, including colitis that is induced by $CD4^+$ T cells undergoing LIP. Interestingly, studies in our laboratory show that in *Gimap5$^{sph/sph}$* mice, the onset of colitis is preceded by a progressive reduction in circulating CD4$^+$ T cells with remaining CD4$^+$ T cells exhibiting a lymphopenia-induced proliferation (LIP) phenotype (CD44^{high} and CD62^{low}) with a large number of cells in S phase([12] and Figure 1). Moreover, CD4$^+$ T cells derived from *Gimap5$^{sph/sph}$* spleen or mesenteric lymph nodes (MLNs) exhibit a higher capacity to produce cytokines, i.e. IFNγ and/or IL-17A following activation of the T cell receptor.

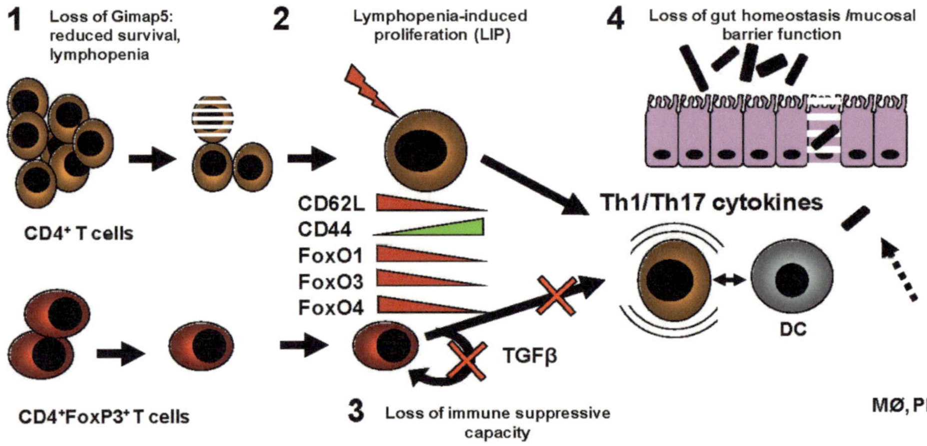

Figure 1. Schematic representation of the events causing colitis in Gimap5-deficient mice. Loss of Gimap5 leads to reduced survival of lymphocytes including CD4$^+$ T cells (1). During lymphopenia, CD4$^+$ T cells undergo LIP exemplified by increased surface expression of CD44 and reduced levels of CD62L (2). Concomitantly, CD4$^+$ T cells exhibit loss of full-length FoxO1, FoxO3 and FoxO4 expression, affecting both immunosuppressive function or Treg cells and the induction of Treg cells in the mesenteric lymph nodes (3). Together these events promote Th17 differentiation and activation of CD4$^+$ T cells in the gut causing inflammation and infiltration of macrophages / neutrophils that further amplify intestinal inflammation (4).

Given the important role of regulatory T cells in immune-mediated sequelae induced by CD4$^+$ T cells undergoing LIP, our laboratory assessed whether the colitis was driven by abnormalities in regulatory T cell development or function. Although relatively normal numbers of Foxp3$^+$ Treg cells are found in 3-week-old mice, a loss of Treg cell numbers is observed by 6 weeks of age particularly in the MLNs[34]. In addition, regulatory T cells in *Gimap5$^{sph/sph}$* mice show a progressive loss of suppressive function. Specifically, whereas Treg cells from 4-week-old *Gimap5$^{sph/sph}$* mice show a slight, but significant reduction in their ability to suppress CD8$^+$ T cell proliferation *in vitro*, Treg cells isolated from 6-week-old *Gimap5$^{sph/sph}$* mice are incapable of suppressing CD8$^+$ T cell proliferation entirely, thus indicating that a progressive impairment in Treg cell survival and function may underlie the colitis development in *Gimap5$^{sph/sph}$* mice. Indeed, colitis can be prevented entirely by injecting wildtype regulatory T cells in 4-week-old *Gimap5$^{sph/sph}$* mice.

Interestingly, the T cell phenotypes in *Gimap5$^{sph/sph}$* mice show striking similarities with those seen in mice deficient in the family of Forkheadbox group O (Foxo) transcription factors.

The family of Foxo transcription factors contain 4 members of which three (Foxo1, Foxo3 and Foxo4) have overlapping patterns of expression and transcriptional activities[35-37]. They play an essential role in the quiescence and survival of $CD4^+$ T cells. Foxo1 expression is critical for maintaining naïve T cell quiescence[38-40]. In addition, Foxo1, 3 expression has been reported to be essential for Treg cell development and function[41, 42]. Specifically, Foxo transcription factors serve a role as coactivators downstream of the TGFβ signaling pathway by interacting with SMAD proteins[43, 44], and directly regulate the induction of a number of Treg cell associated genes, including Foxp3, CTLA-4 and CD25[41, 42]. Indeed immunoblot analysis of $CD4^+$ T (including Treg cells) from *Gimap5*$^{sph/sph}$ mice at various ages, revealed a progressive loss of full-length Foxo1, -3a and -4 proteins, with normal levels at 3 weeks of age, but a complete loss of Foxo-expression in $CD4^+$ T cells from 6-10 week-old *Gimap5*$^{sph/sph}$ mice[34]. The regulation of Foxo3 and Foxo4 protein expression appears to occur at the post-transcriptional level, although the exact mechanism underlying the loss of Foxo-expression remains to be determined. The progressive nature suggest a strong association with the loss of Treg function in *Gimap5*$^{sph/sph}$ over time and link the loss of full-length Foxo expression in *Gimap5*$^{sph/sph}$ lymphocytes with the onset of lymphopenia, impaired lymphocyte proliferation and increased effector function and differentiation into Th17 cells (Figure 1). The detailed mechanistic insight into the loss of immunological tolerance occurring in *Gimap5*$^{sph/sph}$ mice may ultimately provide important leads as to how polyadenylation polymorphisms in GIMAP5 predispose human patients to T1D or SLE.

5. Mutations in hematopietic protein 1; an immunodeficiency resulting in loss of a broad range of immunological functions

Genetic aberrancies causing severe combined immunodeficiency (SCID) are generally rare and associated with a high morbidity and/or mortality. They often present significant challenges in terms of treatment due to the wide variety of immune cells that can be affected. Therefore, besides defining the genetic footprint underlying SCID, a critical challenge lies in obtaining a thorough understanding of the degree of the immunodeficiency presented by specific mutations in genes, including defining the types of immune cells affected and functional aberrancies observed. Our laboratory previously identified a germline mutant, designated *Lampe2*, which exhibited impaired NK as well as $CD8^+$ T cell cytoloytic effector function as determined by the *in vivo* cytotoxicity assay described above (Figure 2a). The G1 pedigrees of these germline mutants were selected to establish a homozygote colony used for genetic analysis and further phenotypic characterization. The mutation behaved as strictly recessive, in that normal cytolytic effector functions were observed in heterozygote mutant mice. Further characterization of 6-week-old homozygote *Lampe2* mutants revealed markedly reduced numbers of $CD8^+$ T, $CD4^+$ T and B cell populations, and a slight reduction in NK cells (Figure 2b). In contrast, an increase in the number of macrophages in the spleen was observed (Figure 2b). Notably, upon necropsy, the liver exhibited white patches at the periphery (Figure 2c) which upon histological analysis revealed large areas of necrosis and significant hematopoietic infiltrate and inflammation (Figure 2d-e).

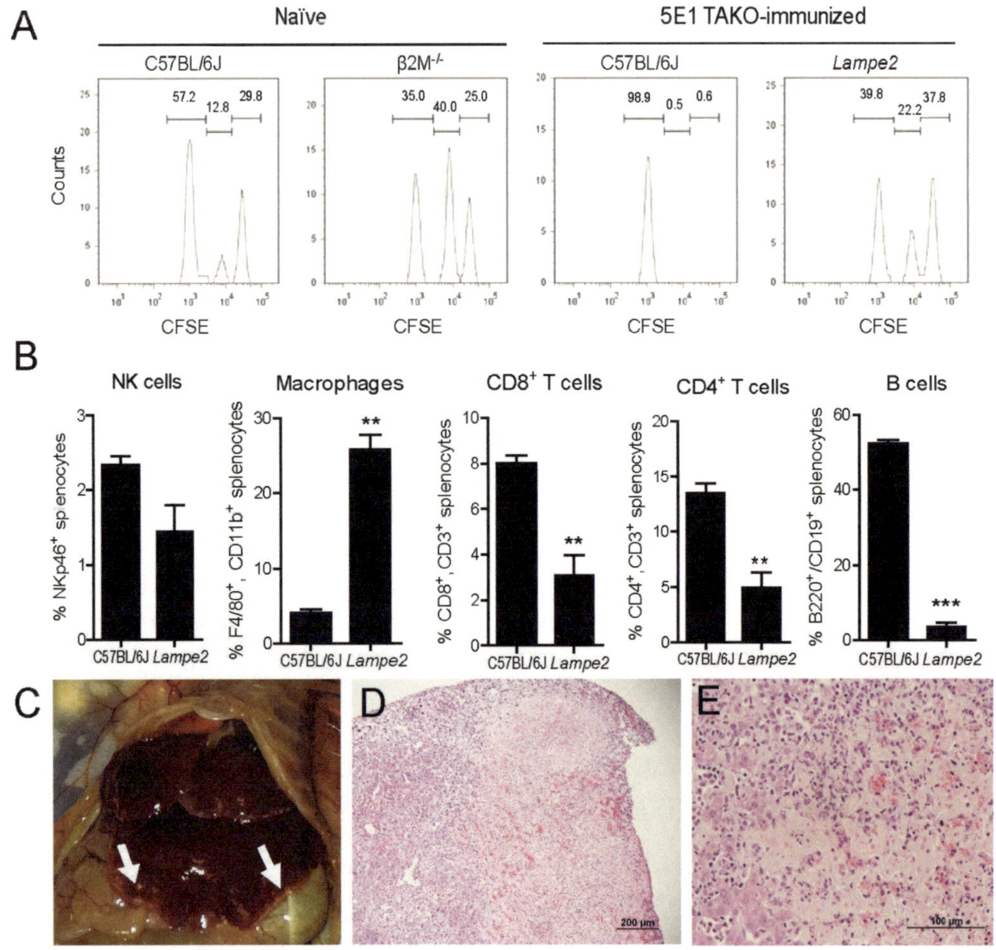

Figure 2. Impaired NK and CD8+ T cell function and development of liver injury in *Lampe2* mice. (**a**) Reduced clearance of CFSE labeled β-2m-deficient and antigen-specific target splenocytes in *Lampe 2* germline mutants compared to C57BL/6J control mice *in vivo*. 48 hours after transfer, blood samples were collected and analyzed for the presence of wildtype splenocytes (low-CFSE) and Kb-deficient splenocytes (medium-CFSE). The percentage killing is calculated from the ratio between β-2m-deficient and C57BL/6J cells administered to β2m-deficient and control naïve C57BL/6J recipients. (**b**) The percentage of NK cells, macrophages, CD8+ T cells, CD4+ T cells and B cells in C57BL/6J and homozygote *Lampe2* mutant mice. (n > 3) (**c**) Macroscopic and histological analysis of *Lampe2* livers.*= P<0.05; **= P<0.01; ***= P<0.001

To identify the causative mutation in *Lampe2* mice, we performed coarse mapping by crossing *Lampe2* C57BL/6J homozygotes males with 129S1/SvImJ females. The resulting F1 offspring were intercrossed to generate a F2 and a total of 24 offspring (6 *Lampe2* mutant- and 18 wildtype-phenotypes) were analyzed for both phenotype and genotype. Genotyping was performed using a genome wide custom-made 353-SNP map distinguishing C57BL/6J and 129S1/SvImJ genetic backgrounds. Coarse mapping revealed a single peak with a LOD score of ~5.86 for SNP rs13482738 located on the distal end of chromosome 15 (Figure 3A).

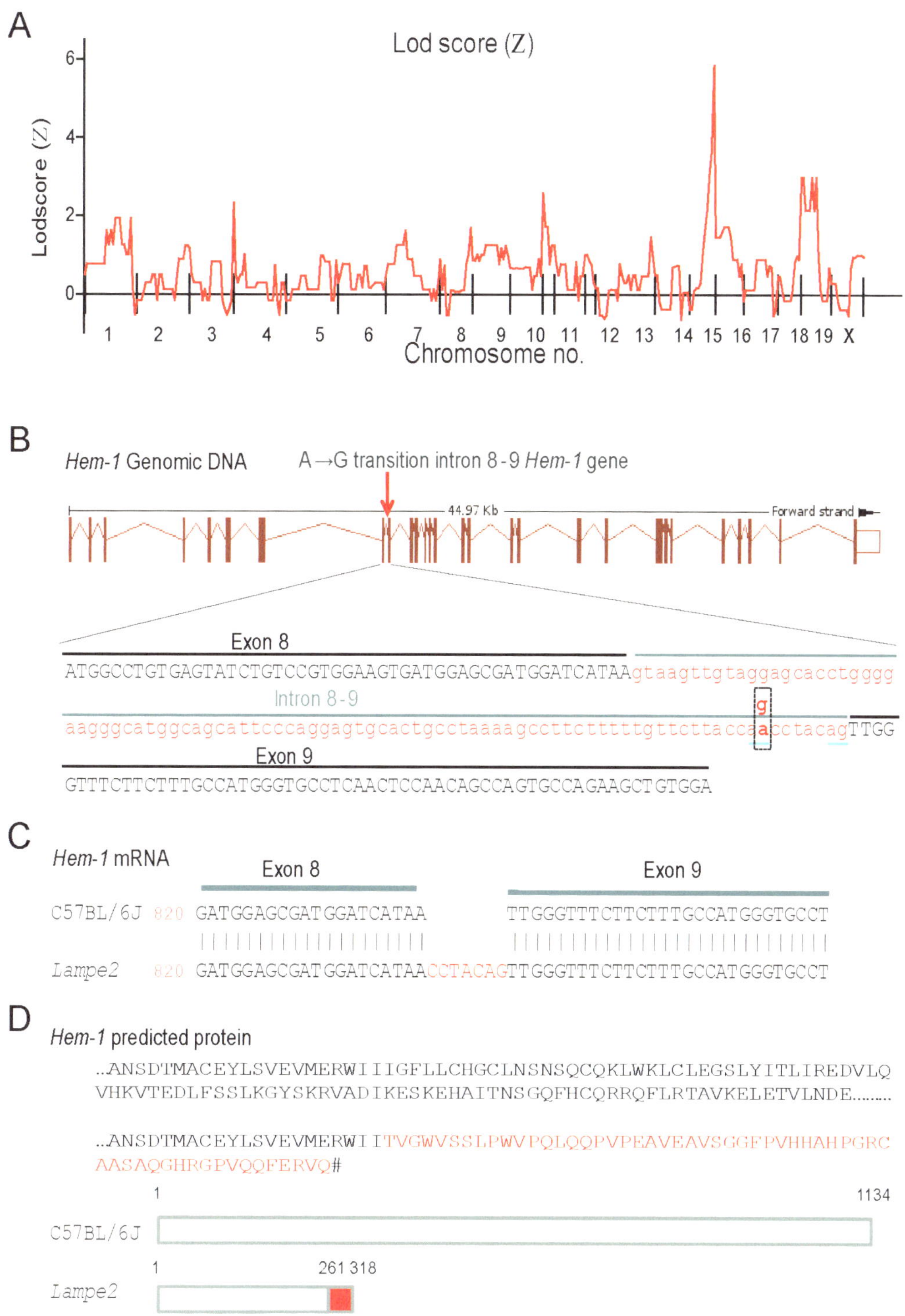

Figure 3. Coarse mapping and identification of the causative mutation in *Lampe2* mice. (a) Low-resolution mapping of the *Lampe2* mutation based on twenty-four mice and a panel of 353 SNPs

covering the entire genome. The *Lampe2* phenotype was linked to the distal site of chromosome 15. (**b,c**), the A→G intronic mutation causes a new acceptor splice site (**b**) resulting in the inclusion of 7 nucleotides intronic sequence into mature *Hem1* transcript as determined by sequencing of *Hem1* cDNA (**c**). (**d**) At the protein level, the mutation is predicted to cause a frameshift at amino acid 261 with alternative coding and premature stop at amino acid 318, resulting in a largely truncated Hem1 protein.

The critical region was defined by proximal marker rs6285067 (at position 95.14 Mb) and the distal end of chromosome 15 (at 103.40 Mb) and consisted of ~8.26 Mb genomic DNA containing 242 annotated genes. Among the annotated genes, *Hematopoietic protein 1* (*Hem-1* aka *NCK associated protein 1 like* or *Nckap1l*) presented a clear candidate gene, in that a previously reported ENU germline mutant carrying a point mutation (referred to as the NBT.1 mutation) causing a premature stop and absence of protein expression, exhibits striking similar phenotypes compared to *Lampe2* mice. Specifically, these mice exhibit lymphopenia with a reduced number of peripheral $CD4^+$ T, $CD8^+$ T and B cells, and on the other hand showed marked expansion of myeloid cells, including neutrophils and macrophages. Moreover, the liver phenotype in mice carrying the NBT.1 mutation bears high resemblance with the liver phenotype observed in *Lampe2* mice—i.e. the occurrence of whitish liver margins and large areas of inflammation. *Hem1* is a member of the Hem family of cytoplasmic adaptor molecules predominantly and is expressed exclusively in hematopoietic cells, including T and B cells, macrophages, DCs and granulocytes. Hem1 plays a critical role in the reorganization of actin cytoskeleton and as such, affects a wide variety of immune functions, including chemotaxis/migration, adhesion, formation of an immune synapse and phagocytosis. Sequencing of *Hem1* exons, including 50 bps of proximal/distal intronic sequence, was performed by Sanger sequencing methodology using genomic DNA and revealed a single A→G point mutation in intron 8-9 located 6 nucleotides upstream of the exon 9 acceptor splice site (Figure 3b). The A→G intronic nucleotide change potentially presented a new acceptor splice site and indeed sequencing of *Hem-1* mRNA isolated from spleen showed the inclusion of 7 intronic nucleotides in mRNA derived from *Lampe2* mice (Figure 3c). At the protein level, the inclusion of intronic nucleotides is predicted to result in a frame-shift and alternative coding following residue 261, and a premature stop at amino acid 381 resulting in a largely truncated Hem-1 protein in *Lampe2* mice (Figure 3d). Given the similarities of the *Lampe2* and *NBT.1* mutant phenotypes and the predicted severe impact of the *Lampe2* mutation on Hem-1 protein expression, we concluded that the mutation in Hem-1 caused the observed phenotypes in *Lampe2* mice (hereafter referred to as $Hem1^{lampe2}$).

Hem1 is part of the Wiskott-Aldrich syndrome protein family Verprolin-homologous protein (WAVE) protein complex in hematopoietic cells regulating cell mobility and intracellular processes requiring rearrangement of the cytoskeleton following immuno-receptor activation, including B and T cell, chemokine and innate immune receptors such as Toll-like receptors. Specifically, receptor triggering causes activation of Rho family of Guanosine triphosphatases (GTPases) such as CDC42, RhoA and Rac ultimately resulting in the activation of downstream adaptor complexes involved in the regulating of actin (de)polymerization. For hematopoietic cells, the adaptor complexes Wiskott-Aldrich syndrome protein (WASP) and WAVE are particularly important for the control of actin polymerization[45-48]. The hematopoietic cell-specific WAVE complex consists of a pentameric subunit complex including, Sra-1 (Specifically Rac-associated protein-1), Hem1, Abi (Abelson interactor 1 or 2), WAVE, and HSPC300 (Hematopoietic stem/progenitor cell protein 300)[49]. Under non-stimulated conditions, the WAVE complex is inactive, but following immunoreceptor activation, GTP-bound Rac binds the pentameric complex presumably through Sra1[49]. In addition, this complex requires binding of phophatidylinositol (3,4,5) triphosphate (PIP_3) interaction and phosphorylation by kinases[50], including Abl kinase and Mitogen-activated protein kinases[51]. Ultimately, this results in a conformational change revealing the WAVE-specific VCA (Verprolin-homology, Cofillin-homology, and acidic) region and allow interaction with the actin-regulatory complex (Arp2/3), ultimately converting monomeric actin (G-actin) into filamentous actin (F-actin). Interestingly, the absence of individual subunit components often causes the degradation of all components of the WAVE complex resulting in aberrant actin polymerization. The consequences of deregulated actin polymerization in hematopoietic cells are wide-ranging and affect broad immunological functions, including but not limited to: 1) leukocyte migration/chemotaxis, 2) loss of immune synapse formation affecting T and B cell receptor signaling (thereby affecting T cell function and development), 3) leukocyte adhesion, and 4) DC-specific phagocytosis and their ability to cross present/prime T cells. As such, mutations in the specific subunit components of the WAVE complex resulting in abnormal gene expression/function cause severe combined immunodeficiencies that stretch beyond lymphocyte populations also affecting granulocyte function and are predicted to correlate with high mortality/morbidity.

6. Implications for human PID

Assessing the immune system using ENU mutagenesis in mice has previously led to important breakthrough discoveries in understanding the genetics in human patients with PID. A prime example is the identification of the *3d* allele—a missense allele of *Unc93b1*, a gene encoding an ER membrane protein with 12 membrane spanning motifs with a previously unknown function. *3d* germline mutants were identified in a screen probing the response of macrophages derived from ENU germline mice to a variety of TLR-ligands. Homozygote *3d* mutant mice were found to be unresponsive to ligands activating endosomal TLRs, but exhibited normal responses to TLRs expressed at the surface[39]. Interestingly, at the same time Casrouge et al. identified two unrelated human patients that presented recurrent infections with Herpes simplex virus-1 (HSV-1) resulting in encephalitis (HSE) and showed remarkable similarities between the phenotypes observed in the human patients and in *3d* mutant mice. Specifically, both patients were unresponsive to endosomal TLR stimulation and showed a high viral susceptibility. Following the identification of the causative mutation in 3d mice as being a missense allele of *Unc93b1*, subsequent sequencing of the human patients indeed revealed aberrant mutations in *UNC93B*[52]. This example highlights the power of ENU mutagenesis and its unbiased approach, by uncovering the function of genes for which a biological function is otherwise difficult to predict.

With regard to the *Gimap5* and *Hem-1* germline mutations described in this chapter, both present examples of genetic mutations leading to severe combined immunodeficiencies. Although limited information is available with regard to genetic mutations causing a null phenotype in human *GIMAP5* or *HEM1*, ample evidence exist that dysregulation of these genes plays an important role in human disease. A previous report suggests that SLE patients were shown to have a trend for lower GIMAP5 mRNA expression in peripheral blood mononuclear cells compared to healthy controls[16]. Moreover, a poly-adenylation mutation in the 3′ region of *GIMAP5*, resulting in minor changes in *GIMAP5* mRNA expression in peripheral blood mononuclear cells, are associated with increased predisposition to SLE and T1D[15, 16, 53]. Thus far, the effect on GIMAP protein expression, specifically in lymphocytes of homozygote/heterozygote carriers for this mutation, remains elusive and warrants further research. Our studies using *Gimap5*-deficient mice point to an important role for Gimap5 in maintaining peripheral immunological tolerance that is intrinsically related to the loss of Gimap5 expression in $CD4^+$ T cells. Thus, research efforts may be directed to a better understanding of GIMAP5 and FOXO protein expression specifically in $CD4^+$ T cells in human patients with SLE or T1D that carry the polyadenylation mutation in GIMAP5.

Finally, perhaps due to its indispensable role in a wide variety of immune pathways, mutations in human HEM1 leading to dysregulated actin polymerization, have thus far not been reported. Nonetheless, over- or under-expression of HEM1 is associated with disease prognosis in leukemia[54]. Specifically, HEM1 overexpression in B-cell chronic lymphocytic leukemia (CLL) is associated with a poor outcome, whereas down-regulation of HEM1 expression in CLL cells rendered tumor cells more susceptible to fludarabine-mediated killing[54]. These findings may indicate the critical role for HEM1 in invasion and/or metastasis of tumor cells from hematopoietic origin.

7. Concluding remarks

A major challenge in the field of genomics is to obtain a comprehensive understanding of the functions of all annotated mammalian genes. Whereas identification and analysis of genome wide SNPs and/or unique nucleotide changes are drastically improved following the development of next generation sequencing technologies, understanding the consequences of such genetic variants remains a major challenge in virtually all biomedical fields. ENU mutagenesis provides one approach that is both powerful and unbiased, uncovering gene function by introducing the sort of genetic abnormalities that can be observed in human patients (e.g. primary immuno-deficiencies). Ultimately, utilization of both forward and reverse genetic approaches will be instrumental in closing the existing phenotype gap and will help us understand the association between identified genetic variants, the implications for protein and biological function, and human disease.

Author details

Kristin Lampe, Siobhan Cashman, Halil Aksoylar and Kasper Hoebe*
Department of Molecular Immunology, Cincinnati Children's Hospital Research Foundation, Cincinnati, Ohio, USA

Acknowledgement

This research was funded by grants from the NIH, including NIH/NIAID RO1 Grant 00426912 and PHS Grant P30 DK078392 (Integrative Morphology Core of the Cincinnati Digestive Disease Research Core Center)

8. References

[1] Visscher PM, Brown MA, McCarthy MI, Yang J. Five years of GWAS discovery. Am J Hum Genet 2012; 90:7-24.
[2] Bousfiha A, Picard C, Boisson-Dupuis S, Zhang SY, Bustamante J, Puel A, et al. Primary immunodeficiencies of protective immunity to primary infections. Clin Immunol 2010; 135:204-9.
[3] Poltorak A, He X, Smirnova I, Liu MY, Van Huffel C, Du X, et al. Defective LPS signaling in C3H/HeJ and C57BL/10ScCr mice: mutations in Tlr4 gene. Science 1998; 282:2085-8.
[4] Brown SD, Peters J. Combining mutagenesis and genomics in the mouse--closing the phenotype gap. Trends Genet 1996; 12:433-5.
[5] Sheridan R, Lampe K, Shanmukhappa SK, Putnam P, Keddache M, Divanovic S, et al. Lampe1: an ENU-germline mutation causing spontaneous hepatosteatosis identified through targeted exon-enrichment and next-generation sequencing. PLoS One 2011; 6:e21979.
[6] Justice MJ, Noveroske JK, Weber JS, Zheng B, Bradley A. Mouse ENU mutagenesis. HumMolGenet 1999; 8:1955-63.
[7] Beutler B, Du X, Xia Y. Precis on forward genetics in mice. NatImmunol 2007; 8:659-64.
[8] Krebs P, Barnes MJ, Lampe K, Whitley K, Bahjat KS, Beutler B, et al. NK-cell-mediated killing of target cells triggers robust antigen-specific T-cell-mediated and humoral responses. Blood 2009; 113:6593-602.
[9] Kawane K, Ohtani M, Miwa K, Kizawa T, Kanbara Y, Yoshioka Y, et al. Chronic polyarthritis caused by mammalian DNA that escapes from degradation in macrophages. Nature 2006; 443:998-1002.
[10] Okabe Y, Kawane K, Akira S, Taniguchi T, Nagata S. Toll-like receptor-independent gene induction program activated by mammalian DNA escaped from apoptotic DNA degradation. JExpMed 2005; 202:1333-9.

---

* Corresponding Author

[11] Tsukumo S, Yasutomo K. DNaseI in pathogenesis of systemic lupus erythematosus. ClinImmunol 2004; 113:14-8.
[12] Barnes MJ, Aksoylar H, Krebs P, Bourdeau T, Arnold CN, Xia Y, et al. Loss of T cell and B cell quiescence precedes the onset of microbial flora-dependent wasting disease and intestinal inflammation in Gimap5-deficient mice. J Immunol 2010; 184:3743-54.
[13] Nitta T, Nasreen M, Seike T, Goji A, Ohigashi I, Miyazaki T, et al. IAN family critically regulates survival and development of T lymphocytes. PLoSBiol 2006; 4:e103.
[14] Schwefel D, Frohlich C, Eichhorst J, Wiesner B, Behlke J, Aravind L, et al. Structural basis of oligomerization in septin-like GTPase of immunity-associated protein 2 (GIMAP2). Proc Natl Acad Sci U S A 2010; 107:20299-304.
[15] Wong V, Saunders A, Hutchings A, Pascall J, Carter C, Bright N, et al. The autoimmunity-related GIMAP5 GTPase is a lysosome-associated protein. self/Nonself 2010; 1:9.
[16] Hellquist A, Zucchelli M, Kivinen K, Saarialho-Kere U, Koskenmies S, Widen E, et al. The human GIMAP5 gene has a common polyadenylation polymorphism increasing risk to systemic lupus erythematosus. J Med Genet 2007; 44:314-21.
[17] Shin JH, Janer M, McNeney B, Blay S, Deutsch K, Sanjeevi CB, et al. IA-2 autoantibodies in incident type I diabetes patients are associated with a polyadenylation signal polymorphism in GIMAP5. Genes Immun 2007; 8:503-12.
[18] Hornum L, Romer J, Markholst H. The diabetes-prone BB rat carries a frameshift mutation in Ian4, a positional candidate of Iddm1. Diabetes 2002; 51:1972-9.
[19] Jacob HJ, Pettersson A, Wilson D, Mao Y, Lernmark A, Lander ES. Genetic dissection of autoimmune type I diabetes in the BB rat. NatGenet 1992; 2:56-60.
[20] Macmurray AJ, Moralejo DH, Kwitek AE, Rutledge EA, Van Yserloo B, Gohlke P, et al. Lymphopenia in the BB rat model of type 1 diabetes is due to a mutation in a novel immune-associated nucleotide (Ian)-related gene. Genome Res 2002; 12:1029-39.
[21] Ramanathan S, Poussier P. BB rat lyp mutation and Type 1 diabetes. ImmunolRev 2001; 184:161-71.
[22] van den Brandt J, Fischer HJ, Walter L, Hunig T, Kloting I, Reichardt HM. Type 1 diabetes in BioBreeding rats is critically linked to an imbalance between Th17 and regulatory T cells and an altered TCR repertoire. J Immunol 2010; 185:2285-94.
[23] Cousins L, Graham M, Tooze R, Carter C, Miller JR, Powrie FM, et al. Eosinophilic bowel disease controlled by the BB rat-derived lymphopenia/Gimap5 gene. Gastroenterology 2006; 131:1475-85.
[24] Podolsky DK. Inflammatory bowel disease. N Engl J Med 2002; 347:417-29.
[25] Xavier RJ, Podolsky DK. Unravelling the pathogenesis of inflammatory bowel disease. Nature 2007; 448:427-34.
[26] Lees CW, Barrett JC, Parkes M, Satsangi J. New IBD genetics: common pathways with other diseases. Gut 2011.
[27] Mackay IR. Clustering and commonalities among autoimmune diseases. J Autoimmun 2009; 33:170-7.
[28] Khoruts A, Fraser JM. A causal link between lymphopenia and autoimmunity. Immunol Lett 2005; 98:23-31.

[29] King C, Ilic A, Koelsch K, Sarvetnick N. Homeostatic expansion of T cells during immune insufficiency generates autoimmunity. Cell 2004; 117:265-77.

[30] Krupica T, Jr., Fry TJ, Mackall CL. Autoimmunity during lymphopenia: a two-hit model. Clin Immunol 2006; 120:121-8.

[31] Aranda R, Sydora BC, McAllister PL, Binder SW, Yang HY, Targan SR, et al. Analysis of intestinal lymphocytes in mouse colitis mediated by transfer of CD4+, CD45RBhigh T cells to SCID recipients. J Immunol 1997; 158:3464-73.

[32] Powrie F, Correa-Oliveira R, Mauze S, Coffman RL. Regulatory interactions between CD45RBhigh and CD45RBlow CD4+ T cells are important for the balance between protective and pathogenic cell-mediated immunity. J Exp Med 1994; 179:589-600.

[33] Powrie F, Leach MW, Mauze S, Caddle LB, Coffman RL. Phenotypically distinct subsets of CD4+ T cells induce or protect from chronic intestinal inflammation in C. B-17 scid mice. Int Immunol 1993; 5:1461-71.

[34] Aksoylar HI, Lampe K, Barnes MJ, Plas DR, Hoebe K. Loss of Immunological Tolerance in Gimap5-Deficient Mice Is Associated with Loss of Foxo in CD4+ T Cells. J Immunol 2012; 188:146-54.

[35] Anderson MJ, Viars CS, Czekay S, Cavenee WK, Arden KC. Cloning and characterization of three human forkhead genes that comprise an FKHR-like gene subfamily. Genomics 1998; 47:187-99.

[36] Biggs WH, 3rd, Cavenee WK, Arden KC. Identification and characterization of members of the FKHR (FOX O) subclass of winged-helix transcription factors in the mouse. Mamm Genome 2001; 12:416-25.

[37] Furuyama T, Kitayama K, Shimoda Y, Ogawa M, Sone K, Yoshida-Araki K, et al. Abnormal angiogenesis in Foxo1 (Fkhr)-deficient mice. J Biol Chem 2004; 279:34741-9.

[38] Kerdiles YM, Beisner DR, Tinoco R, Dejean AS, Castrillon DH, DePinho RA, et al. Foxo1 links homing and survival of naive T cells by regulating L-selectin, CCR7 and interleukin 7 receptor. NatImmunol 2009; 10:176-84.

[39] Ouyang W, Beckett O, Flavell RA, Li MO. An essential role of the Forkhead-box transcription factor Foxo1 in control of T cell homeostasis and tolerance. Immunity 2009; 30:358-71.

[40] Tothova Z, Gilliland DG. FoxO transcription factors and stem cell homeostasis: insights from the hematopoietic system. Cell Stem Cell 2007; 1:140-52.

[41] Kerdiles YM, Stone EL, Beisner DL, McGargill MA, Ch'en IL, Stockmann C, et al. Foxo transcription factors control regulatory T cell development and function. Immunity 2010; 33:890-904.

[42] Ouyang W, Beckett O, Ma Q, Paik JH, DePinho RA, Li MO. Foxo proteins cooperatively control the differentiation of Foxp3+ regulatory T cells. Nat Immunol 2010; 11:618-27.

[43] Gomis RR, Alarcon C, He W, Wang Q, Seoane J, Lash A, et al. A FoxO-Smad synexpression group in human keratinocytes. Proc Natl Acad Sci U S A 2006; 103:12747-52.

[44] Seoane J, Le HV, Shen L, Anderson SA, Massague J. Integration of Smad and forkhead pathways in the control of neuroepithelial and glioblastoma cell proliferation. Cell 2004; 117:211-23.

[45] Park H, Chan MM, Iritani BM. Hem-1: putting the "WAVE" into actin polymerization during an immune response. FEBS Lett 2010; 584:4923-32.

[46] Park H, Staehling-Hampton K, Appleby MW, Brunkow ME, Habib T, Zhang Y, et al. A point mutation in the murine Hem1 gene reveals an essential role for Hematopoietic protein 1 in lymphopoiesis and innate immunity. J Exp Med 2008; 205:2899-913.

[47] Thrasher AJ, Burns SO. WASP: a key immunological multitasker. Nat Rev Immunol 2010; 10:182-92.

[48] Williams DA, Tao W, Yang F, Kim C, Gu Y, Mansfield P, et al. Dominant negative mutation of the hematopoietic-specific Rho GTPase, Rac2, is associated with a human phagocyte immunodeficiency. Blood 2000; 96:1646-54.

[49] Gautreau A, Ho HY, Li J, Steen H, Gygi SP, Kirschner MW. Purification and architecture of the ubiquitous Wave complex. Proc Natl Acad Sci U S A 2004; 101:4379-83.

[50] Lebensohn AM, Kirschner MW. Activation of the WAVE complex by coincident signals controls actin assembly. Mol Cell 2009; 36:512-24.

[51] Danson CM, Pocha SM, Bloomberg GB, Cory GO. Phosphorylation of WAVE2 by MAP kinases regulates persistent cell migration and polarity. J Cell Sci 2007; 120:4144-54.

[52] Casrouge A, Zhang SY, Eidenschenk C, Jouanguy E, Puel A, Yang K, et al. Herpes simplex virus encephalitis in human UNC-93B deficiency. Science 2006; 314:308-12.

[53] Lim MK, Sheen DH, Kim SA, Won SK, Lee SS, Chae SC, et al. IAN5 polymorphisms are associated with systemic lupus erythematosus. Lupus 2009; 18:1045-52.

[54] Joshi AD, Hegde GV, Dickinson JD, Mittal AK, Lynch JC, Eudy JD, et al. ATM, CTLA4, MNDA, and HEM1 in high versus low CD38 expressing B-cell chronic lymphocytic leukemia. Clin Cancer Res 2007; 13:5295-304.

Chapter 10

Mutagenesis in Plant Breeding for Disease and Pest Resistance

Petra Kozjak and Vladimir Meglič

Additional information is available at the end of the chapter

http://dx.doi.org/10.5772/50332

1. Introduction

Food production and food security faces several challenges such as climate change and expanding human growth, the competition of food and non-food uses, and decreasing area of arable land. The role of plant breeding in providing sustainable food production is to enable stable yields with lower inputs of fertilizers, energy and water use, to produce safe and quality food and to meet the demand of a projected raise in human population and livestock production. World population is projected to reach 10 billion by 2100 (United Nations, 2011) with the trend of changing diet towards higher quality food. Mutagenesis could be one of the solution to challenges facing the agriculture. Mutation breeding has substantially contributed the countries' economies and to conservation of biodiversity by stopping gene erosion. Improvement of crop production regarding pest and disease management is one of the main goals in agricultural breeding. Pathogens cause huge yield losses in the agriculture every year with large economic losses and damage to ecosystems. Disease outbreaks pose threats to global food security causing global yield loss of 16% (Oerke, 2006). Actual losses due to pests (weeds, animal pests and pathogens) range from 26-29% for sugar beet, barley, soybean, wheat and cotton, to 31-40% for maize, potato and rice (Oerke, 2006). The actual loss is referring to the losses sustained despite protection measures applied. Plant parasitic nematodes cause crop losses up to 125 US dollars annually (Chitwood, 2003). The constant challenge in plant breeding is to deal with the overcome disease and pest resistance and the development of new aggressive strains of pathogens such as fungi *Puccinia striiformis*, a causal agent of wheat yellow rust. The advances in molecular technology and in recent findings in cloning of disease resistance (*R*) genes allow the improvement of crop disease resistance by applying traditional breeding, genomic approaches, transgenic deployment and mutagenesis tools for enhancing disease and pest resistance. Using radiation breeding, traits for yield, quality, taste and disease and pest resistance have been improved in cereals, legumes, cotton, peppermint, sunflowers, peanut, grapefruit, sesame, banana and cassava. Basic scientific research has substantially benefited from mutagenesis. Using *in vitro* mutagenesis, a considerable progress in understanding the evolution of molecular mechanisms of resistance was achieved.

2. Disease and pest resistance in plants

Plants encounter numerous beneficial and harmful organisms (pathogens) in the environment and use different strategies and mechanisms to cope with in order to survive and reproduce successfully. Basal resistance is referring to the constitutive defence provided by pre-existing physical and chemical barriers in order to disable penetration of pathogen to the host-cell. Another aspect of basal resistance is the recognition of microbial surfaces by cell surface receptors that trigger immune response and offer broad-spectrum resistance. This non-specific resistance is called pathogen associated molecular pattern (PAMP)-triggered immunity (PTI) (Jones & Dangl, 2006). There is an evidence of structural similarity of cell-surface receptors, usually receptor-like kinases, between plants and animals (Nurenberger et al., 2004). The term PAMP is referring to small conserved molecules secreted on the surface of a class of microbes. In bacteria, well characterized PAMPs are: i) flagellin, which is a major structural protein essential for bacteria motility (Ramos et al., 2004), ii) lipopolysaccharides (LPS), a component of the cell wall of Gram-negative bacteria, and iii) peptidoglycan (PGN), a polymer forming the cell wall common to all bacteria (Akira & Takeda, 2004; Janeway & Medzhitov, 2002). In fungi, well characterized PAMPs are chitins, mannans and proteins (Cohn et al., 2001; Holt et al., 2003; Parker, 2003). PTI immune system exist in all higher plants (Boller & He, 2009). For example, homologues of *Arabidopsis FLS2* gene, coding for LRR receptor-like kinase, were found in all sequenced plants. Apart from structural conservation of *FLS2* gene there is proven functionality between different species. Rice *FLS2* gene is functional in *Arabidopsis fls2* mutant, thus suggesting conservation of associated signalling pathways (Takai et al., 2008). During the co-evolution of interplay between successful plant defence and pathogen attack, plant evolved rapid defence responses, involving programmed cell death during hypersensitive response. The response is mediated through R proteins that are either directly involved in the recognition of pathogen effectors or act as a guardian for the modification of plant proteins. Higher level of defence is able to detect specific pathogen effectors and is referred to effector-triggered immunity (ETI). Recent advances in understanding plant immunity suggest that basal resistance and race-specific resistance (ETI) evolve simultaneously as an answer to selection pressure on both actors. Natural selection drives the pathogen to avoid resistance either by evolving the existent effector gene or by acquiring additional effectors. This new effector put the selection pressure on host plant to evolve new *R* gene alleles. The co-evolution of plant defence and pathogen attacks are the result of constant selection pressure that occur across spatial and temporal scales (Ravensdale et al., 2011). In PTI immunity system there is an evidence of molecular evolutionary conservation in structure and functions across kingdoms borders (Medzhitov & Janeway, 1997; Imler & Hoffmann; 2001), however the evidence of existence of ETI in animals is missing. ETI enables the detection of pathogen-specific effectors by protein receptors coded by *R* genes in every single cell in contrast to invertebrate animals that have circulating system, which constitutes to important distinction between plant and animal innate immune systems (Ausubel, 2005). The major players in expressing ETI are plant *R* and pathogen *Avr* genes. Unlike PTI, which is expressed in all plants of a given species, ETI is often expressed in some but not all genotypes within a plant species against pathogen race specific effectors. Although ETI response is fast and effective, plant can also detect pathogens through basal immune system.

2.1. *R* genes

For most proteins coded by *R* genes there are characteristic, conserved, structural domains. In general, we can divide R proteins according to the mode of resistance, to race-specific and race-non-specific. According to structural motif, they can be divided into five classes (Hammond-Kosack & Parker, 2003). In the first class, there are serin/threonin kinases such as *Pto* gene at tomato conferring resistance to bacteria *Pseudomonas syringae*. All other R proteins, combined in four classes, have leucine rich repeat domain and are distinguished by the localization of these domains. R proteins of second class are transmembrane receptors with extracellular LRR domain (*Cf* gene family in tomato), while R proteins of third class have extracellular LRR domain connected to kinase domain (*Xa21* gene at rice). *R* genes belonging to the fourth and fifth group code for intracellular proteins with NBS and LRR domain. LLR domain is important for ligand binding and the recognition of pathogen effectors (Young, 2000). The C- and N-terminal end of LRR domain are proposed to have distinct functions, the C-terminal end is responsible for the ligand recognition and important for determining R-Avr specifity, while N-terminal end is responsible for activation of further signal transduction (Inohara & Nunez, 2003; Tanabe et al., 2004; Chen et al., 2004). Structural similarities between NBS-LRR proteins of different species and taxa confirm the conservation of basic mechanism of defence against pathogens during the evolution and diversification (Moffet et al., 2002). Although R proteins share similar structure at the amino acid level, they clearly differentiate at the nucleotide level. For example, the level of amino acid hop (*Humulus lupulus* L.) RGA sequences compared to cloned *R* genes of evolutionary distant plants such as *Arabidopsis* is mainly restricted to the presumed functional domain (Kozjak et al., 2009).

2.2. Interplay between plant defence and pathogen attack

There are few models describing the interaction between pathogen avirulence (*Avr)* molecules called effectors and *R* proteins that are differing in the mode of action (direct or indirect).

2.2.1. Gene-for-gene

Gene for gene concept is based on direct physical interaction between plant *R* gene and corresponding pathogen avirulence *Avr* gene (Flor, 1955). Examples of such interactions have been described in tomato, where *Pto* interacts with *AvrPto* gene product of

Pseudomonas syringae (Scofield et al. 1996), in rice-rice blast pathosystem, where *Pi-ta* interacts with Avr-Pita (Jia et al., 2000) and in *Arabidopsis*, where RRS1 protein interacts with *Avr-PopP2* gene product of *Ralstonia solanacearum* (Bernoux et al. 2008).

2.2.2. Guard hypothesis

Alternatively, the guard hypothesis is based on the assumption that R proteins act as guards on host target proteins (guardee) and are a part of protein complex. This hypothesis predicts R proteins to be part of surveillance machinery and suggests indirect interaction between R proteins and corresponding *Avr* gene products. R proteins are activated by the modifications of host targets of corresponding pathogen effector (van der Biezen & Jones, 1998; Dangl & Jones, 2001). Two scenarios are proposed for indirect interactions (Figure 1). The Guard model was proposed to explain how the single *R* gene product perceives multiple effectors (Jones & Dangl, 2006) (Figure 1). The first experimental evidence shown for RPM1-mediated disease resistance to *P. syringae* revealed that RPM1 signalling cascade is activated by a protein component RIN4 which also needs to be activated by the phosphorylation in the presence of *AvrB* or *AvrRpm1* (Mackey et al., 2002, 2003). In the absence of effectors, RPM1 is negatively regulated by the RIN4 and stays in inactive form (Mackey et al. 2003). Axtell & Staskawicz (2003) demonstrated that RIN4 has a dual role and acts as a negative regulator of RPS2 activation conferring resistance to *P. syringae* expressing AvrRpt2. In contrast to RIN4 phosphorylation, for the activation of RPM1 signalling pathway, RPS2 activity requires the AvrRpt2-mediated disappearance of RIN4.

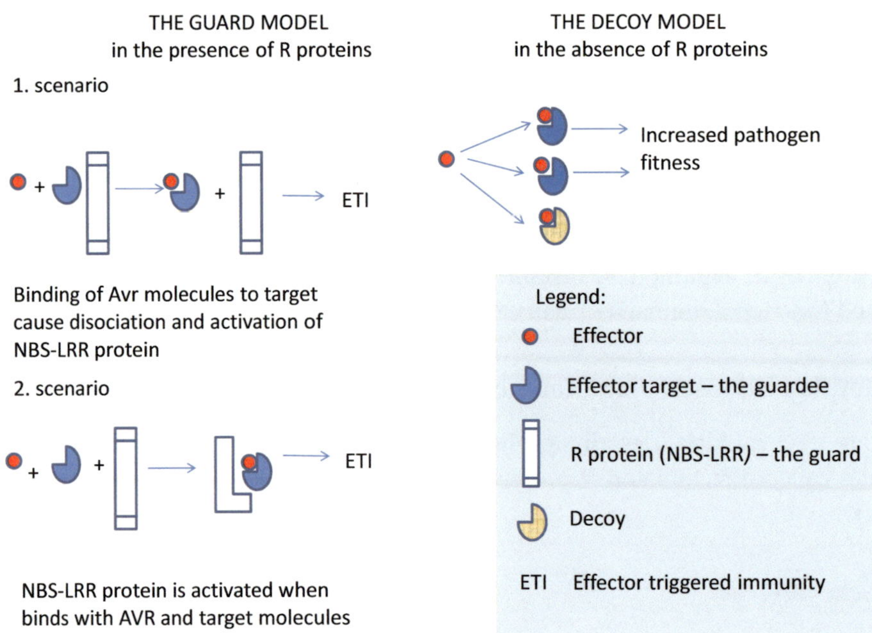

Figure 1. Schematic presentation of Guard and Decoy model

2.2.3. *The Decoy model*

The physical nature of the R-Avr interaction has big impact on the evolution of these proteins (Ravensdale et al., 2011). Effector target and plant guardee are under opposing selection pressures. First, in the absence of *R* gene product, the binding affinity of guardee should decrease in order to avoid detection and modification of a guardee. Just opposite, in the presence of functional *R* gene product, the selection pressure is put on guardee to enhance pathogen detection by improved interactions (van der Hoorn & Kamoun, 2008). This opposite pressure lead to unstable situation that could be released by the host protein that mimics the effector target without contributing to pathogen fitness. This host protein is termed as a "decoy" and is specialized in attracting effector. Difference between the Decoy and Guard models is that in the Decoy model, the pathogen fitness does not benefit from the absence of R protein (van der Hoorn & Kamoun, 2008) (Figure 1).

The Decoy model was proposed just recently and has to be experimentally proven, however few well-studied effector-perception mechanisms support this model. Tomato Pto interacts with avrPto to trigger the resistance to *P. syringae*, with the associated NB-LRR Prf protein that is necessary to trigger further defences. Prf protein acts as a guard on Pto. In addition to Pto, AvrPto binds to different receptor kinase targets, including FLS2 in *Arabidopsis* and LeFLS2 in tomato to block plant immune responses. AvrPto contributes to virulence on tomato even in the absence of Pto (Chang et al. 2000) but not on Arabidopsis lacking FLS2 (Xiang et al., 2008). On *fls* mutants, AvrPto no longer contributes to virulence (Xaing et al., 2008). It has been proposed that Pto competes with FLS2 for AvrPto binding (Zhou & Chai, 2008; Zipfel & Rathjen 2008). In this case, Pto acts as a decoy. Since AvrPto inhibits multiple kinases, Pto could evolve by mimicking one of them by losing some of the structural properties or by duplication and subsequent divergent evolution (Tian et al., 2007, van der Hoorn & Kamoun, 2008). Both of the models, Guard and Decoy, are not necessarily excluding each other since "guardee" may evolve into the "decoy".

2.2.4. *Co-evolution of plant resistance and pathogen virulence*

The co-evolution of host-pathogen interaction is driven by different factors, such as environmental conditions, population size and pathogen dispersal mechanisms that put the selective pressure on each other across space and time. Plant defences against pathogen attacks are dynamic processes that involve regulation of many defence components on the cellular level. NBS-LRR genes take a part in network with other components of signal transduction, since most proteins act as a complex with other components. During the defence, multiple organelles are included in the recognition and signalling mechanisms. The intracellular trafficking of pathogen effectors, mRNA and R proteins between the cytoplasm and nucleus is crucial for successful immune responses (Deslandes & Rivas, 2011). There has been evidence that effectors modulate transcriptional machinery by activation or repression suggesting the involvement of defence associated loci through changes of chromatin (van der Burg & Takken, 2009). The co-evolution of other components is prerequisite for optimal functioning, which is seen as different quantitative characteristics among species (Jones &

Dangl, 2006). This is the case of *Bs2* gene from pepper carrying resistance against bacteria *Xanthomonas sp.*, which is functional in many species within the *Solanaceae* family but not outside the family. Similarly, in *Arabidopsis* some traits may not be relevant to non-brassicaceous plants. Diversity for the virulence (or specialization) and the host resistance is dependent on the reproductive strategies of the host (out crossing or inbreed) and geographical distribution. Host populations can represent distinct groups regarding disease resistance. Ravensdale et al. (2011) analysed host resistance in flax against flax rust resistance and found that resistance structure within populations varied from nearly monomorphic to highly polymorphic, having at least 18 different resistance phenotypes. He concludes that temporal and spatial variation of disease resistance between populations puts stronger selection pressure or drift on the evolution of resistance than on the gene flow. The ZIGZAG model, proposed by Jones & Dangl (2006), illustrates the quantitative output of the plant immune system that can be presented in four phases. In phase 1, plants detect pathogen effectors or PAMPs and trigger PAMP triggered immunity (PTI). In phase 2, pathogen interfere with PTI, in phase 3, an effector is recognized by R protein activating effector triggered immunity (ETI) and in phase 4, pathogen evolve new effectors to suppress ETI thus putting the selection pressure on new R protein alleles in plants.

2.2.4.1. Development and evolution of R genes

R genes develop by different natural mutagenesis mechanisms such as: i) recombination, ii) tandem or segmental duplication gene events, iii) unequal crossing-over, iv) point mutation and v) selection pressure from the environment (Meyers et al. 2005). *R* genes and analogs of *R* genes (RGAs) have strong tendency for clustering in plants (Meyers et al., 1998; Gebhardt and Valkonen, 2001; Mutlu et al., 2006; Di Gaspero et al,. 2007). NBS-LRR genes are unevenly distributed and usually organised in clusters including pseudogenes (Meyers *et al.* 1999). Pseudogenes are assumed to be the source of higher variation than in coding genes and offer a potential reservoir for the *R* gene evolution, so the polymorphism detected in non-coding area of genome is rather expected (Calenge et al., 2005). Recombination is often in closely related and physically close *R* genes (Meyers et al., 2003; Baumgarten et al., 2003) however, in *R* gene cluster of soybean and lettuce a phenomena of suppressed recombination was observed (Kanazin et al. 1996; Meyers et al., 1998). Genome analyses of *Arabidopsis* shows translocation events of NBS-LRR genes by genomic duplications at distant, probably random locations in the genome, these mutations are called ectopic mutations (Meyers et al., 2003; Baumgarten et al., 2003; Leister, 2004). At some loci gene families expand by tandem duplications, doubled sequences are accumulating mutations, which increase the complexity of *R* gene sequences. Comparative sequence analyses of different plant species of *Arabidopsis* (Meyers et al., 2003), tomato (Seah et al., 2007), wild potato (Kuang et al., 2005), wheat (Wicker et al., 2007), rice (Dai et al., 2010), soybean (Innes et al., 2008) and common bean (David et al., 2009) suggest that *R* gene follow different evolution path. Assuming that *R* genes evolve as response to selection pressure of pathogens and changing environment, Kuang et al. (2004; 2005) proposed two evolutionary categories: type I, include genes of frequent sequence exchange among paralogs and type II include slowly evolving genes with the accumulation of single amino acid substitutions.

Although most of *R* genes are dominantly inherited, there are recessive genes that confer race non-specific resistance such as *Mlo* gene at barley against *Erysiphe graminis* (Buschges et al., 1997), *RRS-e* gene at Arabidopsis against different races of *Ralstonia solanacearum* (Deslandes et al., 2002) and *Xa5* at rice against *Xanthomonas oryzae* pv. *oryzae* (Iyer & McCouch, 2004).

3. Breeding for disease and pest resistance

Commonly used methods for improving elite cultivars for disease and pest resistance combine traditional breeding methods (hybridization, selection, and introduction), alternative methods (tissue culture) and mutagenesis using forward and reverse genetic techniques (Figure 2). The induction of mutations for crop improvement is termed mutation breeding. To identify genes and its function two main approaches are employed: forward and reverse genetic techniques. The term forward genetics is used for identifying (cloning) gene, while the term reverse genetics is used to reveal gene function by analyzing phenotypic effects of a gene with known sequence. With the establishment divison for Agriculture & Biotechnology at the International Atomic Energy Agency (IAEA) and the Food and Agriculture Organization, more than 2000 varieties have been released that derived from either direct mutant or crosses between mutants (Ahloowalia et al., 2004). Most of these varieties are improved for increased yield and enhanced quality (improved processing quality, increased stress tolerance,...). Improved characteristics have been released in more than 175 varieties and plant species.

3.1. Classical breeding

The most effective approach to prevent disease outbreak is to cultivate resistant varieties. Transferring genes through conventional transfer process may be hampered by the vertifolia effect that refers to the loss of horizontal resistance during the breeding for vertical resistance (Van der Plank, 1963). A frequent problem associated with *R* genes is their short-term efficacy. Disease resistance of genetically uniform lines with single source of resistance is often defeated by new pathogen races when cultivated large-scale and long-term. This was the case with rice carrying only *Xa4* gene against bacterial blight across several Asian countries (reviewed in Kameswara Rao et al., 2002). Planting a mixture of cultivars would reduce the disease incidence, but intensive mechanization of crop production and modern markets demand uniform crops.

3.1.1. Map-based cloning

Map-based cloning is an approach to identify *R* gene and determine the sequence of a gene using molecular markers. We distinguish two different types of mapping: i) genetic mapping based on the classical techniques using pedigree or breeding of recombinant phenotypes and ii) physical mapping based on the use of biotechnological techniques (genetic fingerprinting) to determine the order and spacing between markers or genes. Linkage map is a genetic map presenting genes in lineage order and distance in between in centimorgans (cM). Mapping Quantitative Trait Loci (QTLs) is effective approach for studying plant disease resistance. The first step in map-based cloning is to place molecular markers that lie near a gene of interest and co-segregate with proposed gene without recombination. It has been shown that soybean cist nematode resistance, rice blast resistance and black mold resistance in tomato, grey leaf spot and common rust in maize are under the control of QTL (Wang et al., 1994, Robert et al., 2001; Concibido et al., 2004; Danson et al., 2008). The second step is to clone the gene by chromosome walking and sequencing the gene. Determination of QTLs is important for studying epistatic interactions and race specifity. More than 35 QTLs in rice were found near *R* genes for resistance to blast (Ballini et al., 2008; Fukuoka & Okuno, 1997; Tabien et al., 2000, 2002). Identification of markers linked to QTL facilitates the targeting of recessive alleles, which can be masked by epistasis in the specific environment (Joshi & Nayak, 2010).

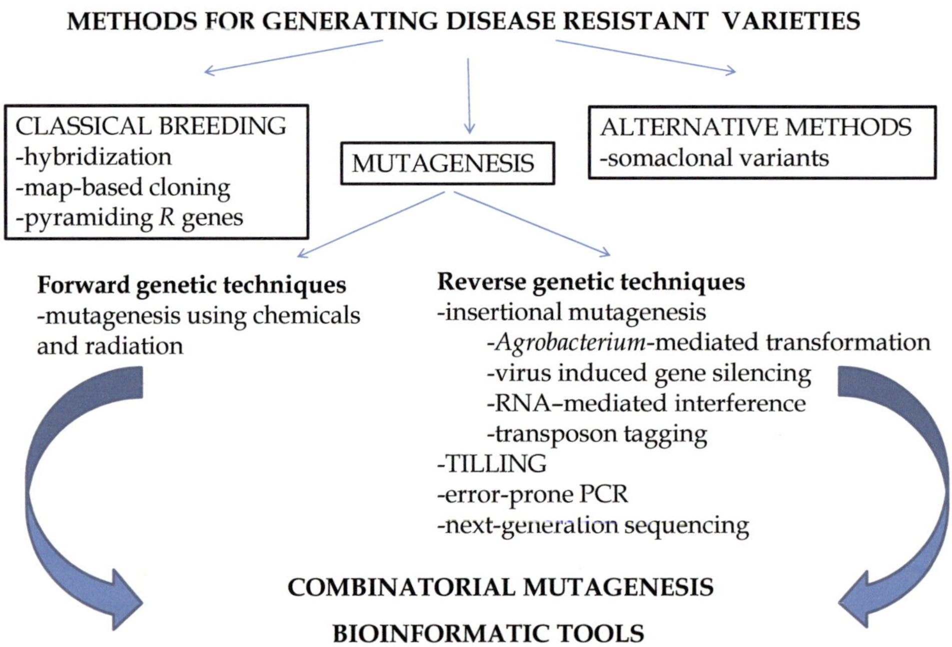

Figure 2. Schematic presentation of most commonly used methods for pest and disease resistance breeding in crops

3.1.2. Pyramiding R genes by marker assisted selection (MAS)

In order to avoid breakdown of resistance conferred by single *R* gene, pyramiding multiple *R* genes in genetically uniform lines presents an alternative. The idea of pyramiding *R* genes into crops is to construct sufficiently large pools of *R* genes that correspond to all avirulence genes in pathogen populations of specific regions. The probability of pathogen to break resistance to two or even more genes is much lower than to single gene. Advantages of pyramiding genes in single genotype are: i) more effective control of insect resistant to single toxin that may be controlled by a second toxin, ii) lower probability of evolving resistance to two independent actions through selection of one toxin, and iii) a single effector cannot break resistance to binding to immunologically different targets (Gahan et al. 2005). The problem of introducing several genes by classical breeding is the transfer of undesirable traits that need to be removed by backcrossing. Gene pyramiding by classical breeding is also difficult due to the dominance and epistatic effects (Singh et al., 2001), but the identification of molecular markers linked to resistance genes or loci ease the identification of desired plants. The selection of desirable phenotype by molecular markers is termed marker assisted selection (MAS). MAS-based gene pyramiding is an analogue approach to classical breeding but less time consuming and relying on the use of molecular markers that speed up the selection procedures. Using sequence tagged sites (STS) markers, MAS based gene pyramiding and marker-aided backcrossing procedures several genes have been successfully transferred in elite rice cultivars (Huang et al. 1997; Singh et al. 2001). In common wheat, three leaf rust resistance genes *Lr41, Lr42, Lr43* were successfully pyramided as well (Cox et al. 1993).

3.2. Mutagenesis

The discovery of x-rays inducing mutations in *Drosophila melanogaster* presents the beginning of mutation breeding in plants. The term mutagenesis applies to methods used for the induction of random or site directed mutations in plant DNA to create new valuable traits in well-adopted cultivars. According to the FAO/IAEA database there are 320 cultivars with improved disease resistance using mutagenic agents that were obtained as direct mutant or derived from hybridization with mutant or from progeny (for example by self-fertilization). Induced mutations have been used to improve economically important crops such as wheat, barley, rice, cotton, peanut, banana etc. Disease and pest resistance in commercial crops was improved mostly in cereals (rice, barley, maize, wheat) and legumes (bean, green pea).

Spontaneous mutations occur at low frequencies, one in a million per gene. If two independent mutations are necessary in recessive alleles to obtain resistant phenotype, the frequency lowers to 10^{-18} per nucleotide (Gressel & Levy, 2006). Mutagenesis is used to accelerate spontaneous mutations in driven evolution. Using chemical mutagen (EMS) in *Arabidopsis* about ten mutations per gene were recorded among 192 genes in 3,000 M_2 plants examined (Greene et al., 2003) with an average of 720 mutations in single M_2 plant (Till et al., 2003). For the improvement of disease resistance, the induction of spontaneous mutations is applied by different mutagenesis approaches: virus induced gene silencing, RNA-mediated interference, *Agrobacterium*-mediated insertional mutagenesis, radiation and chemical mutagenesis and with combined approaches such as Targeting Induced Local Lesions in Genome (TILLING). For the identification of mutants, different methods have been developed through years, that include: i) high resolution melting techniques (HRM), ii) protein truncation test that detect mutants from the terminatioin of mRNA translation, iii) single-strand conformation polymorphism (SSCP) for the detection of frameshift mutations, nonsense and missense mutations, iv) Southern hybridization for detecting large mutations (deletion, insertions, rearrangements), v) denaturing gradient gel electrophoresis (DGGE), vi) DNA microarray, vii) single and multiparallel DNA sequencing, viii) TILLING for the detection of mutations in large exon-rich amplicons and ix) PCR based detection technique. Novel sequencing approaches based on Sequence Candidate Amplicons in Multiple Parallel Reactions are now most commonly used in genomic analyses of gene expression and regulation modes, including the production of genetic maps. The new generation machines (Illumina Genome Analyser, ABI SOLiD, Roche 454) are capable of producing millions of DNA sequences in a single run. The advantage of multiparallel sequencing using pooling strategy is the identification of rare mutations that are distinguishable from background sequencing errors.

3.2.1. Induced mutagenesis by chemical or physical mutagens

Most mutagenic populations are generated by treating seeds with radiation or chemical mutagens. Physical mutagens are X-rays, Gamma rays, alpha particles, UV and radioactive decays. Irradiation usually cause large mutations (large-scale deletions of DNA), while chemical mutagens usually cause point mutations. Fast neutrons are high-energy thermal neutrons produced by nuclear fission. They induce broad range of deletions and chromosomal changes and are often accompanied by gamma radiation. The major fast neutron bombardment technique is Delete-a-gene, a knockouts gene system for plants (Li et al., 2002; Li & Zhang, 2002; Rogers et al., 2009). Delete-a-gene combines fast neutron radiation of seeds and identification of mutants by PCR using two specific primers for targeted locus and shortened PCR extension time to suppress the amplification of a wild type gene. Delete-a-gene is used as alternative method to insertional methods in cases when we do not have well characterized transposons or when the genes are placed in tandem duplication and we cannot inactivate them at the same time in order to observe mutant phenotype (Li et al. 2002). It can be applied to all plants since no transformation and tissue culture is needed. The carbon ions with high linear energy transfer (LET) has been proven very successful for the induction of base substitutions or small deletions/insertions in *Arabidopsis* that can be determined by single nucleotide polymorphism (SNP) detection system which is beneficial for forward genetics and plant breeding (Kazama et al., 2011).

Mutations induced by chemical mutagens are point mutations and are less damaging (not lethal) than large rearrangements.. The advantage of chemical mutagenesis is that is can provide loss- and gain– of - function of genes. There are various chemical mutagens used for generating variability, such as sodium azide, ethyl methanesulphonate (EMS), methyl methanesulphonate (MMS), hydrogen fluoride (HF), diethyl sulphate, hydroxylamine and N-methyl-N-nitrosourea (MNU). Most commonly used mutagen in creating TILLING populations in maize, rice, *Brassica* sp., pea, barley, wheat, soya and cucumber is ethyl methanesulphonate (EMS) (reviewed in Kurowska et al. 2011).

Induced mutagenesis by chemical or radiation mutagens have advantages over insertional methods, since mutagens introduce random changes throughout genome and can generate variety of mutations within a single plant. Comparing to other methods, it is applicable to all crops and it does not demand the establishment of species-specific protocols for transformation and regeneration.

3.2.2. *Agrobacterium-mediated transformation*

Plant transformation technologies employ physical incorporation of foreign DNA into the host genome by different approaches, directly or indirectly. The indirect methods include transformations using *Agrobacteria tumefaciens* or *Agrobacteria rhizogenes*, while direct approaches include protoplast transformation and biolistic or microprojectile bombardment. *Agrobacterium* mediated insertional mutagenesis rely on a natural process of transferring T-DNA as a short segment of *Agrobacterium* plasmid to plant genome when infected by the *Agrobacterium*. The main transgenic crops improved for disease and pest resistance are soybean, maize, rapeseed, cotton, wheat, potato and rice (GMO Compass, 2012). Most of the transgenic research has been focused on virus resistance. In the past, it was believed that monocots are not amenable to *Agrobacterium* mediated transformation, but a successful transformation of wheat (Cheng et al., 1997), maize (Ishida et al. 1996) and barley (Tingay et al., 1997) was reported. Nevertheless, the transformation efficiency in monocots is still unsatisfactory. *Agrobacterium*-mediated transformation is fast and efficient method for introducing genetic material into the host cell and is preferable to many other insertional methods, since it introduce single copies of gene construct using highly efficient vectors that enhance virulence gene expression. However, some crops express hypersenstitive response during inoculation. Alternative transformation methods that exclude tissue culture steps are called *in planta* transformation that allow circumvent the transformation constraints in some monocots. *In planta* transformation, transgenes are delivered into apical meristem of differentiated seed embryo in the form of naked DNA or from *Agrobacterium*. Transgene is injected into the floral tissue of a plant using a needleless-injection device. Once the tissue is transformed, it is removed from the plant and regenerated separately *in vitro*. It has been successfully applied in mulberry, cotton, soybean, and rice (reviewed in Keshamma, 2008).

With sequencing plant genomes, such as *Arabidopsis*, rice and poplar, many genes were identified but their function and localization needs to be proven experimentally *in vivo*. A modified version of *Agrobacterium*-based transformation is Fast Agro-mediated Seedling Transformation (FAST). It offers a transient transformation assay that can take a week, starting from sowing seeds to protein analysis (Li et al. 2009). The advantages of these methods are in addition to time saving also minimal handling with seedlings, high transformation efficiency and big potential for automated high-throughput analysis. This system was applied in *Arabidopsis* to examine the biochemical activity of gene product; it's localization as well as protein-protein interactions. The limitations are in non-expression in different tissues and the need for biological compatible species. It may not be useful for studying disease resistance gene functionality since the co-cultivation with *Agrobacterium* could induce host disease resistance defences. Necrotic responses have been reported in several crops. The defense reaction in grapes was triggered by elevated levels of auxin produced by wild-type T-DNA (Deng et al. 1995), while in tomato, resistance responses were triggered by flagellin (Felix et al. 1999).

3.2.3. Insertional mutagenesis using transposon

Transposon mutagenesis is used when plants are not susceptible to *Agrobacterium*-mediated transformation. Transposons are mobile elements able to move within genome and exist in several copies within the wild-type genome. In order to distinguish novel insertion events from wild type transposon, foreign sequences are introduced into transposon construct. Alternatively, transposon is transferred between different species. Comparing to T-DNA insertional mutagenesis, transposon insertion is more unstable, so different systems are developed to generate more stable transposon insertions (Twyman and Kohli, 2003). The most common is two-component transposon system. One component consists of Activator (Ac) mobile element that includes its own transposase for mobility within the genome, while the second component is lacking the transposase gene, named Dissociation (Ds) element. For the incorporation into host cell, both components are necessary but Ac element can be eliminated by further crossing in order to disable Ds element to move. A transposon tagging is a method of cloning genes whose function is not known. The first step is to identify mutant plants with changed phenotype for a specific trait because of insertion of transposon, truncation and inactivation of a gene. The genomic library is then generated from selected mutants and screened for transposable element. Any clone containing the element will also contain the mutated gene adjacent to the transposon that can be further sequenced and analysed. By transposon tagging, the first cloned *R* genes were isolated from maize, *Hm1* gene that conferrs resistance fungal pathogen *Cochliobolus carbonum* (Johal & Briggs, 1992), *Cf-9* gene from tomato against fungus *Cladosporium fulvum* (Jones et al., 1994) and *N* gene from tobacco conferring resistance to tobacco mosaic virus (TMV) (Dinesh-Kumar et al., 1995). Targeted tranposon tagging is a choice when we target single gene, while for isolating a group of genes a modified method, non-targeting transposon tagging, is an alternative (Gierl & Saedler, 1992).

3.2.4. Insertional mutagenesis using RNAi silencing

The phenomena of RNA silencing were discovered as a side effect during the plant transformation, in which the transgene and homologous endogene were silenced after the successful transformation. RNA silencing is a natural mechanism of wild *R* gene regulations that can be exploited in molecular breeding. This regulatory mechanism provides defence systems by destroying foreign nucleic acids of different nature. In *Arabidopsis*, *RPP5* locus contains structurally unrelated genes combining *RPP4* gene that confers resistance to downy mildew *Peronspora parasitica* and *SNC1* gene against multiple pathogens (van der Biezen et al., 2002). Small RNAs are generated from *RPP5* locus that could be a gradient form for generation of double-stranded small RNAs involved in RNA silencing (Nakano et al. 2006; Kasschau et al., 2007). It has been shown that RNA silencing may reduce fitness costs for expressing multiple *R* genes in the absence of pathogen and offers the possibility to express broad-spectrum resistance (Eckardt, 2007). Disadvantage of using RNA silencing is that it has very variable success in different crops and its time consuming due to the vector construction and transformation of a plant. One of the first commercial outcome using RNA silencing was transgenic papaya with resistance to Papaya ringspot virus (Fuchs and Gonsalves, 2007). Destroying RNAs of viruses can also be achieved by using artificial short RNAs called miRNA. Apart from conferring resistance to viruses, miRNA has broader application for resistance to other pathogens as well. RNA silencing was induced in tobacco plants transformed with constructs against root-knot nematode gene that showed effective resistance (Fairbairn et al., 2007). Baum et al. (2007) identified 14 genes at western corn rootworm larvae that are destroyed by the dsRNA. Transforming maize with dsRNA genes gave protection similar to *Bacillus thuringiensis* transgene. Example of miRNA contributing to bacteria resistance is miRNA from *Arabidopsis* against *P. syringae* (Navarro et al., 2006). It has also been shown that *Arabidopsis* gene silencing is involved in specific resistance to funghi of the *Verticillium* genus (Ellendorff et al., 2009).

3.2.5. Virus-induced gene silencing (VIGS)

Virus-induced gene silencing is based on cloning 200-1300 bp long plant gene cDNA in RNA of a virus and incorporates it into plant genome by *Agrobacterium*-mediated transformation. It is applicable in monocot and dicot species. The advantage is that several homologues genes are targeted by single construct, but the phenotype is transient and mutations are not inherited.

In plants, manifestation of a pathway is termed as post-transcriptional gene silencing (VIGS). VIGS was used to unravel tobacco genes involved in *N* gene mediated defence pathway conferring resistance to tobacco mosaic virus (TMV). Three genes *Rar1, EDS1* and *NPR1/NIM1* were recognised to play an important role in signalling pathway aginst TMV, since the infection with TMV occurs in the presence of *N* gene if these genes are silenced (Liu et al., 2004). In *Arabidopsis*, silencing genes *rar1*, *hsp90* and *ndr1* in functional analysis of RPS2-dependent resistance demonstrated their involvement in expressing disease resistance caused by *P. syringae* (Cai et al., 2006). Using VIGS, the role of Hsp90 was proven in *I-2*-mediated resistance pathway against fungus *Fusarium oxysporum* in tomato (de la Fuente van Bentem et al., 2005) and *Mla13*-mediated resistance against fungus *Blumeria graminis* in barley (Hein et al., 2005). Although VIGS has been employed in important findings, the main disadvantage is the inability to employ vectors in certain varieties due to the natural resistance against those vectors.

3.2.6. Targeting induced local lesions in genome

Targeting Induced Local Lesions in Genome (TILLING) was developed as an alternative to insertional mutagenesis. The strategy was described by McCallum et al. (2000), who describes three main steps: i) treatment of seeds with mutagen and development of M_1 and M_2 generation and creation of DNAs pools of M2 plants, ii) detection of mutations (PCR amplification of specific fragments, heteroduplex formation and identification of heteroduplex using DHPLC, cleavage by specific endonucleases, high-throughput sequencing, identification of mutant plants and determination of mutations) and iii) analysis of mutant phenotype (Kurowska et al., 2011). TILLING is a non-transgenic strategy for providing large spectrum of mutations (point mutations, small insertion, truncation and deletions) that can be applied when the sequence of gene is known and the methodology of detection of SNPs has been developed (Colbert et al., 2001). Advantages of TILLING over T-DNA insertions are in smaller population needed for the saturation mutagenesis (5,000 M_1 plants for TILLING compared to 360,000 M_1 plants for T-DNA mutagenesis) due to higher frequency of mutations (Østergaard & Yanofsky, 2004; Alonso & Ecker, 2006). This method provides allelic series of mutants, including knockouts. There is no need to have fully sequenced genome of the studied species, the sequences can be retrieved from gene databases (GenBank) and homologs identified by the BLAST search. Nevertheless, the search for evolutionary distant species should be done for amino acid rather than nucleotide queries. Bioinformatics analyses are necessary during all steps in TILLING strategy, from the determination of a gene to the determination of allele impact on protein function (Kurowska et al. 2011). TILLING is used mainly for basic research, the potential for commercial purposes still need to be established.

Since the invention of the method, many modifications have been developed such as Eco-TILLING (Comai et al., 2004) and individual TILLING (iTILLING) (Bush & Krysan, 2010). The difference between Tilling and iTILLING is that in Tilling DNA is polled from M_2 plants, while in iTILLING, DNA is isolated from pooled seeds collected in bulks of M_1 plants, which is cheaper and less time consuming. In iTILLING the detection of mutations is based on high-resolution melt-curve analysis of PCR products to reveal mutations of interest. Eco-TILLING is efficient method for cataloguing natural polymorphisms (SNPs, small insertions and deletions) in wild populations. It is cost effective because only one individual per haplotype is sequenced and it is applicable to all species. Eco-TILLING is used for identification of resistance genes to novel diseases and in discovering disease resistance gene variation. Allelic variants of *eIF4E* and *eIF(iso)4E* genes in *Capsicum* species that are involved in elimination of RNA viruses were identified as valuable source for resistance to RNA viruses (Ibiza & Nuez, 2010). Using Eco-TILLING different allele variants were examined in barley at *Mlo* and *mla* genes that are involved in resistance to powdery mildew (Mejlhede et al., 2006). The powdery mildew resistance gene *mlo* is a single copy gene that encodes protein involved in cell wall process. Using EcoTILLING it was possible to identified 11 *mlo* mutants. The *Mla* region combines several classes of genes with defence responses. More than 20 alleles of *Mla* locus have been identified from wild barley in Israel (Jahoor & Fischbeck 1987 & 1993, Kintzios et al., 1995).

3.2.7. Error-prone PCR

Error-prone PCR is a method for generating mutants in order to analyse the relationship between gene sequence, structure and function of protein (Pritchar et al. 2005). It uses imperfect PCR to enhance natural error rate of polymerase to generate beneficial mutations in directed evolutional experiment. Imperfect PCR reaction conditions reduce the fidelity of *Taq* polymerase to generate randomized nucleotide sequences, which is called gene shuffling. This method was used to study protein interaction of RIN4 and RPS2 association in *Arabidopsis* conferring resistance to *P. syringae* (Day et al., (2005). Association of RIN4 and RPS2 was previously confirmed *in planta* (Axtell & Staskawicz, 2003). Day et al. (2005) identified two distinct regions in RIN4 protein as key determinants for RPS2 regulation in *Arabidopsis*.

3.2.8. Alternative methods

There are also unconventional ways of producing mutants. Plant tissue culture can be used as a source for generating variability in regenerants called somaclonal variability. It can be of genetic or epigenetic nature. Genetic variability is caused by mutations or other changes in DNA (changes in ploidy, structural changes in DNA, activation of transposon and chimera rearregement) and is heritable, while epigenetic variability is caused by temporary phenotypic changes (rejuvenation). Seven wheat cultivars having some degree of resistance to *Bipolaris sorokiniana*, *Magnaporthe grisea* or *Xanthomonas campestris* pv. *undulosa* (Xcu) provide somaclonal variation for disease resistance (Mehta & Angra, 2000). The stability of somaclonal variants must be examined through several generations in order to distinguish from epigenetic changes, which is the reason for lower utility.

4. Conclusions

One of the main goals of future agriculture is to achieve durable and broad-spectrum resistance. *R* genes do not always provide durable resistance due to the evolution of pathogen that break host ETI immune system (Jones and Dangl, 2006). Mutagenesis enables the identification of wild *R* genes or the creation of novel *R* genes. Induced mutagenesis offers many benefits to agriculture, especially when there is no reliable source of resistance found in the nature that makes it impossible to introduce to susceptible cultivars by hybridization. Understanding defence responses offer the possibility to introduce new combinations of alleles from ancestral varieties into modern crop. Genetic characteristics of pea (*Pisum sativum* L.) recessive *er1* gene show similarities to *Arabidopsis*, barley and tomato resistance to powdery mildew, which is caused by the loss-of-function of MLO gene family members. An *er1* resistance line was produced by the induction of mutagenesis using alkylating diethyl sulphate that carry a single nucleotide polymorphism in the PsMLO and lead to the premature termination of translation (Pavan et al., 2011). Genetic variation in basal resistance of ancestral plants can be exploited to provide more durable disease protection, which was proven successful with introduction of *WKS1* gene from ancestral wheat *Triticum turgidum* to commercial wheat variety (Fu et al., 2009). *WKS1* gene confers partial and temperature-dependent resistance to stripe rust *Puccinia striiformis* (Fu et al., 2009). In barley, *Mlo* locus comprises different recessive alleles that confer resistance to broad spectrum of fungal pathogen causing powdery mildew and offer durable resistance.

Although mutagenesis is valuable toll to researcher, mutants produced by genetic transformation (GMOs) cause some public concerns, especially in Europe. On the other hand, mutagenized plants are much more acceptable to consumers, breeders, environmentalist and governments. New findings of trans-generation memory of mutants, improved either by mutagenesis or transformation, open the debate if should mutagenized crops also be considered as GMO. Each mutation or transgenic event cause a stress to the cell or organism and lead to altered expression of genes. Stress event is memorized through several generations that can be explained by epigenetic modifications, although the changes are decreasing with each new generation (Molinier et al. 2006). The environmental factors lead to changes in physiology and in genome flexibility that can be transferred to next, untreated generations. Batista et al. (2008) summarized genes whose expression was altered by microarray analysis into six groups: 1) genes involved in signalling pathways and stress, 2) genes involved in regulatory pathways as second messengers, 3) gene involved in reactive oxygen species (ROS) network; 4) genes implicated in protein modification, 5) genes encoding transcription factors and 6) genes encoding for retrotransposons. They pointed out that similar phenotype does not necessarily mean similar expression profile.

Future assignments of mutation breeding are:

- to speed the use of mutations and the release of commercial varieties,
- to establish public mutant gene banks,
- to maintain free access of mutant varieties to global agriculture,
- to apply next generation sequencing techniques for evaluation of wild genetic variation of entire genomes of a population,
- to improve bioinformatics tools. The distinction between true SNPs and sequencing errors still remain problem that can be solved by programmers,
- to improve combined traits such as tolerance to abiotic and biotic stress. *R* genes play part of signalling pathway involved in many metabolic processes, so the change in disease resistance may affect other traits as well,
- to introduce mutants in organic breeding and
- to transfer findings from basic research of plant disease resistance mechanisms to other organisms and research fields, as is the use of RNA silencing in human chemotherapy.

Author details

Petra Kozjak and Vladimir Meglič
Agricultural Institute of Slovenia, Ljubljana, Slovenia

5. References

Ahloowalia, B.S.; Maluszynski, M. & Nichterlein, K. (2004). Global impact of mutation-derived varieties. *Euphytica*, Vol. 135, pp. 187-204

Akira, S. & Takeda, K. (2004). Toll-like receptor signalling. *Nature Reviews Immunology*, Vol. 4, pp. 499-511

Alonso, J.M. & Ecker, J.R. (2006). Moving forward in reverse: genetic technologies to enable genome-wide phenomic screens in Arabidopsis. *Nature Reviews Genetics*, Vol. 7, pp. 524–536

Ausubel, F.M. (2005). Are innate immune signaling pathways in plants and animals conserved? *Nature Immunology*, Vol. 6, pp. 973-979

Axtell, M.J. & Staskawicz, B.J. (2003.) Initiation of RPS2-specified disease resistance in *Arabidopsis* is coupled to the AvrRpt2-directed elimination of RIN4. *Cell*, Vol. 112, pp. 369-377

Ballini, E.; Morel, J.B.; Droc, G.; Price, A.; Courtois, B.; Notteghem, J.L. & Tharreau, D. (2008). A genome-wide meta-analysis of rice blast resistance genes and quantitative trait loci provides new insights into partial and complete resistance. *Molecular Plant-Microbe Interactions*, Vol. 21, pp. 859-868

Batista, R.; Saibo N.; Lourenço, T. & Oliveira, M.M. (2008). Microarray analyses reveal that plant mutagenesis may induce more transcriptomic changes than transgene insertion. *Proceedings of the National Academy of Sciences*, Vol. 105, pp. 3640-3645

Baum, J.A.; Bogaert, T.; Clinton, W.; Heck, G.R.; Feldmann, P.; Ilagan, O.; Johnson, S.; Plaetinck, G.; Munyikwa, T.; Pleau, M.; Vaughn, T. & Roberts, J. (2007). Control of coleopteran insect pests through RNA interference. *Nature Biotechnology*, Vol. 25, pp. 1322–1326

Baumgarten, A.; Cannon, S.; Spangler, R. & May, G. (2003). Genome-Level Evolution of Resistance Genes in *Arabidopsis thaliana*. *Genetics*, Vol. 165, pp. 309-319

Bernoux, M.; Timmers, T.; Jauneau A.; Briere, C.; de Wit, P.; Marco, Y. & Deslandes, L. (2008). RD19, an *Arabidopsis* cysteine protease required for RRS1-R-mediated resistance, is relocalized to the nucleus by the *Ralstonia solanacearum* Popp2 effector. *Plant Cell*, Vol. 20, pp. 2252-2264

Boller, T. & He, S.Y. (2009). Innate immunity in plants: an arms race between pattern recognition receptors in plants and effectors in microbial pathogens. *Science*, Vol. 324, pp. 742-744

Bush S.M. & Krysana P.J. (2010). iTILLING: A personalized approach to the identification of induced mutations in *Arabidopsis*. Plant Physiology, Vol. 154, pp. 25-35

Cai, X.Z.; Xu, Q.F.; Wang, C.C. & Zheng, Z. (2006). Development of a virus-induced gene silencing system for functional analysis of the RPS2-dependent resistance signalling pathways in Arabidopsis. *Plant Molecular Biology*, Vol. 62, pp. 223–232

Calenge, F.; Van der Linden, C.G.; Van de Weg, E.; Schouten, H.J.; Van Arkel, G.; Denancé, C. & Durel, C.E. (2005). Resistance gene analogues identified through the NBS-profiling method map close to major genes and QTL for disease resistance in apple. *Theoretical and Applied Genetics*, Vol. 110, pp. 660-668

Chang, J.H.; Rathjen, J.P.; Bernal, A.J.; Staskawicz, B.J. & Michelmore, R.W. (2000). AvrPto enhances growth and necrosis caused by *Pseudomonas syringae* pv. *tomato* in tomato lines lacking either *Pto* or *Prf*. *Molecular Plant-Microbe Interactions*, Vol. 13, pp. 568–571

Chen, C.M.; Gong Y.; Zhang M. & Chen J.J. (2004) Reciprocal cross talk between Nod2 and TAK1 signaling pathways. *The Journal of Biological Chemistry*, Vol. 279, pp. 25876-25882

Cheng, M.; ,Fry, J.E.; Pang, S.; Zhou, H.; Hironaka, C.M.; Duncan, D.R.; Conner, T.W.; Wan, Y. (1997). Genetic transformation of wheat mediated by Agrobacterium tumefaciens. *Plant Physiology*, Vol. 115, pp. 971-980

Chitwood, D. (2003). Research on plant-parasitic nematode biology conducted by the United States Department of Agriculture-Agricultural Research Service. *Pest Management Science*, Vol. 59, pp. 748-753

Cohn, J; Sessa, G. & Martin, G.B. (2001). Innate immunity in plants. *Current Opinion in Immunology*, Vol. 13, pp. 55–62

Colbert, T.; Till, B.J.; Tompa, R.; Reynolds, S.; Steine, M.N.; Yeung, A.T.; McCallum, C.M.; Comai, L. & Henikoff, S. (2001). High Throughtput Screening for Induced Point Mutations. *Plant Physiology*, Vol. 126, pp. 480-484

Comai, L.; Young, K.; Till, B.J.; Reynolds, S.H.; Greene, E.A.; Codomo, C.A.; Enns, L.C.; Johnson, J.E.; Burtner, C.; Odden, A.R. & Henikoff, S. (2004). Efficient discovery of DNA polymorphisms in natural populations by Ecotilling. *Plant Journal*, Vol. 37, pp. 778–786

Concibido, V.C.; Diers, B.W. & Arelli, P.R. (2004). A decade of QTL mapping for cyst nematode resistance in soybean. *Crop Science*, Vol. 44, pp. 1121-1131

Cox, T.S.; Raupp, W.J. & Gill, B.S. (1993). Leaf rust-resistance genes *Lr41*, *Lr42* and *Lr43* transferred from *Triticum tauschii* to common wheat. *Crop Science*, Vol. 34, pp. 339-343

Dai, L.; Wu, J.; Li, X.; Wang, X.; Liu, X.; Jantasuriyarat, C.; Kudrna, D.; Yu, Y.; Wing, R.A.; Han, B.; Zhou, B. & Wang, G.L. (2010). Genomic structure and evolution of the *Pi2/9* locus in wild rice species. *Theoretical and Applied Genetics*, Vol. 121, pp. 295-309

Dangl, J.L. & Jones, J.D.G. (2001). Plant pathogens and integrated defence responses to infection. *Nature*, Vol. 411, pp. 826-833.

Danson, J.; Lagat, M.; Kimani, M. & Kuria A. (2008). Quantitative trait loci (QTLs) for resistance to gray leaf spot and common rust diseases of maize. *African Journal of Biotechnology*, Vol. 7, pp. 3247-3254

David, P.; Chen, N.W.G.; Pedrosa-Harand, A.; Thareau, V.; Sevignac, M.; Cannon, S.B.; Debouck, D.; Langin, T. & Geffroy, V. (2009). A Nomadic Subtelomeric Disease Resistance Gene Cluster in Common Bean. *Plant Physiology*, Vol. 151, pp. 1048-1065

Day, B.; Dahlbeck, D.; Huang, J.; Chisholm, S.T.; Li, D. & Staskawicz, B.J. (2005). Molecular basis for the RIN4 negative regulation of RPS2 disease resistance. *Plant Cell*, Vol. 17, pp. 1292-1305

de la Fuente van Bentem, S.; Vossen, J.H.; de Vries, K.J.; van Wees, S.; Tameling, W.I.; Dekker, H.L.; de Koster, C.G.; Haring, M.A.; Takken, F.L. & Cornelissen B.J. (2005). Heat shock protein 90 and its co-chaperone protein phosphatase 5 interact with distinct regions of the tomato *I-2* disease resistance protein. *Plant Journal*, Vol. 43, pp. 284–298

Deng, W.; Pu, X.; Goodman, R.N.; Gordon, M.P. & Nester, E.W. (1995). T-DNA genes responsible for inducing a necrotic response on grape vines. *Molecular Plant–Microbe Interactions*, Vol. 8, pp. 538–548

Deslandes, L. & Rivas, S. (2011). The plant Nucleus. A true arena for the fight between plants and pathogens. *Plant Signalling & Behavior*, Vol. 6, pp. 42-48

Deslandes, L.; Olivier, J.; Theulieres, F.; Hirsch, J.; Feng, D.X.; Bittner-Eddy, P.; Beynon, J. & Marco Y. (2002). Resistance to *Ralstonia solanacearum* in *Arabidopsis thaliana* is conferred by the recessive RRS1-R gene, a member of a novel family of resistance genes. *Proceedings of the National Academy of Sciences USA*, Vol. 99, pp. 2404-2409

Di Gaspero, G.; Cipriani, G.; Adam-Blondon, A.F. & Testolin R. (2007). Linkage maps of grapevine displaying the chromosomal locations of 420 microsatellite markers and 82 markers for *R*-gene candidates. *Theoretical and Applied Genetics*, Vol. 114, pp. 1249-1263.

Dinesh-Kumar, S.P.; Whitham, S.; Choi, D.; Hehl, R.; Corr, C. & Baker, B. (1995.) Transposon tagging of tobacco mosaic virus resistance gene N: its possible role in the *TMV*-N-mediated signal transduction pathway. *Proceedings of the National Academy of Sciences USA*, Vol. 92, pp. 4175–4180

Eckardt, N.A. (2007). Positive and Negative Feedback Coordinate Regulation of Disease Resistance Gene Expression. *The Plant Cell*, Vol. 19, pp. 2700-2702

Ellendorff, U.; Fradin, E.F.; de Jonge, R. & Thomma, B.P.H.J. (2009). RNA silencing is required for *Arabidopsis* defence against *Verticillium* wilt disease. *Journal of Experimental Botany*, Vol. 60, pp. 591-602.

Fairbairn, D.J.; Cavalloro, A.S.; Bernard, M.; Mahalinga-Iyer, J.; Graham, M.W. & Botella, J.R. (2007). Host-delivered RNAi: an effective strategy to silence genes in plant parasite nematodes. *Planta*, Vol. 226, pp. 1525–1533

Felix, G.; Duran, J.D.; Volko, S. & Boller, T. (1999). Plants have a sensitive perception system for the most conserved domain of bacterial flagellin. *Plant Journal*, Vol. 18, pp. 265–276

Flor H.H. (1955). Host-parasite interaction in flax rust - its genetics and other implications. *Phytopathology*, Vol. 45, pp. 680–685

Fu, D.; Uauy, C.; Distelfeld A.; Blechl A.; Epstein L.; Chen X.; Sela H.; Fahima T. & Dubcovsky J. (2009). A kinase-START gene confers temperature-dependent resistance to wheat stripe rust. *Science*, Vol. 323, pp. 1357-1360

Fuchs, M. & Gonsalves, D. (2007). Safety of virus-resistant transgenic plants two decades after their introduction: lessons from realistic field risk assessment studies. *Annual Review of Phytopathology*, Vol. 45, pp. 173–202

Fukuoka, S. & Okuno, K. (1997). QTL analysis for field resistance to rice blast using RFLP markers. *Rice Genetics Newsletter*, Vol. 14, p. 99

Gahan, L.J.; Ma, Y.T.; Coble, M.L.M.; Gould, F.; Moar, W.J. & Heckel, D.G. (2005). Genetic basis of resistance to Cry1Ac and Cry2Aa in *Heliothis virescens* (Lepidoptera: Noctuidae). *Journal of Economic Entomology*, Vol. 98, pp. 1357–1368

Gebhardt, C. & Valkonen, J.P. (2001). Organization of genes controlling disease resistance in the potato genome. *Annual Review of Phytopathology*, Vol. 39, pp. 79-102

Gierl, A. & Saedler, H. (1992). Plant-transposable elements and gene tagging. *Plant Molecular Biology*, Vol. 19, pp. 39-49.

GMO Compass. (2012). Information on genetically modified organisms. (20.2.2012) Available from: http://www.gmo-compass.org

Greene, E.A.; Codomo, C.A.; Taylor, N.E.; Henikoff, J.G.; Till, B.J.; Reynolds, S.H.; Enns L.C.; Burtner, C.; Johnson, J.E.; Odden, A.R.; Comai, L. & Henikoff, S. (2003). Spectrum of chemically induced mutations from a large-scale reverse-genetic screen in *Arabidopsis*. *Genetics*, Vol. 164, pp. 731–740

Gressel J. & Levy A. (2006). Agriculture: The selector of improbable mutations. *Proceedings of the National Academy of Sciences*, Vol. 103, pp. 12215-12216

Hammond-Kosack, K.E. & Parker J.E. (2003). Deciphering plant-pathogen communication: fresh perspective for molecular resistance breeding. *Current Opinion in Biotechnology*, Vol. 14, pp. 177-193

Hein, I.; Barciszewska-Pacak, M.; Hrubikova, K;, Williamson, S.; Dinesen, M.; Soenderby, I.E.; Sundar, S.; Jarmolowski, A.; Shirasu, K. & Lacomme, C. (2005). Virus-induced gene silencing-based functional characterization of genes associated with powdery mildew resistance in barley. *Plant Physiology*, Vol. 138, pp. 2155–2164

Holt, B.F.; Hubert, D.A. & Dangl, J.L. (2003). Resistance gene signalling in plants – complex similarities to animal innate immunity. *Current Opinion in Immunology*, Vol.15, pp. 20–25

Huang, N.; Angeles, E.R.; Domingo, J.; Magpantay, G.; Singh, S.; Zhang, G.; Kumaravadivel, N.; Bennett, J. & Khush G.S. (1997). Pyramiding bacterial blight resistance genes in rice: marker assisted selection using RFLP and PCR. *Theoretical and Applied Genetics*, Vol. 95, pp. 313-320

Ibiza, V.P.; Cañizares, J. & Nuez, F. (2010). EcoTILLING in Capsicum species: searching for new virus resistances. *BMC Genomics*, Vol.11: 631

Imler, J.L. & Hoffmann, J.A. (2001). Toll receptors in Innate immunity. *Trends in Cell Biology*, Vol. 11, pp. 304-311

Innes, R.W.; Ameline-Torregrosa, C.;Ashfield, T.; Cannon, E.; Cannon S.B.; Chacko B.; Chen N.W.G.; Couloux A.; Dalwani A., Denny R. *et al.* (2008). Differential Accumulation of Retroelements and Diversification of NB-LRR Disease Resistance Genes in Duplicated Regions following Polyploidy in the Ancestor of Soybean. *Plant Physiology*, Vol. 148, pp. 1740-1759

Inohara, N. & Nunez, G. (2003). Nods: Intracellular proteins involved in inflammation and apoptosis. *Nature Review of Immunology*, Vol. 3, pp. 371-382

Ishida, Y.; Saito, H.; Ohta, S.; Hiei, Y.; Komari, T. & Kumashiro, T. (1996). High efficiency transformation of maize (*Zea mays* L.) mediated by *Agrobacterium tumefaciens*. *Nature Biotechnology*, Vol. 14, pp. 745-750

Iyer, A.S.; McCouch, S.R. (2004). The rice bacterial blight resistance gene *Xa5* encodes a novel form of disease resistance. *Molecular Plant-Microbe Interactions*, Vol. 17, pp. 1348-1354

Jahoor, A. & Fischbeck, G. (1993). Identification of new genes for mildew resistance of barley at the *Mla* locus in lines derived from *Hordeum spontaneum*. *Plant Breeding*, Vol. 110, pp. 116—122

Jahoor, A. & Fischbeck, G. (1987). Genetic studies of resistance to powdery mildew in barley lines derived from *Hordeum spontaneum* collected from Israel. *Plant Breeding*, Vol. 99, pp. 265—273

Janeway,C.A.Jr. & Medzhitov, R. (2002). Innate immune recognition. *Annual Review of Immunology*, Vol. 20, pp. 197-216

Jia, Y.; McAdams, S.A.; Bryan G.T.; Hershey H.P. & Valent B. (2000). Direct interaction of resistance gene and avirulence gene products confers rice blast resistance. *EMBO Journal*, Vol. 19, pp. 4004–4014

Johal G.S. & Briggs S.P. (1992). Reductase activity encoded by the *HM1* disease resistance gene in maize. *Science*, Vol. 258, pp. 985-987

Jones, J. D. & Dangl, J. L. (2006). The plant immune system. *Nature*, vol. 444, pp. 323-329

Jones, D.A.; Thomas, C.M.; Hammond-Kosack, K.E.; Balint-Kurti, P.J. & Jones J.D.G. (1994). Isolation of the tomato *Cf-9* gene for resistance to *Cladosporium fulvum* by transposon tagging. *Science*, Vol. 266, pp. 789-793

Joshi, R.K. & Nayak, S. (2010). Gene pyramiding-A broad spectrum technique for developing durable stress resistance in crops. *Biotechnology and Molecular Biology Review*, Vol. 5, pp. 51-60

Kameswara Rao, K.; Lakshminarasu, M. & Jena, K.K. (2002). DNA markers and marker-assisted breeding for durable resistance to bacterial blight disease in rice. *Biotechnology Advances*, Vol. 20, pp. 33-47

Kanazin, V.; Marek, L.F. & Shoemaker, R.C. (1996). Resistance gene analogs are conserved and clustered in soybean, Proceedings of National Academy of Science USA, Vol. 93, pp. 11746-11750

Kasschau, K.D.; Fahlgren, N.; Chapman, E.J.; Sullivan, C.M.; Cumbie, J.S.; Givan, S.A.; & Carrington, J.C. (2007). Genome-wide profiling and analysis of *Arabidopsis* siRNAs. PLoS Biology 5(3): e57

Kazama, Y.; Hirano, T.; Saito, H.; Liu, Y.; Ohbu, S.; Hayashi, Y.; Abe, T. (2011). Characterization of highly efficient heavy-ion mutagenesis in Arabidopsis thaliana. *BMC Plant Biology*, Vol. 11:161

Keshamma, E.; Rohini, S.; Rao, K. S.; Madhusudhan, B. & Udaya Kumar, M. (2008). Tissue Culture-independent *In Planta* Transformation Strategy: an *Agrobacterium tumefaciens*-Mediated Gene Transfer Method to Overcome Recalcitrance in Cotton (Gossypium hirsutum L.). *The Journal of Cotton Science*, Vol. 12, pp. 264–272

Kintzios, S.; Jahoor, A. & Fischbeck, G. (1995). Powdery mildew resistance genes *Mla29* and *Mla32* in *H. spontaneum* derived winter barley lines. *Plant Breeding*, Vol. 114, pp. 265—266

Kozjak, P.; Jakšc, J. & Javornik, B. (2009). Isolation and sequence analysis of NBS-LRR disease resistance gene analogues from hop *Humulus lupulus* L. *Plant Science*, Vol. 176, pp. 775-782

Kuang, H.; Wei, F.; Marano, M.; Wirtz, U.; Wang, X.; Liu, J.; Shum, W.; Zaborsky, J.; Tallon, L.; Rensink, W. *et al.* (2005). The R1 resistance gene cluster contains three groups of independently evolving, type I R1 homologues and shows substantial structural variation among haplotypes of *Solanum demissum*. *Plant Journal*, Vol. 44, pp. 37-51

Kuang, H.; Woo, S.S.; Meyers, B.C.; Nevo, E. & Michelmore, R.W. (2004). Multiple genetic processes result in heterogeneous rates of evolution within the major cluster disease resistance genes in lettuce. *Plant Cell*, Vol. 16, pp. 2870-2894

Kurowska, M.; Daszowska-Golec, A.; Gruszka, D.; Marzec, M.; Szurman M., Szarejko, I. & Maluszynski, M. (2011). TILLING: a shortcut in functional genomics. *Journal of Applied Genetics*, Vol. 52, pp. 371-390

Leister, D. (2004). Tandem and segmental gene duplication and recombination in the evolution of plant disease resistance gene. *Trends in Genetics*, Vol. 20, pp. 116-122

Li, J.F.; Park, E.; von Arnim, A.G. & Nebenführ, A. (2009) The FAST technique: A simplified Agrobacterium-based transformation method for transient gene expression analysis in seedlings of Arabidopsis and other plant species. *Plant Methods* 5:6

Li, X & Zhang, Y. (2002). Reverse genetics by fast neutron mutagenesis in higher plants. *Functional & Integrative Genomics*, Vol.2, pp. 254–258

Li, X.; Lassner, M. & Zhang Y. (2002). Deletagene: a fast neutron deletion mutagenesis-based gene knockout system for plants. *Comparative and Functional Genomics*, Vol. 3, pp. 158-160

Liu, Y.; Nakayama, N.; Schiff, M.; Litt, A.; Irish, V.F. & Dinesh-Kumar, S.P. (2004). Virus induced gene silencing of a DEFICIENS ortholog in Nicotiana benthamiana. *Plant Molecular Biology*, Vol. 54, pp. 701-711.

Mackey, D.; Belkhadir, Y.; Alonso, J.M.; Ecker, J.R. & Dangl, J.L. (2003). Arabidopsis RIN4 is a target of the type III virulence effector AvrRpt2 and modulates RPS2-mediated resistance. *Cell*, Vol. 112, pp. 379-389

Mackey, D.; Holt, B.F.3rd; Wiig, A. & Dangl JL. (2002). RIN4 interacts with *Pseudomonas syringae* type III effector molecules and is required for RPM1-mediated resistance in *Arabidopsis*. *Cell*, Vol. 108, pp. 743-754

McCallum, C.M.; Comai, L.; Greene, E.A. & Henikoff, S. (2000). Targeting Induced Local Lesions IN Genomes (TILLING) for plant functional genomics. *Plant Physiology*, Vol. 123, pp. 439-442

Medzhitov, R. & Janeway, C. (1997). Innate immunity: the virtues of a nonconal system of recognition. *Cell*, Vol. 91, pp. 295-298

Mehta, Y.R. & Angra, D.C. (2000). Somaclonal variation for disease resistance in wheat and production of dihaploids through wheat x maize hybrids. *Genetics and Molecular Biology*, Vol. 23, pp. 617-622

Mejlhede, N.; Kyjovska, Z.; Backes, G.; Burhenne, K.; Rasmussen, S.K. & Jahoor, A. (2006). EcoTILLING for the identification of allelic variation in the powdery mildew resistance genes mlo and Mla of barley. *Plant Breeding*, Vol. 125, pp. 461-467

Meyers, B.C.; Kaushik, S. & Nandety, R.S. (2005). Evolving disease resistance genes. *Current Opinion in Plant Biology*, Vol.8, pp. 129-134

Meyers, B.C.; Kozik, A.; Griego, A.; Kuang, H.; Michelomeore, R.W. (2003). Genome-wide analysis of NBS-LRR-encoding genes in *Arabidopsis*. *Plant Cell*, Vol. 15, pp. 809-834

Meyers, B.C.; Dickerman, A.W.; Michelmore, R.W.; Sivaramakrishnan, S.; Sobral, B.W. & Young, N.D. (1999). Plant disease resistance genes encode members of an ancient and diverse protein family within the nucleotide-binding superfamily. *Plant Journal*, Vol. 20, pp. 317-332

Meyers, B.C.; Chin, D.B.; Shen, K.A.; Sivaramakrishnan, S.; Lavelle, D.O.; Zhang, T. & Michelmore, R.W. (1998). The major resistance gene cluster in letucce is highly duplicated and spans several megabases. *Plant Cell*, Vol. 10, pp. 1817-1832

Moffett, P.; Farnham, G.; Peart, J. & Baulcombe, D.C. (2002). Interaction between domains of a plant NBS-LRR protein in disease resistance-related cell death. *EMBO Journal*, Vol. 21, pp. 4511-4519

Molinier, J.; Ries, G.; Zipfel, C. & Hohn, B. (2006). Transgeneration memory of stress in plants. *Nature*, Vol. 442, pp. 1046–1049

Mutlu, N.; Miklas, P.N. & Coyne, D.P. (2006). Resistance gene analog polymorphism (RGAP) markers co-localize with disease resistance genes and QTL in common bean. *Molecular Breeding*, Vol. 17, pp. 127-135

Nakano, M.; Nobuta, K.; Vemaraju, K.; Tej, S.S.; Skogen, J.W. & Meyers, B.C. (2006). Plant MPSS databases: Signature-based transcriptional resources for analyses of mRNA and small RNA. *Nucleic Acids Research*, Vol. 34, pp. D731–D735

Navarro, L.; Dunoyer, P.; Jay, F.; Arnold, B.; Dharmasiri, N.; Estelle, M.; Voinnet, O. & Jones J.D.G. (2006). A plant miRNA contributes to antibacterial resistance by repressing auxin signaling. *Science*, Vol. 312, pp. 436–439

Nürnberger, T.; Brunner, F.; Kemmerling B. & Piater, L. (2004). Innate immunity in plants and animals: striking similarities and obvious differences. *Immunological Reviews*, Vol. 198, pp. 249-266

Oerke, E.C. (2006). Crop losses to pests. *Journal of Agricultural Science*, Vol. 144, pp. 31-43

Østergaard, L. & Yanofsky, M.F. (2004). Establishing gene function by mutagenesis in *Arabidopsis thaliana*. *Plant Journal*, Vol. 39, pp. 682–696

Parker, J.E. (2003). Plant recognition of microbial patterns. *Trends Plant Science*, Vol. 8, pp. 245–247

Pavan, S.; Schiavulli, A.; Appiano, M.; Marcotrigiano, A.R.; Cillo, F.; Visser, R.G.; Bai, Y.; Lotti, C. & Ricciardi, L. (2011). Pea powdery mildew er1 resistance is associated to loss-of-function mutations at a MLO homologous locus. *Theoretical and Applied Genetics*, Vol. 123, pp. 1425-1431

Pritchard, L., Corne, D., Kell, D., Rowland, J. & Winson, M. (1995). A general model of error-prone PCR. *Journal of Theoretical Biology*, Vol. 234, pp. 497-509

Ramos, H.C.; Rumbo, M. & Sirard, J.C. (2004). Bacterial flagellins: mediators of pathogenicity and host immune responses in mucosa. *Trends Microbiology*, Vol. 12, pp. 509-517

Ravensdale, M.; Nemri, A.; Thrall, P.H.; Ellis, J. G. & Dodds, P. N. (2011). Co-evolutionary interactions between host resistance and pathogen effector genes in flax rust disease. *Molecular Plant Pathology*, Vol. 12, pp. 93–102

Robert, V.J.M. (2001). Marker assisted introgression of black mold resistance QTL alleles from wild *Lycopersicon cheesmanii* to cultivated tomato (*L. esculentum)* and evaluation of QTL phenotypic effects. *Molecular Breeding*, Vol. 8, pp. 217-223

Rogers, C.; Wen, J.; Chen, R. & Oldroyd, G. (2009). Deletion-based reverse genetics in *Medicago truncatula*. *Plant Physiology*, Vol. 151, pp. 1077–1086

Scofield, S.R.; Tobias, C.M.; Rathjen, J.P.; Chang, J.H.; Lavelle, D.T.; Michelmore, R.W. & Staskawicz, B.J. (1996). Molecular Basis of Gene-for-Gene Specifity in Bacterial Speck Disease of Tomato. *Science*, Vol. 274, pp. 2063-2065

Seah, S.; Telleen, A.C. & Williamson, V.M. (2007). Introgressed and endogenous *Mi-1* gene clusters in tomato differ by complex rearrangements in flanking sequences and show sequence exchange and diversifying selection among homologues. *Theoretical and Applied Genetics*, Vol. 114, pp. 1289-1302

Singh, S.; Sidhu, J. S.; Huang, N.; Vikal, Y.; Li, Z.; Brar, D. S.; Dhaliwal, H. S. & Khush, G. S. (2001). Pyramiding three bacterial blight resistance genes (xa5, xa13 and Xa21) using marker-assisted selection into indica rice cultivar PR106. *Theoretical and Applied Genetics*, Vol. 102, pp. 1011-1015

Tabien, R.E.; Li, Z.; Paterson, A.H.; Marchetti, M.A.; Stansel, J.W. & Pinson, S.R.M. (2002). Mapping QTL for field resistance to the rice blast pathogen and evaluating their individual and combined utility in improved varieties. *Theoretical Applied Genetics*, Vol. 105, pp. 313-324

Tabien, R.E.; Li., Z.; Paterson, A.H.; Marchetti, M.A.; Stansel, J.W.; Pinson, S.R.M. & Park, W.D. (2000). Mapping of four major rice blast resistance genes from 'Lemont' and 'Teqing', and evaluation of their combinatorial effect for field resistance. Theoretical Applied Genetics, Vol. 101, pp. 1215- 1225

Takai, R.; Isogai, A.; Takayama, S. & Che, F.S. (2008). Analysis of flagellin perception mediated by flg22 receptor OsFLS2 in rice. *Molecular Plant-Microbe Interactions*, Vol. 21, pp. 1635-1642

Tanabe, T.; Chamaillard, M.; Ogura, Y.; Zhu, L.; Qiu, S.; Masumoto, J.; Ghosh, P., Moran, A., Predergast, M.M; Tromp, G.; Williams, C.J.; Inohara, N. & Núñez, G. (2004). Regulatory regions and critical residues of NOD2 involved in muramyl dipeptide recognition. *EMBO Journal*, Vol. 7, pp. 1587–1597

Tian, M.; Win, J.; Song, J.; Van der Hoorn, R.A.L.; Van der Knaap, E. & Kamoun, S. (2007). A *Phytophthora infestans* cystatin-like protein interacts with and inhibits a tomato papain-like apoplastic protease. *Plant Physiology*, Vol. 143, pp. 364–377

Till, B.J.; Reynolds, S.H.; Greene, E.A.; Codomo, C.A.; Enns, L.C.; Johnson, J.E.; Burtner, C.; Odden, A.R.; Young, K.; Taylor, N.E.; Henikoff, J.G.; Comai, L. & Henikoff, S. (2003). Large-scale discovery of induced point mutations with high-throughput TILLING. *Genome Research*, Vol. 13, 524–530

Tingay, S.; McElroy, D.; Kalla, R.; Fieg, S.; Wang, M.; Thornton, S. & Brettell, R. (1997). Agrobacterium tumefaciens-mediated barley transformation. *Plant Journal*, Vol. 11, pp. 1369-1376

Ton, J.; D'Alessandro, M.; Jourdie, V.; Jakab, G.; Karlen, D.; Held, M.; Mauch-Mani, B. & Turlings, T.C. (2007). Priming by air-borne signals boosts direct and indirect resistance in maize. *Plant Journal*, Vol. 49, pp. 16–26

Twyman, R.M. & Kohli, A. (2003). Genetic modification: Insertional and transposon mutagenesis. In: Thomas B, Murphy DJ, Murray B (eds) *Encyclopedia of Applied Plant Sciences*. Elsevier Science, London UK, pp. 369-377

United Nations. (2011). World Population to reach 10 billion by 2100 if Fertility in all Countries Converges to Replacement Level. World Population Prospects: The 2010 Revision, Press Release (3 May 2011). Retrieved from: http://esa.un.org/wpp/Other-Information/Press_Release_WPP2010.pdf

Van der Biezen, E.A.; Freddie, C.T.; Kahn, K.; Parker, J.E. & Jones, J.D.G. (2002). *Arabidopsis* RPP4 is a member of the RPP5 multigene family of TIR-NB-LRR genes and confers downy mildew resistance through multiple signalling components. *Plant Journal*, Vol. 29, pp. 439–451

Van der Biezen, E.A. & Jones, J.D.G. (1998). The NB-ARC domain: a novel signaling motif shared by plant disease resistance gene products and regulators of cell death in animals. *Current Biology*, Vol. 8, pp. 226-227

Van der Burg, H.A. & Takken, F.L. (2009). Does chromatin remodelling mark systemic acquired resistance? *Trends in Plant Science*, Vol. 14, pp. 286-294

Van der Hoorn, R.A.L. & Kamoun S. (2008). From Guard to Decoy: A new model for Perception of Plant Pathogen Effectors. *The Plant Cell*, Vol. 20, pp. 2009-2017

Van der Plank, J.E. (1963). Plant Diseases: Epidemics and Control. Academic Press, New York & London, 349 p.

Wang, G.L.; Mackill, D.J.; Bonman, J.M.; McCouch, S.R.; Champoux, M.C. & Nelson R.J. (1994). RFLP mapping of genes conferring complete and partial resistance to blast in a durably resistance rice cultivar. *Genetics*, Vol. 136, pp. 1421-1434

Wicker, T.; Yahiaoui, N. & Keller, B. (2007). Contrasting rates of evolution in Pm3 loci from three wheat species and rice. *Genetics*, Vol. 177, pp. 1207-1216

Xiang, T.; Zhong, N.; Zou, Y.; Wu, Y.; Zhang, J.; Xing, W.; Li, Y.; Tang, X.; Zhu, L.; Chai, J. & Zhou, J.M. (2008). *Pseudomonas syringae* effector *AvrPto* blocks innate immunity by targeting receptor kinases. *Current Biology*, Vol. 18, pp. 74–80.

Young, N.D. (2000). The genetic architecture of resistance. *Current Opinion in Plant Biology*, Vol. 3, pp. 285-290

Zhou, J.M. & Chai, J. (2008). Plant pathogenic bacteria type III effectors subdue host responses. *Current Opinion in Microbiology*, Vol. 11, pp. 179–185

Zipfel, C. & Rathjen, J.P. (2008). Plant Immunity: *AvrPto* targets the frontline. *Current Biology*, Vol. 18, pp. R218–R220

Chapter 11

Mutagenesis and Temperature-Sensitive Little Machines

María Pertusa, Hans Moldenhauer, Sebastián Brauchi,
Ramón Latorre, Rodolfo Madrid and Patricio Orio

Additional information is available at the end of the chapter

http://dx.doi.org/10.5772/50333

1. Introduction

In mammals a class of ion channels able to sense a wide range of temperatures (0-60 °C) has evolved. These molecular thermodynamic machines called thermo Transient Receptor Potential (thermoTRP) are spread through the different TRP channel subfamilies having members inside the TRPM (melastatin) subfamily, where TRPM2, TRPM3, TRPM4 and TRPM5 are heat-activated, whereas TRPM8 is activated by cold. The TRPV (vanilloid) subfamily contains four thermoTRP channels (TRPV1, TRPV2, TRPV3 and TRPV4), which are all activated by heat; and TRPA1 (ankyrin) channel which is activated by noxious cold (reviewed in [28, 107], Figure 1). More recently, a member of TRPC (canonical) subfamily, TRPC5, was identified as a cold receptor in the temperature range 37-25 °C [1].

Located in cutaneous nerve endings of thermoreceptors and nociceptors, and because extreme temperatures produce discomfort and pain, thermoTRP channels are involved in nociception and can be activated by a long list of other noxious stimuli such as low pH and irritant chemicals [2].

What characterizes these channels is their exquisite temperature sensitivity. Thermodynamic analyses reveal that thermoTRP channels undergo large enthalpy changes (ΔH) that account for their high temperature sensitivity [3-8]. For example, the enthalpy change between close and open in TRPV1 and TRPM8 involves ΔHs of ~100 kcal/mol and ~-60 kcal/mol, respectively [3, 5]. It is obvious that in order to make the closed-open reaction reversible these enthalpy changes must be accompanied by large entropy (ΔS) changes. These activation enthalpies are 3-5 times the enthalpy change for voltage- or ligand-dependent channel gating (ΔH ~20 kcal/mol; [108]). Actually, Yao et al. [7] pointed out that in the case of TRPV1, the ΔH involved in the closed-open transition is equivalent to an electrical energy moving 71 unit charges across 60 mV!

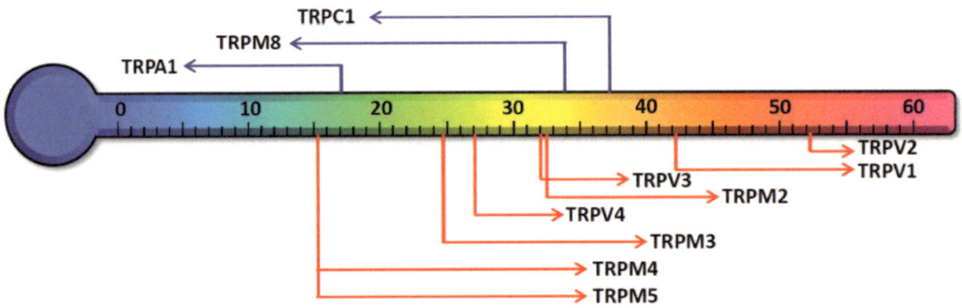

Figure 1. Schematic representation of the eleven mammalian thermoTRPs indicating their reported temperature sensitivity.

Enthalpy changes of this magnitude mean large structural rearrangements of the channel-forming protein. Just consider that the ΔH of the thermal denaturation of the ribonuclease (RNase A), a 124 residues protein is 57 kcal/mol [9]. Since protein denaturation involves a change from a molecule with a very well defined structure to a random coil, how is it possible that that the TRPV1 channel opening reaction is defined by such a huge enthalpy change leading to Q_{10} ~270 for TRPV1? It is difficult to think that, as in the process of denaturation, these channels undergo global disturbance of their structure, in particular given that channel opening demands a well kept pore structure and responses to voltage and agonists are maintained at all temperatures.

We rather are of the opinion that thermoTRPs might possess specific residues, domains or domain interactions that are specifically affected by temperature in the channel activation pathway. The data available at present strongly suggests that these structure(s) are different from those in charge of determining voltage-dependence or agonist binding [4].

The first goal of this chapter is to present the reader the different genetic and mutagenesis procedures that have been used so far in the quest for finding the "Holy Grail" of thermoTRP channels: the thermal sensor. The situation is at present rather confuse since the molecular determinants for temperature sensitivity in thermoTRP channels have been claimed to be in the N-terminus of TRPV1 and *Drosophila* TRPA1 [10, 11], the pore region of TRPV1 and TRPV3 [12, 13] and in the C-terminus [14-16]. Bona fide components of such a thermal sensor can only be, however, those components capable of appreciably perturbing the enthalpy of the channel. Mutagenesis, as we describe below, has been indispensable in the search for those components.

In this chapter we will also show an analysis of the phenotypes of knockout mice that have been used so far in the study of the physiological role of these exciting temperature-sensitive little machines. We will see that several thermoTRPs are critical molecular components of the thermotransduction machinery in primary sensory neurons of the somatosensory system. We will also see that, in spite of their high temperature dependency, some thermoTRPs are playing roles apparently unrelated to their temperature sensibility. Thus, knockout mice have been of great value in unveiling both expected and unexpected roles for thermo TRP channels.

2. A chimeric approach to search for thermal sensors in thermoTRP channels

Channels are most likely built as modular structures [4, 17-22], and we can hypothesize that temperature sensors need to be contained in these protein modules. The tale about the search for the temperature sensor in thermoTRP channels started with a deletion mutagenesis strategy designed to chop parts of the C-terminal domain. Such approximation renders phenotypes with altered thermal sensitivity where progressive deletions correlate with progressive loss of temperature dependency [15]. Prompted by the work of Vlachova et al [15], the group of Latorre engineered chimeras in which the entire C-terminal domains were swapped between cold (TRPM8) and hot receptors (TRPV1) [16]. The resultant chimeric channels –specially the one carrying TRPM8 C-terminal– inherits the temperature sensitivity of the channel contributing with the C-terminal, however, eliciting a rather small Q10 compared with WT channels. Interestingly, the chimeras were often unable to recover from either activation or deactivation process, this was described as "locking" behavior; extreme voltages or long incubations at different temperatures were needed to recover from that new state. The observations obtained on those chimeric proteins led to the conclusion that the C-terminal has an essential role as a "thermal modulator" of channel gating. This is not surprising considering the proximity of the swapped region to both the bundle crossing [23] and to the PIP_2 regulatory region [24]. Further work on the C-terminal domain unveiled a short region in the C-terminal of TRPV1 able to change the TRPM8 phenotype to that of a heat receptor [14]. Although the studies described above did not unequivocally identify the C-terminal domain as the temperature sensor, these results demonstrated that the PIP_2-, voltage- and thermal-responsive elements are contained in different channel molecular structures [14-16].

Recently, fluorescence resonance energy transfer (FRET) experiments done at the turret, a loop connecting the 5th transmembrane domain with the pore helix and located above the external mouth of the pore, was reported to be involved in temperature sensing. Using FRET in combination with electrophysiological recordings and site directed mutagenesis, Yang et al. [25] showed that conformational rearrangements of the turret are essential for temperature-dependent activation. This result was somewhat supported by results presented by other groups in which pore mutations near the turret region either ablate or severely affect temperature-dependent gating [see high throughput section]. However, the striking results presented by Yang et al. have been severely questioned by the group of Qin who showed that deletions of the entire turret region are not affecting temperature sensitivity. Clearly this controversy has to be solved [26, 27].

This saga continues with a nice blend of the use of an ingenious fast temperature clamp ($>10^5$ °C/s) developed by Qin's laboratory [7], and mutagenesis to unveil structural domains in thermo-TRP channels that confer to these channels their exquisite temperature sensitivity. In this case, temperature jumps were produced using a single emitter laser diode as the heat source. This temperature clamp is able to change the temperature of the bath much faster than the time course of the development of the thermoTRP-induced currents (Figure 2). The technique allowed measuring directly activation and deactivation kinetics of the TRPV1 channel, a thermoTRP channel of the vanilloid family (Reviewed in [28]), as a function of temperature. The results indicated that the reaction path is asymmetrical, with temperature mainly driving the opening reaction while the closing rate is, if anything, sensitive to cold [7].

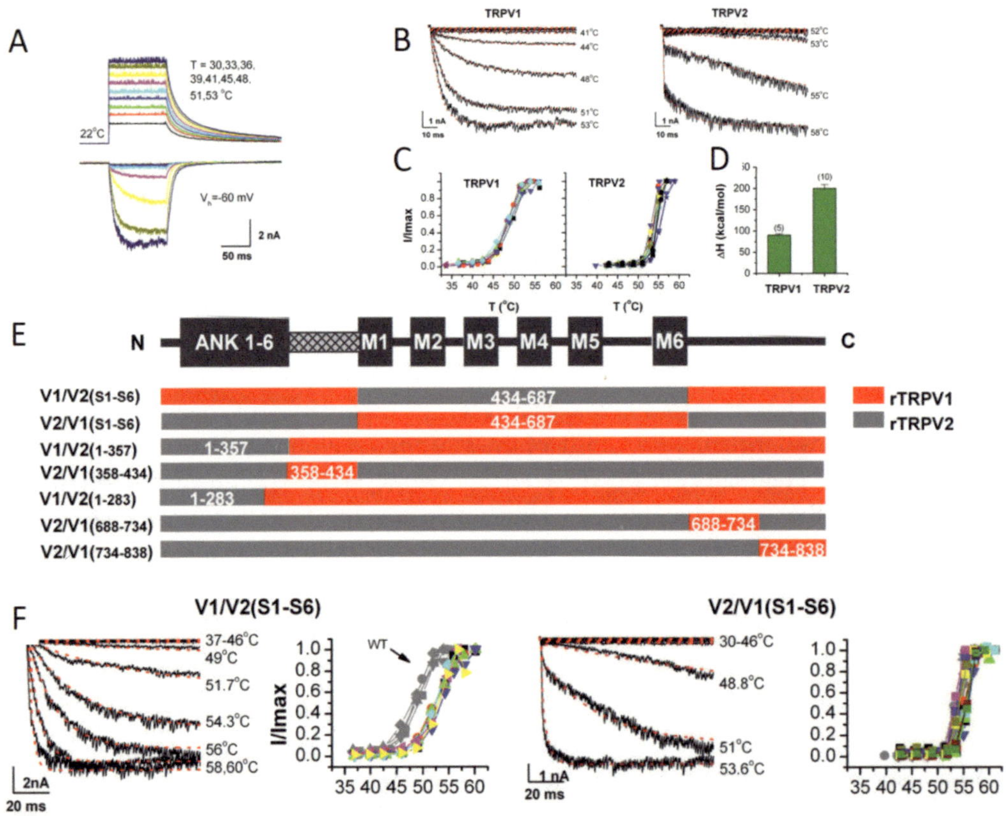

Figure 2. Searching for the molecular determinants of temperature sensitivity in thermoTRPV channels. A. Top. Submillisecond temperature steps generated by infrared laser irradiation. Bottom. TRPV1 channel responses induced by the rapid temperature changes shown in A. **B**. Comparison of the current time course of the TRPV1 currents (left) and TRPV2 (right). **C**. Temperature dependence of the steady state response taken from eperiments like those shown in B. **D**. Activation enthalpies for TRPV1 and TRPV2. **E**. Composition of the chimeric channel proteins used by Yao et al. [13]. **F**. Temperature sensitivity of TRPV1 and TRPV2 channels resides in the N-terminal. Notice that the V1/V2 chimera has a TRPV1 phenotype regarding gating kinetics and enthalpy changes, and that the V2/V1 chimera possesses a TRPV2 phenotype.

In a search for the protein domain(s) involved in thermal sensitivity of TRPV1 channels, Yao et al. [11] used TRPV1 that has an enthalpy change (ΔH) of activation of ~100 kcal/mol, and TRPV2 in which the channel closed-opening reaction involve a ΔH ~200 kcal/mol. A systematic chimeric analysis on TRPV1 and TRPV2 allowed to conclude that temperature sensitivity is associated with the N terminus (Figure 2). In the TRPV1 N-terminus, they further identified N-terminus a fragment of 80 residues that connects the ankyrin repeats to the first transmembrane segment able to transfer to TRPV2 the temperature sensitivity characteristics of TRPV1. Notably, this channel region is precisely the segment missing in the nonfunctional TRPV1b splice variant [29, 30]. Alterations of this region profoundly altered the energetic of thermal sensing in all temperature-sensitive vanilloid receptor homologues (TRPV1-4), while its replacement in temperature-insensitive homologues successfully conferred gain-of-function. It is important to note here that swapping other domains like the C terminus or other domains did not have any effect on the temperature sensitivity. For example, a mutant containing the first 357 amino acids of TRPV2 (i.e., all 6 transmembrane domains and the C terminus belonged to TRPV1) has a TRPV2 phenotype. These results demonstrated that these channels possess localized molecular components for temperature detection.

A chimeric strategy was also used to unveil the thermal-sensing structures in the TRPA1 channels. In this case, the group of David Julius took advantage of the fact that mammalian TRPA1 channels are heat insensitive while the snake TRPA1 version is activated by heat. Through engineering chimeras between mammalian and snake ion channels, the authors turned the mammalian TRPA1 channel into a temperature-sensitive channel identifying the N-terminal region of TRPA1 -within the ankyrin domain of the snake channel- that behave as transferable temperature sensitive modules. The chimeric approximation also suggests that both, sensitivity to chemical stimuli and intracellular calcium dependence, also localize to the N-terminal ankyrin repeat-rich domain [10].

Thus, the current scenario presents the N-terminal region of TRPV1, TRPV2 and TRPA1 as a strong candidate for containing the temperature–sensitive domain with the C-terminal playing a modulatory role. One interesting possibility envisioned by Brauchi et al. [16] was that temperature may affect the interaction between a particular portion of the proximal C-terminal and some other regions of the channel. It may be that the structural arrangements induced by temperature involve an inter-molecular interaction between the proximal C-terminal and specific regions of the N-terminal domain (eg. ANK domains in TRPVs). Such setting would be extremely convenient, because it could explain the large entropy associated to the hydrophobic effect [3, 31] without the necessity of an argument that involves protein unfolding. Lacking ANK domains, this hypothesis plotted for TRPVs and TRPA1 may not be necessarily valid for the case of TRPMs.

3. High throughput mutagenesis and thermoTRP channels

In the field of ion channels the mutagenesis is one of the most powerful tools to understand structural-function relationships. The most common strategy is the replacement of certain residue that one might consider important for channel function by another, and then to perform functional experiments to test our hypothesis of how a particular ion channel gates or transport ions. Depending on primer design you can even to replace 2 or 3 amino acids in a single PCR reaction but what can you do if the protein under study has at least 1000 amino acids? Repeating the single point mutation it is not an option if you must replace a large number of residues by one of the 20 amino acids. The picture becomes more complicated if you wish to test more than one amino acid at each position. This can be an incredible time (and money!) consuming task.

The ideal experimental maneuver would be to perform multiple point mutations of our protein of interest, by means of a technique that would allow you to obtain, for example, 12000 single-point mutations and to test the functional properties of each one of them in one day. Now the good news: that technique exists and is called *high throughput mutagenesis coupled to a cell based assay using a Fluorescent Imaging Plate Reader (FLIPR)*.

The basis of this technique is a massive random mutagenesis in our target cDNA sequence. This is done by means of a PCR reaction using the blend of 2 specific error prone DNA polymerases, and depending on the commercial kit of choice the names could be Mutazyme I and Taq DNA polymerase mutant. The explanation for using a mix of enzymes is that we need the same frequency of mutations in the 4 nucleotides. In the past, the kits used only one enzyme, Mutazymes I and the frequency of mutation in the Cs and Gs was higher than in the As and Ts. The procedure starts by setting up how many mutations the DNA polymerase will introduce per clone, namely the mutation frequency. The whole idea is to obtain 1 mutation per clone, and this can be controlled by using the adequate amount of target DNA and number of PCR reaction cycles. High amounts of DNA give lower frequency of mutations, because one single molecule of DNA has less replication cycles; the same occurs with the PCR cycles in which high amount of DNA and less cycles give a lower mutation frequency.

This procedure yields, however, mutants clones with 2 and 3 mutations, but if those clones present an interesting phenotype we can design the single point mutants to evaluate the contribution of each mutation. Once the random mutagenesis process is finished you are left with a large library with thousand of mutants, with the only caveat that it is probable that you do not have every possible amino acid replacement in each position of a particular region of your protein. If certain mutations were particularly interesting after functional evaluation and are grouped in a well defined region, a second screening can be done, but this time in a saturating way, that means changing every amino acid by one of the other 20. The procedure is the same describe above using the same mutagenesis approach, but with the difference that here we take a particular section of the protein primary structure and change every residue by each of the other 20. To obtain this, it is necessary to take lower concentrations of DNA and to do more than 25 cycles of PCR (frequently 30). Often, to obtain a frequency equal or greater than 20 mutations per Kb, it is necessary to perform many PCR reactions in tandem.

At this point, you have at least 10000 different mutants of your favorite protein, but this is completely useless unless you have a fast method to evaluate the functional properties of each one of them. This is done using the fluorimetric cell-based assay method dubbed FLIPR [32]. The machine is basically a fluorescent plate reader that can stimulate the sample in a specific wavelength and detect the change in the fluorescent emission in other wavelength (this wavelength depends on the fluorescent probe used). A remarkable advantage of the method is that it allows the use of 384-wells plate which makes easy to test the effect of chemical compounds on ion channel function, i.e. an agonist or a blocker. This method makes possible to perform a large number of experiments in a short period of time.

The 384-wells plate have a standardized quantity of mammalian cells like HEK-293 or CHO transfected with the ion channel DNA and each well contains a different mutant clone. The cells in every well are loaded with a fluorescent probe that can report the activity of our ion channel. If you are interested in studying ion channels one of the most important requirements for using this technique is that the channel allows the passage of ions that can be detected by a probe. For example, if you are working with a TRP channel that permeates Ca^{2+} we can use a calcium probe like Fluo-3 or Fluo-4. The technique is particularly suited for thermoTRP since these channels are polymodal receptors that are activated by different agonists and temperature and many of them are Ca^{2+} permeable [12, 13, 33].

This is an unbiased approach, different from making chimeras between different channel-forming proteins and other kind of mutations techniques. Since the mutations are done randomly, the high throughput mutagenesis technique produced mutants that are not biased by our previous knowledge of how ion channels may work. The discoveries using this approach in the field of thermoTRP channels are: a) important amino acid residues involved in temperature sensing; and b) the binding sites for the agonist menthol in mTRPM8 [12, 13, 33] (Figure 3).

For TRPV1, Grandl et al. [13] focused in a mutant library of 4400 clones from which they found 3 mutations that affect the heat response; N628K located in the pore region, adjacent to the pore helix and N652T and Y653T placed in the extracellular loop between the selectivity filter and transmembrane domain 6 (TM6). The three mutations show a decrease in their heat sensitivity as determined by a right shift in the temperature threshold to higher temperatures. The double and triple mutants N652T;Y653T and N628K;N652T;Y653T, have a stronger phenotype than the single-point mutants, with a greater decrease in their heat sensitivity. However, their activation by the agonists capsaicin and 2-Aminoethoxydiphenyl borate (2APB [34, 35]), as well as by voltage and acid pH is the same as in the wild type. As a quality control of the screening, they found 2 residues previously reported: E600V that produces a loss in the pH sensitivity [13] and F489Y that produces a right shift in the EC50 capsaicin activation. This kind of information gives us the confidence that the procedure is trustable since it corroborates the results obtained with other techniques.

In the case of mTRPV3, from a mutant library of 14.000 clones Grandl et al. [12] discovered 3 mutants with a decrease in their heat sensitivity; I644S, N647Y and Y661C all of them located between the pore helix and TM6. The 3 mutants have a normal response to the agonist 2APB, and unaltered ion selectivity. This is important because the mutations are located near to the TRPV3 selectivity filter. These libraries are not saturating (i.e., every amino acid of the channel-forming protein replaced by one of the other 20), for this reason, Grandl et al. [12] made 45 more mutants in the region between TM5 and TM6 finding two other clones in TM6 with an altered temperature phenotype (F654S and L657E). Molecular modeling suggests that the three mutations in TM6 (F654S, L657E and Y661C) are located in a periodic pattern probably aligned on the lipid-facing side of the α-helix[12].

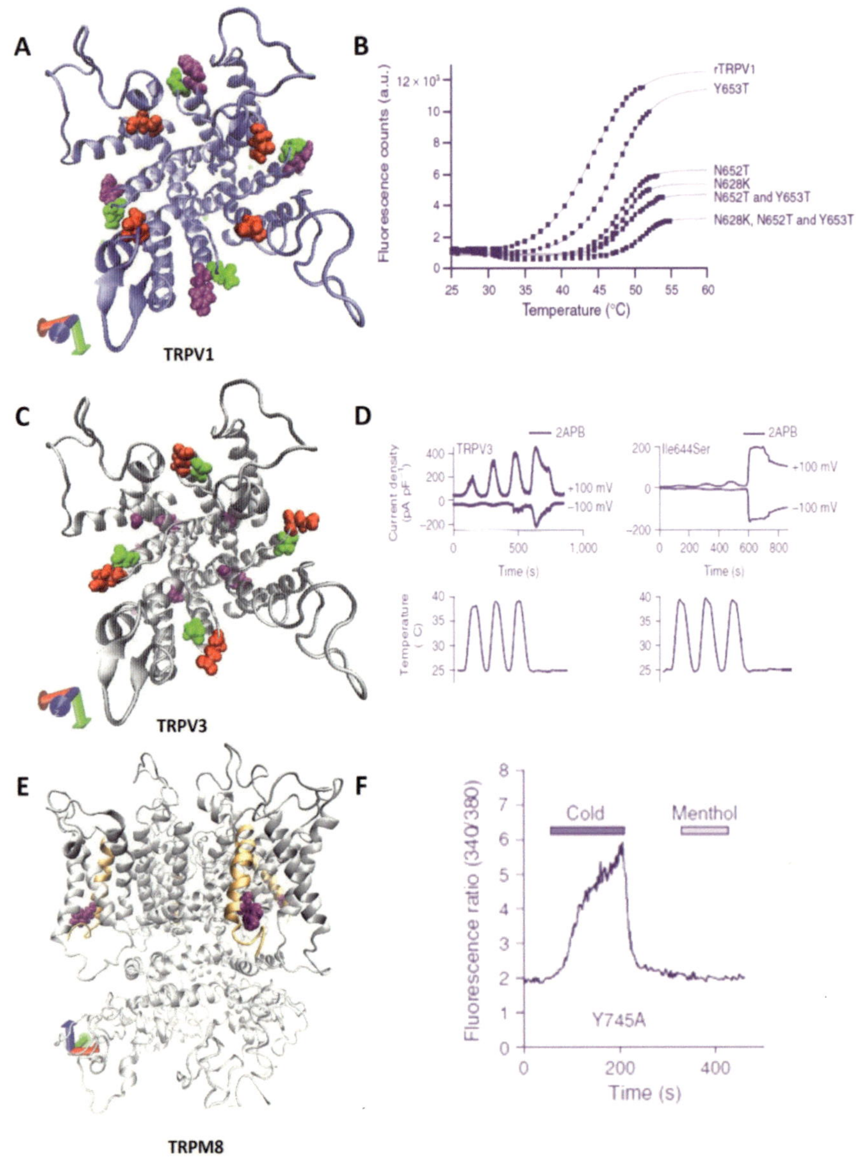

Figure 3. The high throughput screening technique used to identify key residues in thermoTRP channels. **A**. A top view of TRPV1 structure highlighting residues N628 (red), N652 (green), and Y653 (violet). **B**. A graph showing the normalized fluorescence change in response to an increase in temperature. The different mutants have a lower response compared with the wild type channel and the decrease in response is proportional to the number of residues mutated. **C**. Top view of the TRPV3 structure highlighting the residues I644 (red), Y661 (violet),and N647 (green). **D**. Responses to temperature changes of the wild type TRPV3 (left) and the I644S TRPV3 mutant. Notice that despite the fact that the temperature response in the mutant has been obliterated the response to the agonist 2APB is vey similar to that of the wild type channel. **E**. Lateral view of the TRPM8 structure showing the position of the residue Y745. **F**. The experiment shows that although the mutant is still sensitive to cold, it has lost its ability to respond to menthol.

Finally, for TRPM8 the screening random mutagenesis technique was used to search for the residues constituting the menthol binding site [33]. Over a library of 14000 clones, Bandell et al. identified two mutations in TM2, Y745H and L1009R, able to produce a failure in the menthol activation but keeping the other TRPM8 channel properties like voltage-dependence, temperature sensitivity and PIP_2 sensitivity intact. In the case of L1009R mutation, mutations by any other amino acid besides R do not have any effect in the menthol activation showing us the power of this unbiased assay, very helpful when you don't have a crystal structure or an adequate molecular model of the channel.

4. Physiological role of thermoTRP channels

ThermoTRPs channels are key elements in many physiological processes. Powerful mutagenic tools such as knockout mice generation have been intensely used to study the physiological function of thermoTRP channels *in vivo*. In this section, we will summarize the information about the physiological role of thermoTRP channels, obtained from the study of the phenotype of knockout mice.

4.1. TRPA1

Transient Receptor Potential Ankiryn 1 (TRPA1) channel is activated by several pungent agents found in mustard oil, cinnamon and garlic, among others [36, 37]. This Ca^{2+}-permeable nonselective cation channel, expressed in a large subpopulation of nociceptors, was described to be activated by cold temperatures, $<17°C$ [38]. Thus, it was postulated to be the thermoreceptor for noxious cold temperatures in primary sensory neurons. This channel is activated by a large list of irritant substances, and nowadays it is considered the main molecular sensor of noxious chemical stimuli in the somatosensory system [39-41, reviewed in 42]. Two knockout mice of TRPA1 were described in 2006 [23, 39]. These mice were generated by targeted recombination that deleted the exons coding the pore region. Although disruption of the TRPA1 gene abolishes the behavioral responses to chemical activators of the channel, it has no effect in the response or prevalence of cold-sensitive primary neurons [23, 39]. However, there are deficits in the response to mechanical stimuli [23, 43]. Later it was found that like TRPV1 (see below), TRPA1 is actually involved in inflammatory-related hyperalgesia [44, 45]. Most likely, this role is given by its activation or modulation by a great variety of chemical agents, including several inflammatory- or cell damage-related molecules. Although TRPA1 seems to have a minor role in peripheral cold thermotransduction, a study with the knockout mice shows that this channel participates in the cold sensation in visceral nerves [46]. In vagal sensory neurons from the nodose ganglion, $TRPA^{-/-}$ mice have significantly less cold-sensitive neurons than wild type animals. Also the pharmacological profile of cold-evoked responses in neurons from nodose ganglia is compatible with a major role of TRPA1 in their cold sensitivity [46]. However, a more recent and detailed study of the phenotype of $TRPA1^{-/-}$ animals by Karashima and coworkers strongly suggest an important role of TRPA1 channels in noxious cold thermotransduction [47].

4.2. TRPC5

Transient Receptor Potential Canonical 5 (TRPC5) is a Ca^{2+}-permeable nonselective cation channel expressed in the brain and several other tissues, including vascular smooth muscle cells, endothelial cells, adrenal medulla, mammary glands, yolk sac, activated T cells and monocytes and cardiac ventricles in hypertension (reviewed by [48]). Recently, it has been reported that TRPC5 displays a high sensitivity to cooling into the mild cold range [1]. This cold-dependent activation can be potentiated by activation of G_q-linked muscarinic type 1 receptor via carbachol, as well as by PLC activation via extracellular lanthanum [1]. The activation of TRPC5 by low temperatures, together with the fact that this channel is expressed in approximately 30% of mouse primary sensory neurons in culture, makes it a potential candidate to participate in cold sensing.

A TRPC5 deficient mouse was generated in 2009 [49]. However, from the data obtained using this TRPC5 deficient mice, it was difficult to assign a relevant role of TRPC5 in cold sensing. First, behavioural tests showed no differences between wild type and TRPC5$^{-/-}$ mice in various temperature-sensing assays [1]. In contrast, in cultured primary sensory neurons, TRPC5$^{-/-}$ mice displayed a significant reduction in the percentage of cold-sensitive neurons and also an interesting reduction in TRPM8 channels detected by immunohistochemistry, with no changes in the nociceptive markers CGRP, IB_4, NF200, peripherin, or TRPV1 protein. TRPC5-mediated currents could not be measured in this preparation of primary sensory neurons from dorsal root ganglia in wild type and knockout mice. These results suggest that TRPC5 activity in response to cold could be used for other adaptive or regulatory processes, such as localized metabolic changes, local vascular changes, retraction of nerve endings, or initiation of transcriptional programs [1]. Other possible explanation would be that the deletion of TRPC5 results in compensatory replacement by functionally overlapping of other cold transducers. This compensatory replacement, as it has been described for instance in tetrodotoxin-sensitive ion channels in Nav1.8-deficient mice [43], could result in the absence of a clear TRPC5$^{-/-}$ phenotype in behavioural tests, and could explain the avoidance of cold temperatures displayed by TRPC5$^{-/-}$ mice [1]. Further studies are needed to establish the possible role of TRPC5 in cold transduction and the physiological significance of the potentiation of its activity by cold temperatures.

4.3. TRPM2

Transient Receptor Potential Melastatin 2 (TRPM2), is an ion channel permeable to all physiological cations including Ca^{2+}, which activation leads to an increase in intracellular Ca^{2+} concentration and/or membrane depolarization (reviewed by [48]). TRPM2 exhibits a widespread distribution that includes brain, bone marrow, spleen, heart, liver and lung, and also different cell types including immune cells (neutrophils, megakaryocytes, monocytes/macrophages), pancreatic β-cells and, endothelial cells, cardiomyocytes, microglia and neurons [48]. TRPM2 is activated by warm temperatures, with a threshold of ~35°C [50]. This channel is not present in primary sensory neurons and a role in thermosensation has not been reported.

TRPM2 contains a Nudix hydrolase (NudT9-H) domain in its C-terminus, that activates the channel through its binding with the adenosine 5′-diphosphoribose (ADPR)[48]. It has been described that TRPM2 is also activated by H_2O_2 and plays an important physiological role in regulation of the oxidative stress [51]. In addition, this channel can function as an intracellular channel [52, 53].

Knockout mice for these channels were generated by targeted homologous recombination in ES cells, disrupting the third exon of the TRPM2 gene [54]. Analysis of these animals evidenced that they are key regulators of intracellular calcium levels and they participate in signaling cascades related to the function of the immune system [54, 55]. TRPM2 also participates in the control of insulin secretion in pancreatic β cells [56], making TRPM2$^{-/-}$ mice to have an impaired glucose metabolism because of low insulin secretion.

4.4. TRPM3

Transient Receptor Potential Melastatin 3 (TRPM3) is a Ca^{2+}-permeable nonselective cation channel expressed in a range of different tissues, including brain, kidney, endocrine pancreas, ovary , and sensory neurons [57, 58]. It has been reported that this channel is activated by hypotonic cell swelling, D-erytrosphingosine, strong depolarization, removal of extracellular Na^+, and pregnenolone sulphate (reviewed by [59]). TRPM3, like other TRPM channels closely related such as TRPM2, TRPM4, TRPM5 and TRPM8, is a thermosensitive channel activated by heat [60].

Little is yet known about the physiological role of TRPM3 *in vivo*. In addition, the TRPM3 gene encodes several isoforms presenting different biophysical properties [61]. So far, this channel has been involved in the modulation of the secretion of insulin and interleukin-6, promotion of vascular contraction, and thermotransduction [60, 62, 63]. The generation of the deficient TRPM3 mice by homologous recombination by Vriens and colleagues, has been an important step to elucidate the physiological relevance of this channel. TRPM3$^{-/-}$ mice exhibited no obvious deficits in fertility, gross anatomy, body weight, core body temperature, locomotion, or exploratory behaviour. No differences in resting blood glucose were found, suggesting that basal glucose homeostasis is not affected. Thus, TRPM3$^{-/-}$ mice appear generally healthy, with no indications of major developmental or metabolic deficits [60].

The thermosensitivity of TRPM3, its expression in sensory neurons, and the painful effect of the systemic administration of the activator of TRPM3 pregnenolone sulphate, allow to hypothesize about the potential role of TRPM3 in thermotransduction and nociception [60]. The analysis of primary sensory neurons from wild-type and TRPM3$^{-/-}$ animals, showed a reduction in the percentage of heat sensitive neurons, and allowed to establish the existence of four distinct populations of heat-sensitive neurons. The largest population of the heat sensitive neurons responded to both pregnenolone sulphate (PS) and capsaicin, suggesting the coexpression of TRPM3 and TRPV1. The second most abundant population of heat positive neurons responded to PS but not to capsaicin (TRPM3-positive), and a minor fraction responded to capsaicin but not to PS (TRPV1-positive). Finally, a small percentage (2%) of heat-activated neurons was unresponsive to both PS and capsaicin, indicating the existence of a TRPM3- and TRPV1-independent heat-sensing mechanism.

The number of heat-sensitive neurons was reduced in a 25% in TRPM3$^{-/-}$ mice. The subgroup of heat-sensitive neurons responding to PS but not to capsaicin disappeared, whereas an increase was observed in the number of neurons that responded to heat and capsaicin and in the number of heat-positive neurons independent of TRPM3 and TRPV1. In agreement with these observations, the study of the heat responses of sensory neurons from the TRPM3 knockout mice in the presence of a specific antagonist of TRPV1, showed that an important fraction of heat sensitive neurons remained, pointing out the existence of another molecular entity responsible for the heat transduction [60]. The reduction in the population of heat-sensitive neurons in TRPM3$^{-/-}$ is consistent with a strong deficit in the detection of noxious heat stimuli displayed by these animals, as evidenced by prolonged reaction latencies in the tail immersion tests and in hot plate assays, and a reduced avoidance of hot temperatures. In addition, these mice show a significant and specific deficit in the nocifensive responses to TRPM3-activating stimuli, and a strong deficit in the development of inflammation-induced heat hyperalgesia [60]. Taken together, these results establish that TRPM3 works as a chemo- and thermosensor in the somatosensory system, involved in the detection of noxious stimuli.

4.5. TRPM4

Transient Receptor Potential Melastatin 4 (TRPM4) is an ion channel selective for monovalent cations and no permeable to Ca^{2+}. It is inhibited by intracellular free ATP and activated by internal Ca^{2+}. This activation is modulated by ATP, Ca^{2+}-calmodulin, and PKC. PIP_2 also regulates the activity of the channel, by modulating its calcium and voltage sensitivity (reviewed by [64]). It has been described that heat, in the 15-35°C range, modulates the voltage sensitivity of TRPM4, resulting in an increase in current [65]. Despite of TRPM4 being detected in a large numbers of tissues such as heart, pancreas, placenta, and prostate and at lower levels in the kidney, skeletal muscle, liver, intestines, thymus, and spleen [64], it has not been reported in sensory neurons and thus it is very unlikely to have a role in thermosensation.

Knockout mice for TRPM4 were first generated by a Cre-loxP strategy that excised exons 15 and 16 of the gene (containing the first membrane-spanning segment of the protein), replacing them by a PGK-neo^r cassette that was removed by Cre recombinase [66]. A second study used the same strategy to target a region of the gene spanning from exons 3 to 6 [67]. Studies using the TRPM4 deficient mice show that this channel, like TRPM2, plays a role in controlling intracellular calcium levels during mast cell activation and dendritic cell migration [66-68]. It has also been shown that this ion channel helps to limit catecholamine release from chromaffin cells, indicating that TRPM4$^{-/-}$ mice have an increased sympathetic tone and hypertension [69].

4.6. TRPM5

Transient Receptor Potential Melastatin 5 (TRPM5) is a cationic Ca^{2+}-activated channel with an important role in vertebrate taste transduction. This channel was cloned by Perez and coworkers in 2002 [70]. TRPM5 is strongly expressed in taste cells where it is co-expressed with several taste-signaling molecules [70]. The transduction of sweet, bitter and amino acid tastes depends on the activation of G protein-coupled membrane receptors that involve the participation of a common intracellular pathway. These receptors signal through a heterotrimeric G protein that activates phospholipase C β2, whose activation increase inositol 1,4,5-triphosphate (IP_3) levels, inducing Ca^{2+}-release from intracellular stores that finally activates TRPM5 in the basolateral plasma membrane of taste receptor cells.

The first knockout mouse of TRPM5 was developed by Zhang and colleagues in 2003 [71], where exons 15 to 19 were replaced by the PGK-neomycin resistance cassette. This region encodes for the first five transmembrane domains and the pore region of the channel. These knockout animals are undistinguishable from the wild type animals in terms of weight, viability, general behavior and morphology and number of taste cells, taste receptors, and other signaling molecules. Nevertheless sweet, amino acid and bitter taste detection was completely abolished by TRPM5 ablation, with no effects on sour or salty tastes.

Rong and colleagues also generated a TRPM5 knockout mouse [72]. In this case, TRPM5 gene has a deletion of 2.4-kb of the 5′-flanking region of the gene. This deletion includes the promoter and coding region encompassing exons 1 to 4. Using this genetically modified mouse, Talavera and colleagues studied the temperature sensitivity of TRPM5, and they found that this is a heat activated channel [65]. Modulation by temperature of the human perception of different taste modalities is a well known fact, and temperature dependence of TRPM5 appears to contribute to this phenomenon *in vivo*. By using this animal model, Talavera and coworkers demonstrated that heat potentiates the gustatory (chorda tympani) nerve responses to sweet compounds in wild type animals but not in $TRPM5^{-/-}$ mice [65]. Molecular ablation of TRPM5 also eliminates, to a large extent, the responses to natural and artificial sweet compounds recorded in chorda tympani nerve. Interestingly, the responses to umami and bitter tastants were not potentiated by heat in both wild type and knockout animals. The thermal sensitivity of TRPM5 could explain the stronger perceived sweetness of sweet beverages at warmer temperatures in humans. On the other hand, the direct modulation of TRPM5 by temperature could explain why heating or cooling of the tongue can evoke sensations of taste.

4.7. TRPM8

The Transient Receptor Potential Melastatin 8 (TRPM8) channel is the main molecular entity responsible for the transduction of moderate cold in the somatosensory system (reviewed in [73, 74]). This Ca^{2+}-permeable cation channel, identified in 2001 as a protein up-regulated in prostate cancer [75] and cloned by two groups independently in 2002 [76, 77], is activated not only by cold but also by chemical cooling compounds such as menthol and by voltage [3, 6, 76, 77]. This channel is expressed mainly in dorsal root and trigeminal ganglia, but its expression has been detected in several other tissues, where its function is under intense study [73, 74].

Knockout mice lacking functional expression of TRPM8 were generated by three groups independently in 2007 [78-80]. Survival and general appearance of these mutant mice are normal, with no differences in the mean core body temperatures compared to wild type.

However, the cold sensitivity of all these mutants is strongly compromised, especially in the range of innocuous cold temperatures. Bautista and coworkers [78] generated a TRPM8 knockout mice by deletion of the coding region between residues 594 and 661, into the large intracellular N-terminal domain of TRPM8, introducing a stop codon before and frameshift after this segment. The non-functional truncated transcript produced by TRPM8$^{-/-}$ mice allowed to confirm by *in situ* hybridization that the loss of TRPM8 functional protein does not eliminate those neurons that normally express the channel. Behavioral studies revealed that these mice present a profound impairment to discriminate between cold and warm environments. Electrophysiological recordings of single sensory fibers from TRPM8 knockout mice show not only a strongly reduced sensitivity to innocuous cold stimuli in the low-threshold cold sensitive fibers, but also a marked reduction in the responses to temperature reductions in the high-threshold cold sensitive afferent neurons. Cold sensitive C- fibers of these animals present a reduced basal firing rate, indicating that TRPM8 is also important for the generation of the basal action potential firing of these neurons at resting temperatures. The molecular ablation of TRPM8 does not affect the conduction velocity and electrical activation threshold of these fibers. Calcium imaging experiments in cultured sensory neurons from TRPM8$^{-/-}$ mice show a decrease in both number and magnitude of the responses to cold, completing a picture where TRPM8 appears as the main molecular entity responsible for the cold sensitivity, with a role not only in the transduction of innocuous cold but also in the detection of cold in the noxious range.

Mice lacking functional expression of TRPM8 have been also generated by Colburn and coworkers by using homologous recombination [79]. These knockout mice are completely normal regarding general behaviors, lifespan and fertility. TRPM8$^{-/-}$ mice show a complete absence of behavioral responses to systemic chemical activation of TRPM8 by icilin, the strongest synthetic activator described for this channel so far. On the other hand, in a test to evaluate thermal preference, TRPM8$^{-/-}$ mice spend the same fraction of time in cold surfaces than in comfortable warm ones, unlike wild type animals that prefer warm surfaces. The development of neuropathic cold-induced pain was also strongly reduced by the genetic ablation of the channel in these animals. TRPM8 gene disruption also reduces cold sensitivity of primary sensory neurons in culture, studied by calcium imaging [79].

Dhaka and coworkers have also developed a TRPM8 knockout mouse. They used a targeting construct to delete amino acids in the N-terminal region knocked in by using a farnesylated enhanced green fluorescent protein (EGFP-F) followed by an SV40polyA tail in frame with the start codon of the channel [80]. This SV40polyA tail prevents the transcription of TRPM8, and EGFP allows to identify neurons that would express the channel in normal conditions. These TRPM8-deficient mice are completely viable, and they exhibit a strongly reduced avoidance to cold in two-temperature assays and in thermotaxis assays of temperature gradients. Molecular ablation of TRPM8 also reduced the responses to cooling chemicals. Using this mouse and a preparation that allows to record the electrical activity of unitary cold-sensitive nerve endings in the cornea, Parra and colleagues demonstrated that not only the responses to cold but also the ongoing electrical activity of these sensory neurons was proportional to the functional expression level of TRPM8 channels [81]. Altogether, these evidences confirm a critical role of TRPM8 in cold thermosensitivity.

4.8. TRPV1

The Transient Receptor Potential Vanilloid 1 (TRPV1), a Ca^{2+}-permeable nonselective cation channel, was the first thermoTRP channel to be cloned and the first protein to be shown to have such an unusually high dependence on temperature [82]. Its open probability increases with heat with a threshold of around 43°C (in heterologous expression systems), as well as upon exposure to capsaicin (the active compound in chili peppers) or low pH. Given its properties as well as its presence in a population of primary sensory neurons from the dorsal root and trigeminal ganglia, it was hypothesized as the main responsible for heat sensory transduction in mammals. In the year 2000, two groups were able to generate knock-out mice for the TRPV1 gene by targeted homologous recombination [83, 84]. Both works reported absence of heat-, capsaicin- or acid-evoked currents in cultured sensory neurons from TRPV1$^{-/-}$ mice. Moreover Caterina et al. [83] reported an impaired physiological and behavioral response to capsaicin in the knockout animals: while a normal animal would avoid consumption of a capsaicin-containing solution and decrease its body temperature upon capsaicin administration, mice with a disrupted TRPV1 gene lacked these responses. However the most expected prediction was not fulfilled in behavioral tests, as TRPV1$^{-/-}$ animals can detect non-noxious heat as wild type animals do. Instead, a deficit was detected in the knock-out animals regarding their response to noxious heat (>50°C) stimuli, a result that was striking given the threshold of 43°C for channel activation. This result can be explained assuming that the TRPV2 channel (see below) takes the place of TRPV1, however, the heat-responding neurons in TRPV1$^{-/-}$ mice do not express TRPV2 [85].

In addition, what was actually found in TRPV1$^{-/-}$ mice was a deficit in heat hyperalgesia [83, 84]. This process consists in the sensitization of the pain receptors upon persistence of the pain stimulus. The noxious heat in this case, may involve different inflammatory signaling cascades that lead to a sensitization of TRPV1 either by decreasing PIP_2 levels, activation of PKC, activation of PKA, and/or an increase of the number of TRPV1 channels in the plasma membrane [86].

Regarding body temperature (T_B) control, mice lacking this channel have a wider daily T_B cycle compared to wild type animals, with lower T_B during the day (inactivity period), and higher T_B during the night (active phase) [87, 88]. However, they still display a normal heat tolerance, i.e., their body temperature increases as much as in normal mice when exposed to a hot ambient [87]. On the contrary, when the TRPV1 channels of wild type mice are desensitized with capsaicin, these animals show a robust heat-intolerance [87], evidencing that knockout mice give only a partial picture of the role of TRPV1 as a heat sensor in physiological conditions.

4.9. TRPV2

The Transient Receptor Potential Vanilloid 2 (TRPV2) is activated by higher temperatures than TRPV1, with a threshold of ~52°C [89], and by cannabinoids[30]. The role of this channel is still poorly understood and although it is expressed in sensory ganglia no sensory function has yet been established. Instead, it has been associated to the axon outgrowth of developing DRG and motor neurons [90]. Knockout mice were generated by a loxP recombination that disrupted 4 exons comprising the 5th TM segment, the pore loop and the 6th TM segment [91]. These mice have unimpaired responses to heat and mechanical stimuli; however they were reported to be susceptible to perinatal lethality and to have a reduced body weight [91]. When looking at macrophage function [92], it was found that TRPV2 has a critical role in the early stages of phagocytosis. The channel is recruited to nascent phagosomes where it depolarizes the membrane initiating a signal cascade that results in clustering of phagocytic receptors. Moreover, macrophages from TRPV2$^{-/-}$ mice have a diminished motility [92].

4.10. TRPV3

The Transient Receptor Potential Vanilloid 3 (TRPV3) is a Ca^{2+}-permeable cation channel activated by innocuous warm temperatures (≥33°C) [93]. This channel is highly expressed in skin keratinocytes and oral and nasal epithelia, while the expression levels of TRPV3 in brain, sensory ganglia and spinal cord are in general weak [94]. TRPV3 can be activated by a large list of chemicals such as eugenol, thymol, and carvacrol, savory, clove, thyme, camphor, and 2-aminoethoxydiphenyl borate [94, 95]. Heat activation of TRPV3 shows a marked sensitization under repeated heat stimulation, both in recombinant systems and in keratinocytes.

The first knockout mouse for TRPV3 was developed by Moqrich and coworkers [95]. Cultured keratinocytes from these wild type mice show sensitization to repeated heat stimulation, responses and sensitization to camphor stimuli, and blockage by ruthenium red. Keratinocytes from TRPV3$^{-/-}$ mice did not respond to camphor or heat stimuli [95]. The most outstanding results that pointed out TRPV3 as a molecular sensor of the thermal stimuli in the warm range were obtained from behavioral experiments in TRPV3$^{-/-}$ mice. TRPV3-deficient mice showed a reduced tendency to migrate towards warm surfaces and a defect in their responses to noxious heat stimulation [95]. The remaining sensitivity to warm temperatures in TRPV3$^{-/-}$ mice implies that other molecular entities are also involved in innocuous warm thermosensation. On the other hand, if skin keratinocytes participate in the detection of warm temperatures at the surface of the skin, a mechanism to transmit the information from keratinocytes to sensory nerves is needed. In this line, it has been proposed that ATP released from keratinocytes due the activation of TRPV3 channels under warm temperature stimulation, could potentially signal to a variety of P2X- or P2Y-expressing sensory nerve terminals within the epidermis to transmit thermal information [96]. In addition to their role in thermosensation, several rodent strains bearing mutant TRPV3, such as the autosomal dominant DS-Nh (no-hair) mouse and the WBN/Kob-Ht rat, are spontaneously hairless and develop atopic dermatitis-like lesions. These two strains present a point mutation in the S4-S5 linker of TRPV3 (G573S in DS-Nh and G573C in WBN/Kob-Ht mice) [97]. The study of these point mutations, using a recombinant system, showed that a single substitution of the glycine 573 results in a constitutive active channel, insensitive to thermal and chemical stimuli [98]. In addition to these hairless strains, the TRPV3 deficient mice also display a hair abnormality. In the first description of TRPV3 deficient mice, a subtle and temporary hair irregularity in the abdominal area was reported around the third postnatal week [95]. However, in a more exhaustive study about the relationship between TRPV3 and hair morphogenesis, using another TRPV3$^{-/-}$ strain [99], the authors found important phenotypic changes related to abnormalities in skin, hair, and whiskers. The results of this work show that TRPV3-deficient mice displayed a deregulation of TGF-α/EGFR signaling that affect keratinocyte terminal differentiation, affecting hair generation [99].

4.11. TRPV4

Transient Receptor Potential Vanilloid 4 (TRPV4) is a Ca^{2+}-permeable cation channel that was cloned in 2000 by two groups independently [100, 101]. Originally identified as an osmosensitive channel, TRPV4 is expressed in kidney, lung, spleen, testis, tongue, fat cells, keratinocytes, inner and outer hair cells of the organ of Corti, sensory ganglia and central nervous system [100, 102-104]. In heterologous expression systems, TRPV4 show a threshold temperature of activation near to 34°C, a temperature close to the resting temperature of the skin [103], suggesting that TRPV4 could work as a warm detector in physiological conditions. A TRPV4 knockout mouse was developed and described by Suzuki and colleagues in 2003 [104]. TRPV4 gene disruption was achieved by homologous recombination, using a neomycin resistance (PKG-neo) cassette to replace the region encoding exon 4. Behavioral studies show that mice lacking TRPV4 did not display any impairment to sense noxious heat [104], but deficiencies in thermal selection between wild type and knockout mice were evident in temperature selection tests in the range of warm comfortable temperatures [105]. Interestingly, electrophysiological recordings of sensory neurons in the femoral nerve show a decrease in the warmth-dependent electrical activity in TRPV4$^{-/-}$ mice. Moreover, molecular ablation of TRPV4 also increased the latency to escape from hot (from 35 to 50°C) in animals with carrageenan-induced heat hypersensitivity [106], suggesting an important role for TRPV4 in both detection of warm temperatures and thermal hyperalgesia.

In Table 1 we summarize the information obtained so far from the study of the phenotype of knockout mice of thermoTRP channels.

5. Conclusions

Mutagenesis combined with electrophysiological studies have been fundamental in the understanding of the molecular mechanisms involved in the gating of temperature-activated TRP channels. The evidence accumulated from the molecular cloning of TRPV1 channel until nowadays, points out to the thermoTRP channels not only as predominant sensors of thermal and noxious stimuli in the somatosensory system *in vivo*, but also as key players in many other physiological processes.

ThermoTRP channel	Temperature sensitivity	Knockout mice phenotype and physiological role
TRPA1	≤17 °C	Loss of sensitivity to pungent natural compounds, environmental irritants and proalgesic agents. No effects in cold sensation of peripheral receptors. Deficits in cold sensation at vagal sensory neurons. Reduced hyperalgesia to mechanical stimuli. Mediator of inflammatory-related hyperalgesia. Contribution to noxious cold transduction. Cold transductor in visceral nerves.
TRPM8	≤34 °C	Impaired detection of innocuous and noxious cold. Suppression of the ongoing activity of cold thermoreceptor fibers. Strong reduction of cold-sensitive neurons in culture. Main molecular sensor to cold in the somatosensory system.
TRPC5	≤37 °C	No temperature-sensitive behavioural changes. Gain of function in C-cold nociceptors. Reduction in TRPM8 expressing cells and cold-sensitive neurons. Role in the detection and local adaptation to cold temperatures in the peripheral nervous system.
TRPM2	35-45 °C	Impaired immune response. Impaired insulin secretion and glucose metabolism. Ca^{2+} influx through TRPM2 controls signaling cascades responsible for chemokine production. Involved in insulin secretion stimulated by glucose.
TRPM4	15-35 °C	Increased IgE-dependent mast cell activation, impaired chemokine-dependent migration of dendritic cells. Impaired catecholamine release from chromaffine cells leading to increased sympathetic tone and hypertension. Involved in intracellular calcium regulation.
TRPM5	15-35 °C	Impaired detection of sweet, bitter and umami tastants. Key molecular component of taste transduction machinery. Proposed molecular counterpart of the modulation by temperature of the human perception of different taste modalities.
TRPM3	25-40 °C	Impaired detection of noxious heat stimuli. Deficit in the development of heat hyperalgesia. Reduction in the percentage of heat positive neurons in dorsal root and trigeminal ganglia. Chemo- and thermosensor in the somatosensory system, involved in the detection of noxious stimuli in healthy and inflamed tissue.

ThermoTRP channel	Temperature sensitivity	Knockout mice phenotype and physiological role
TRPV4	27-34 °C	Deficiencies in thermal selection in the range of warm comfortable temperatures. Reduction of inflammation-induced heat hypersensitivity. Decreased warmth-dependent electrical activity of primary sensory neurons. Significant role in detection of warm temperatures and thermal hyperalgesia.
TRPV3	32-39 °C	Reduced tendency to migrate towards warm surfaces and a defect in their responses to noxious heat stimulation. Cultured keratinocytes do not respond to camphor or heat stimuli. Heat-activated channel expressed in keratinocytes, with a significant role in thermosensation.
TRPV1	≥ 42 °C	Impaired nociception and pain sensation. Almost normal heat sensation but absence of inflammatory thermal hyperalgesia. Normal heat tolerance. Wider daily body temperature rhythm. Critical mediator of inflammatory-related hyperalgesia.
TRPV2	≥ 52 °C	Impaired innate immunity, leading to augmented perinatal lethality. Normal responses to heat and mechanical stimuli. Plays a pivotal role in chemoatractant-elicited motility of macrophages.

Table 1. The thermoTRP channels. Table summarizing the temperature sensitivity, knockout mice phenotype, and physiological role of mammalian thermoTRP channels.

Author details

María Pertusa and Rodolfo Madrid
Laboratorio de Neurociencia, Departamento de Biología, Facultad de Química y Biología, Universidad de Santiago de Chile, Santiago, Chile

Ramón Latorre* and Patricio Orio
Centro Interdisciplinario de Neurociencia de Valparaíso, Facultad de Ciencias, Universidad de Valparaíso, Valparaíso, Chile

Hans Moldenhauer
Centro Interdisciplinario de Neurociencia de Valparaíso, Facultad de Ciencias, Universidad de Valparaíso, Valparaíso, Chile
Programa de Doctorado en Ciencias mención Neurociencias, Facultad de Ciencias, Universidad de Valparaíso, Valparaíso, Chile

* Corresponding Author

Sebastián Brauchi
Instituto de Fisiología, Universidad Austral de Chile, Valdivia, Chile

Acknowledgement

Authors are supported by Fondecyt Grants 1110430 to R. Latorre; 1110906 to S. Brauchi; 1100983 to R. Madrid; 11090308 to P. Orio; 3110128 to M. Pertusa. R. Madrid thanks the support of Vicerrectoría de Investigación y Desarrollo of the Universidad of Santiago de Chile. H. Moldenhauer is recipient of a Conicyt Ph.D. Fellowship. The Centro Interdisciplinario de Neurociencia de Valparaíso is supported by the Millenium Science Initiative of the Ministry of Economy (Chile).

6. References

[1] Zimmermann, K., et al., *Transient receptor potential cation channel, subfamily C, member 5 (TRPC5) is a cold-transducer in the peripheral nervous system.* Proceedings of the National Academy of Sciences of the United States of America, 2011. 108(44): p. 18114-9.

[2] Basbaum, A.I., et al., *Cellular and molecular mechanisms of pain.* Cell, 2009. 139(2): p. 267-84.

[3] Brauchi, S., P. Orio, and R. Latorre, *Clues to understanding cold sensation: thermodynamics and electrophysiological analysis of the cold receptor TRPM8.* Proc Natl Acad Sci U S A, 2004. 101(43): p. 15494-9.

[4] Latorre, R., et al., *ThermoTRP channels as modular proteins with allosteric gating.* Cell calcium, 2007. 42(4-5): p. 427-38.

[5] Liu, B., K. Hui, and F. Qin, *Thermodynamics of heat activation of single capsaicin ion channels VR1.* Biophysical journal, 2003. 85(5): p. 2988-3006.

[6] Voets, T., et al., *The principle of temperature-dependent gating in cold- and heat-sensitive TRP channels.* Nature, 2004. 430(7001): p. 748-54.

[7] Yao, J., B. Liu, and F. Qin, *Kinetic and energetic analysis of thermally activated TRPV1 channels.* Biophysical journal, 2010. 99(6): p. 1743-53.

[8] Zakharian, E., C. Cao, and T. Rohacs, *Gating of transient receptor potential melastatin 8 (TRPM8) channels activated by cold and chemical agonists in planar lipid bilayers.* J Neurosci, 2010. 30(37): p. 12526-34.

[9] Tanford, C., *Protein denaturation.* Advances in protein chemistry, 1968. 23: p. 121-282.

[10] Cordero-Morales, J.F., E.O. Gracheva, and D. Julius, *Cytoplasmic ankyrin repeats of transient receptor potential A1 (TRPA1) dictate sensitivity to thermal and chemical stimuli.* Proceedings of the National Academy of Sciences of the United States of America, 2011. 108(46): p. E1184-91.

[11] Yao, J., B. Liu, and F. Qin, *Modular thermal sensors in temperature-gated transient receptor potential (TRP) channels.* Proceedings of the National Academy of Sciences of the United States of America, 2011. 108(27): p. 11109-14.

[12] Grandl, J., et al., *Pore region of TRPV3 ion channel is specifically required for heat activation.* Nature neuroscience, 2008. 11(9): p. 1007-13.

[13] Grandl, J., et al., *Temperature-induced opening of TRPV1 ion channel is stabilized by the pore domain.* Nature neuroscience, 2010. 13(6): p. 708-14.
[14] Brauchi, S., et al., *Dissection of the components for PIP2 activation and thermosensation in TRP channels.* Proceedings of the National Academy of Sciences of the United States of America, 2007. 104(24): p. 10246-51.
[15] Vlachova, V., et al., *Functional role of C-terminal cytoplasmic tail of rat vanilloid receptor 1.* The Journal of neuroscience : the official journal of the Society for Neuroscience, 2003. 23(4): p. 1340-50.
[16] Brauchi, S., et al., *A hot-sensing cold receptor: C-terminal domain determines thermosensation in transient receptor potential channels.* J Neurosci, 2006. 26(18): p. 4835-40.
[17] Schreiber, M., A. Yuan, and L. Salkoff, *Transplantable sites confer calcium sensitivity to BK channels.* Nature neuroscience, 1999. 2(5): p. 416-21.
[18] Jiang, Y., et al., *Structure of the RCK domain from the E. coli K+ channel and demonstration of its presence in the human BK channel.* Neuron, 2001. 29(3): p. 593-601.
[19] Choe, S., *Potassium channel structures.* Nature reviews. Neuroscience, 2002. 3(2): p. 115-21.
[20] Zagotta, W.N., et al., *Structural basis for modulation and agonist specificity of HCN pacemaker channels.* Nature, 2003. 425(6954): p. 200-5.
[21] Long, S.B., E.B. Campbell, and R. Mackinnon, *Voltage sensor of Kv1.2: structural basis of electromechanical coupling.* Science, 2005. 309(5736): p. 903-8.
[22] Murata, Y., et al., *Phosphoinositide phosphatase activity coupled to an intrinsic voltage sensor.* Nature, 2005. 435(7046): p. 1239-43.
[23] Nieto-Posadas, A., A. Jara-Oseguera, and T. Rosenbaum, *TRP channel gating physiology.* Curr Top Med Chem, 2011. 11(17): p. 2131-50.
[24] Rohacs, T., *Phosphoinositide regulation of non-canonical transient receptor potential channels.* Cell Calcium, 2009. 45(6): p. 554-65.
[25] Yang, F., et al., *Thermosensitive TRP channel pore turret is part of the temperature activation pathway.* Proceedings of the National Academy of Sciences of the United States of America, 2010. 107(15): p. 7083-8.
[26] Yang, F., et al., *Reply to Yao et al.: Is the pore turret just thermoTRP channels' appendix?* Proceedings of the National Academy of Sciences, 2010. 107(32): p. E126-E127.
[27] Yao, J., B. Liu, and F. Qin, *Pore turret of thermal TRP channels is not essential for temperature sensing.* Proceedings of the National Academy of Sciences of the United States of America, 2010. 107(32): p. E125; author reply E126-7.
[28] Latorre, R., C. Zaelzer, and S. Brauchi, *Structure-functional intimacies of transient receptor potential channels.* Q Rev Biophys, 2009. 42(3): p. 201-46.
[29] Lu, G., et al., *TRPV1b, a functional human vanilloid receptor splice variant.* Molecular pharmacology, 2005. 67(4): p. 1119-27.
[30] Neeper, M.P., et al., *Activation properties of heterologously expressed mammalian TRPV2: evidence for species dependence.* The Journal of biological chemistry, 2007. 282(21): p. 15894-902.

[31] Clapham, D.E. and C. Miller, *A thermodynamic framework for understanding temperature sensing by transient receptor potential (TRP) channels.* Proceedings of the National Academy of Sciences of the United States of America, 2011. 108(49): p. 19492-7.

[32] Sullivan, E., E.M. Tucker, and I.L. Dale, *Measurement of [Ca2+] using the Fluorometric Imaging Plate Reader (FLIPR).* Methods in molecular biology, 1999. 114: p. 125-33.

[33] Bandell, M., et al., *High-throughput random mutagenesis screen reveals TRPM8 residues specifically required for activation by menthol.* Nat Neurosci, 2006. 9(4): p. 493-500.

[34] Chung, M.K., et al., *2-aminoethoxydiphenyl borate activates and sensitizes the heat-gated ion channel TRPV3.* The Journal of neuroscience : the official journal of the Society for Neuroscience, 2004. 24(22): p. 5177-82.

[35] Hu, H.Z., et al., *2-aminoethoxydiphenyl borate is a common activator of TRPV1, TRPV2, and TRPV3.* The Journal of biological chemistry, 2004. 279(34): p. 35741-8.

[36] Bandell, M., et al., *Noxious cold ion channel TRPA1 is activated by pungent compounds and bradykinin.* Neuron, 2004. 41(6): p. 849-57.

[37] Bautista, D.M., et al., *Pungent products from garlic activate the sensory ion channel TRPA1.* Proceedings of the National Academy of Sciences of the United States of America, 2005. 102(34): p. 12248-52.

[38] Story, G.M., et al., *ANKTM1, a TRP-like channel expressed in nociceptive neurons, is activated by cold temperatures.* Cell, 2003. 112(6): p. 819-29.

[39] Bautista, D.M., et al., *TRPA1 mediates the inflammatory actions of environmental irritants and proalgesic agents.* Cell, 2006. 124(6): p. 1269-82.

[40] Macpherson, L.J., et al., *An ion channel essential for sensing chemical damage.* The Journal of neuroscience : the official journal of the Society for Neuroscience, 2007. 27(42): p. 11412-5.

[41] Andersson, D.A., et al., *Transient receptor potential A1 is a sensory receptor for multiple products of oxidative stress.* The Journal of neuroscience : the official journal of the Society for Neuroscience, 2008. 28(10): p. 2485-94.

[42] Venkatachalam, K. and C. Montell, *TRP channels.* Annu Rev Biochem, 2007. 76: p. 387-417.

[43] Zimmermann, K., et al., *Sensory neuron sodium channel Nav1.8 is essential for pain at low temperatures.* Nature, 2007. 447(7146): p. 855-8.

[44] Petrus, M., et al., *A role of TRPA1 in mechanical hyperalgesia is revealed by pharmacological inhibition.* Molecular pain, 2007. 3: p. 40.

[45] Fernandes, E.S., et al., *A distinct role for transient receptor potential ankyrin 1, in addition to transient receptor potential vanilloid 1, in tumor necrosis factor alpha-induced inflammatory hyperalgesia and Freund's complete adjuvant-induced monarthritis.* Arthritis and rheumatism, 2011. 63(3): p. 819-29.

[46] Fajardo, O., et al., *TRPA1 channels mediate cold temperature sensing in mammalian vagal sensory neurons: pharmacological and genetic evidence.* The Journal of neuroscience : the official journal of the Society for Neuroscience, 2008. 28(31): p. 7863-75.

[47] Karashima, Y., et al., *TRPA1 acts as a cold sensor in vitro and in vivo.* Proceedings of the National Academy of Sciences of the United States of America, 2009. 106(4): p. 1273-8.

[48] Jiang, L.H., N. Gamper, and D.J. Beech, *Properties and therapeutic potential of transient receptor potential channels with putative roles in adversity: focus on TRPC5, TRPM2 and TRPA1.* Current drug targets, 2011. 12(5): p. 724-36.

[49] Riccio, A., et al., *Essential role for TRPC5 in amygdala function and fear-related behavior.* Cell, 2009. 137(4): p. 761-72.

[50] Togashi, K., et al., *TRPM2 activation by cyclic ADP-ribose at body temperature is involved in insulin secretion.* The EMBO journal, 2006. 25(9): p. 1804-15.

[51] Takahashi, N., et al., *Roles of TRPM2 in oxidative stress.* Cell Calcium, 2011. 50(3): p. 279-87.

[52] Patel, S. and R. Docampo, *In with the TRP channels: intracellular functions for TRPM1 and TRPM2.* Science signaling, 2009. 2(95): p. pe69.

[53] Sumoza-Toledo, A. and R. Penner, *TRPM2: a multifunctional ion channel for calcium signalling.* The Journal of physiology, 2011. 589(Pt 7): p. 1515-25.

[54] Yamamoto, S., et al., *TRPM2-mediated Ca2+influx induces chemokine production in monocytes that aggravates inflammatory neutrophil infiltration.* Nature medicine, 2008. 14(7): p. 738-47.

[55] Knowles, H., et al., *Transient Receptor Potential Melastatin 2 (TRPM2) ion channel is required for innate immunity against Listeria monocytogenes.* Proceedings of the National Academy of Sciences of the United States of America, 2011. 108(28): p. 11578-83.

[56] Uchida, K., et al., *Lack of TRPM2 impaired insulin secretion and glucose metabolisms in mice.* Diabetes, 2011. 60(1): p. 119-26.

[57] Harteneck, C. and G. Schultz, *TRPV4 and TRPM3 as Volume-Regulated Cation Channels*, in *TRP Ion Channel Function in Sensory Transduction and Cellular Signaling Cascades*, W.B. Liedtke and S. Heller, Editors. 2007: Boca Raton (FL).

[58] Staaf, S., et al., *Dynamic expression of the TRPM subgroup of ion channels in developing mouse sensory neurons.* Gene expression patterns : GEP, 2010. 10(1): p. 65-74.

[59] Nilius, B. and T. Voets, *A TRP channel-steroid marriage.* Nature cell biology, 2008. 10(12): p. 1383-4.

[60] Vriens, J., et al., *TRPM3 is a nociceptor channel involved in the detection of noxious heat.* Neuron, 2011. 70(3): p. 482-94.

[61] Oberwinkler, J., et al., *Alternative splicing switches the divalent cation selectivity of TRPM3 channels.* The Journal of biological chemistry, 2005. 280(23): p. 22540-8.

[62] Naylor, J., et al., *Pregnenolone sulphate- and cholesterol-regulated TRPM3 channels coupled to vascular smooth muscle secretion and contraction.* Circulation research, 2010. 106(9): p. 1507-15.

[63] Wagner, T.F., et al., *Transient receptor potential M3 channels are ionotropic steroid receptors in pancreatic beta cells.* Nature cell biology, 2008. 10(12): p. 1421-30.

[64] Guinamard, R., M. Demion, and P. Launay, *Physiological roles of the TRPM4 channel extracted from background currents.* Physiology, 2010. 25(3): p. 155-64.

[65] Talavera, K., et al., *Heat activation of TRPM5 underlies thermal sensitivity of sweet taste.* Nature, 2005. 438(7070): p. 1022-5.

[66] Vennekens, R., et al., *Increased IgE-dependent mast cell activation and anaphylactic responses in mice lacking the calcium-activated nonselective cation channel TRPM4.* Nature immunology, 2007. 8(3): p. 312-20.

[67] Barbet, G., et al., *The calcium-activated nonselective cation channel TRPM4 is essential for the migration but not the maturation of dendritic cells.* Nature immunology, 2008. 9(10): p. 1148-56.

[68] Shimizu, T., et al., *TRPM4 regulates migration of mast cells in mice.* Cell Calcium, 2009. 45(3): p. 226-32.

[69] Mathar, I., et al., *Increased catecholamine secretion contributes to hypertension in TRPM4-deficient mice.* The Journal of clinical investigation, 2010. 120(9): p. 3267-79.

[70] Perez, C.A., et al., *A transient receptor potential channel expressed in taste receptor cells.* Nature neuroscience, 2002. 5(11): p. 1169-76.

[71] Zhang, Y., et al., *Coding of sweet, bitter, and umami tastes: different receptor cells sharing similar signaling pathways.* Cell, 2003. 112(3): p. 293-301.

[72] Damak, S., et al., *Trpm5 null mice respond to bitter, sweet, and umami compounds.* Chemical senses, 2006. 31(3): p. 253-64.

[73] Babes, A., et al., *TRPM8, a Sensor for Mild Cooling in Mammalian Sensory Nerve Endings.* Curr Pharm Biotechnol, 2011. 12(1): p. 78-88.

[74] Latorre, R., et al., *A cool channel in cold transduction.* Physiology, 2011. 26(4): p. 273-85.

[75] Tsavaler, L., et al., *Trp-p8, a novel prostate-specific gene, is up-regulated in prostate cancer and other malignancies and shares high homology with transient receptor potential calcium channel proteins.* Cancer Res, 2001. 61(9): p. 3760-9.

[76] McKemy, D.D., W.M. Neuhausser, and D. Julius, *Identification of a cold receptor reveals a general role for TRP channels in thermosensation.* Nature, 2002. 416(6876): p. 52-8.

[77] Peier, A.M., et al., *A TRP channel that senses cold stimuli and menthol.* Cell, 2002. 108(5): p. 705-15.

[78] Bautista, D.M., et al., *The menthol receptor TRPM8 is the principal detector of environmental cold.* Nature, 2007. 448(7150): p. 204-8.

[79] Colburn, R.W., et al., *Attenuated cold sensitivity in TRPM8 null mice.* Neuron, 2007. 54(3): p. 379-86.

[80] Dhaka, A., et al., *TRPM8 is required for cold sensation in mice.* Neuron, 2007. 54(3): p. 371-8.

[81] Parra, A., et al., *Ocular surface wetness is regulated by TRPM8-dependent cold thermoreceptors of the cornea.* Nat Med, 2010. 16(12): p. 1396-9.

[82] Caterina, M.J., et al., *The capsaicin receptor: a heat-activated ion channel in the pain pathway.* Nature, 1997. 389(6653): p. 816-24.

[83] Caterina, M.J., et al., *Impaired nociception and pain sensation in mice lacking the capsaicin receptor.* Science, 2000. 288(5464): p. 306-13.

[84] Davis, J.B., et al., *Vanilloid receptor-1 is essential for inflammatory thermal hyperalgesia.* Nature, 2000. 405(6783): p. 183-7.

[85] Woodbury, C.J., et al., *Nociceptors lacking TRPV1 and TRPV2 have normal heat responses.* The Journal of neuroscience : the official journal of the Society for Neuroscience, 2004. 24(28): p. 6410-5.

[86] Zhang, X. and P.A. McNaughton, *Why pain gets worse: the mechanism of heat hyperalgesia.* The Journal of general physiology, 2006. 128(5): p. 491-3.

[87] Szelenyi, Z., et al., *Daily body temperature rhythm and heat tolerance in TRPV1 knockout and capsaicin pretreated mice.* The European journal of neuroscience, 2004. 19(5): p. 1421-4.

[88] Garami, A., et al., *Thermoregulatory phenotype of the Trpv1 knockout mouse: thermoeffector dysbalance with hyperkinesis.* The Journal of neuroscience : the official journal of the Society for Neuroscience, 2011. 31(5): p. 1721-33.
[89] Caterina, M.J., et al., *A capsaicin-receptor homologue with a high threshold for noxious heat.* Nature, 1999. 398(6726): p. 436-41.
[90] Shibasaki, K., et al., *TRPV2 enhances axon outgrowth through its activation by membrane stretch in developing sensory and motor neurons.* The Journal of neuroscience : the official journal of the Society for Neuroscience, 2010. 30(13): p. 4601-12.
[91] Park, U., et al., *TRP vanilloid 2 knock-out mice are susceptible to perinatal lethality but display normal thermal and mechanical nociception.* The Journal of neuroscience : the official journal of the Society for Neuroscience, 2011. 31(32): p. 11425-36.
[92] Link, T.M., et al., *TRPV2 has a pivotal role in macrophage particle binding and phagocytosis.* Nature immunology, 2010. 11(3): p. 232-9.
[93] Xu, H., et al., *TRPV3 is a calcium-permeable temperature-sensitive cation channel.* Nature, 2002. 418(6894): p. 181-6.
[94] Xu, H., et al., *Oregano, thyme and clove-derived flavors and skin sensitizers activate specific TRP channels.* Nature neuroscience, 2006. 9(5): p. 628-35.
[95] Moqrich, A., et al., *Impaired thermosensation in mice lacking TRPV3, a heat and camphor sensor in the skin.* Science, 2005. 307(5714): p. 1468-72.
[96] Mandadi, S., et al., *TRPV3 in keratinocytes transmits temperature information to sensory neurons via ATP.* Pflugers Archiv : European journal of physiology, 2009. 458(6): p. 1093-102.
[97] Asakawa, M., et al., *Association of a mutation in TRPV3 with defective hair growth in rodents.* The Journal of investigative dermatology, 2006. 126(12): p. 2664-72.
[98] Xiao, R., et al., *The TRPV3 mutation associated with the hairless phenotype in rodents is constitutively active.* Cell Calcium, 2008. 43(4): p. 334-43.
[99] Cheng, X., et al., *TRP channel regulates EGFR signaling in hair morphogenesis and skin barrier formation.* Cell, 2010. 141(2): p. 331-43.
[100] Liedtke, W., et al., *Vanilloid receptor-related osmotically activated channel (VR-OAC), a candidate vertebrate osmoreceptor.* Cell, 2000. 103(3): p. 525-35.
[101] Strotmann, R., et al., *OTRPC4, a nonselective cation channel that confers sensitivity to extracellular osmolarity.* Nature cell biology, 2000. 2(10): p. 695-702.
[102] Chung, M.K., et al., *TRPV3 and TRPV4 mediate warmth-evoked currents in primary mouse keratinocytes.* The Journal of biological chemistry, 2004. 279(20): p. 21569-75.
[103] Guler, A.D., et al., *Heat-evoked activation of the ion channel, TRPV4.* The Journal of neuroscience : the official journal of the Society for Neuroscience, 2002. 22(15): p. 6408-14.
[104] Suzuki, M., et al., *Impaired pressure sensation in mice lacking TRPV4.* The Journal of biological chemistry, 2003. 278(25): p. 22664-8.
[105] Lee, H., et al., *Altered thermal selection behavior in mice lacking transient receptor potential vanilloid 4.* The Journal of neuroscience : the official journal of the Society for Neuroscience, 2005. 25(5): p. 1304-10.

[106] Todaka, H., et al., *Warm temperature-sensitive transient receptor potential vanilloid 4 (TRPV4) plays an essential role in thermal hyperalgesia.* The Journal of biological chemistry, 2004. 279(34): p. 35133-8.

[107] [107] Patapoutian, A., et al., *ThermoTRP channels and beyond: mechanisms of temperature sensation.* Nature reviews. Neuroscience, 2003. 4(7): p. 529-39.

[108] [108] Hille, B., *Ion Channels of Excitable Membranes*. 3rd ed. 2001, Sunderland, MA, USA: Sinauer Associates Inc.